Travel and Conflict in the Early Modern World

This edited collection examines the meeting points between travel, mobility, and conflict to uncover the experience of travel – whether real or imagined – in the early modern world. Until relatively recently, both domestic travel and voyages to the wider world remained dangerous undertakings. Physical travel, whether initiated by religious conversion and pilgrimage, diplomacy, trade, war, or the desire to encounter other cultures, inevitably heralded disruption: contact zones witnessed cultural encounters that were not always cordial, despite the knowledge acquisition and financial gain that could be reaped from travel. Vast compendia of travel such as Hakluyt's *Principla Navigations, Voyages and Discoveries*, printed from the late sixteenth century, and Prévost's *Histoire Générale des Voyages* (1746–1759) underscored European exploration as a marker of European progress, and in so doing showed the tensions that can arise as a consequence of interaction with other cultures. In focusing upon language acquisition and translation, travel and religion, travel and politics, and imaginary travel, the essays in this collection tease out the ways in which travel was both obstructed and enriched by conflict.

Gábor Gelléri is Lecturer in French at Aberystwyth University.

Rachel Willie is Reader in Early Modern Literary Studies at Liverpool John Moores University.

Routledge Studies in Renaissance and Early Modern Worlds of Knowledge

Series Editors:
Harald E. Braun
(University of Liverpool, UK)
Emily Michelson
(University of St Andrews, UK)

This series explores Renaissance and Early Modern Worlds of Knowledge (c.1400–c.1700) in Europe, the Americas, Asia and Africa. The volumes published in this series study the individuals, communities and networks involved in making and communicating knowledge during the first age of globalization. Authors investigate the perceptions, practices and modes of behaviour which shaped Renaissance and Early Modern intellectual endeavour and examine the ways in which they reverberated in the political, cultural, social and economic sphere.

The series is interdisciplinary, comparative and global in its outlook. We welcome submissions from new as well as existing fields of Renaissance Studies, including the history of literature (including neo-Latin, European and non-European languages), science and medicine, religion, architecture, environmental and economic history, the history of the book, art history, intellectual history and the history of music. We are particularly interested in proposals that straddle disciplines and are innovative in terms of approach and methodology.

The series includes monographs, shorter works and edited collections of essays. The Society for Renaissance Studies (http://www.rensoc.org.uk) provides an expert editorial board, mentoring, extensive editing and support for contributors to the series, ensuring high standards of peer-reviewed scholarship. We welcome proposals from early career researchers as well as more established colleagues.

17 Travel and Conflict in the Early Modern World
Edited by Gábor Gelléri and Rachel Willie

For more information about this series, please visit: https://www.routledge.com/Routledge-Studies-in-Renaissance-and-Early-Modern-Worlds-of-Knowledge/book-series/ASHSER4043

Travel and Conflict in the Early Modern World

Edited by
Gábor Gelléri
and Rachel Willie

Routledge
Taylor & Francis Group
LONDON AND NEW YORK

First published 2021
by Routledge
2 Park Square, Milton Park, Abingdon, Oxon OX14 4RN

and by Routledge
52 Vanderbilt Avenue, New York, NY 10017

Routledge is an imprint of the Taylor & Francis Group, an informa business

British Library Cataloguing-in-Publication Data
A catalogue record for this book is available from the British Library

Library of Congress Cataloging-in-Publication Data
Names: Gelléri, Gábor, 1976- editor. | Willie, Rachel, editor.
Title: Travel and conflict in the early modern world / edited by Gábor Gelléri and Rachel Willie.
Description: Abingdon, Oxon; New York, NY: Routledge, 2021. | Includes bibliographical references and index.
Identifiers: LCCN 2020025814 (print) | LCCN 2020025815 (ebook) | ISBN 9780367524210 (hardback) | ISBN 9781003057871 (ebook)
Subjects: LCSH: Travel--History. | Voyages and travels. | International relations--History. | Cultural relations--History.
Classification: LCC G156 .T29 2021 (print) | LCC G156 (ebook) | DDC 910.9/032--dc23
LC record available at https://lccn.loc.gov/2020025814
LC ebook record available at https://lccn.loc.gov/2020025815

ISBN: 978-0-367-52421-0 (hbk)
ISBN: 978-1-003-05787-1 (ebk)

Typeset in Sabon
by MPS Limited, Dehradun

Contents

List of figures

Acknowledgements

This edited collection had its genesis in the *Travel and Conflict in the Medieval and Early Modern World* conference, held at Bangor University in September 2015. We would like to thank the Modern Humanities Research Association, the Royal Historical Society, the Society for Renaissance Studies, the Aberystwyth-Bangor Institute for Medieval and Early Modern Studies (and affiliated departments) for their generous support in organizing this event. We are especially grateful to Andrew Hiscock and Rhun Emlyn who were part of the conference committee, and to Claire Jowitt, Sue Niebrzydowski, Raluca Radulescu, Björn Weiller, and Helen Wilcox for their support. We also thank all who attended for the stimulating conversations that arose.

This volume could not have come together without our friends and family. We would like to thank Julie Duran-Gelléri for helping with translation and Andrew and Val Willie for assisting with transcription. Liz Oakley-Brown generously shared her work with us, Eva Johanna Holmberg kindly commented upon a draft of our introduction, and Harald Braun and Emily Michelson proved patient and very helpful series editors. Finally, the editors would like to thank all the scholars who contributed to this volume, and Max Novick and Michael Greenwood at Routledge, without whom this book would not have been possible.

A note to the reader

When quoting from primary sources, we have maintained original spelling. Due to citing texts in several languages that have varying orthographic conventions, we have decided not to follow contemporary practice in silently amending u/v and i/j. Unless otherwise stated in individual chapters, English translations are the contributors' own. With the exception of Marie-Christine Gomez-Géraud's essay, in which the argument is predicated upon linguistic expression, we have only included the English translations of texts that originally appeared in other languages.

In 1582, much of Catholic Western Europe adopted the Gregorian Calendar, and Western Protestant nations gradually followed suit. England, along with its Scandinavian neighbours, did not subscribe to this calendar until the eighteenth century and instead used the Julian Calendar, which leads to a discrepancy in the dating of the days of the year. Consequently, when days of the year are referenced, we have adopted the calendar most appropriate to the chapter to maintain consistency with source materials. Hence, chapters dealing primarily with Italian sources have followed the practice of modernizing the days of the year, but we have maintained the original date for English source material. In accordance with current scholarly practice, we have silently amended the calendar year to begin on January 1 and not March 25 as was the custom in much of early modern Europe. For example, the Battle of Pavia took place on 24 February 1525, not 24 February 1524/25.

List of abbreviations

ARSI	Archivum Romanum Societatis Iesu
ASMa	Archivio di Stato Mantova, Archivio Gonzaga, Affari Esteri
ASMi, AS, PE	Archivio di Stato Milano, Archivio Sforzesco, Carteggio Estero, Potenze Estere
ASMo, ASE, CA	Archivio di Stato Modena, Archivio Segreto Estense, Carteggio Ambasciatori
ASS	Archivio di Stato Siena
BNCF	Biblioteca Nazionale Centrale di Firenze
LP	J. S. Brewer, J. Gairdner and R. H. Brodie, eds. *Letters and Papers, Foreign and Domestic, of the Reign of Henry VIII, 1509–1547.* 21 vols. London, 1862–1932
OED	Oxford English Dictionary
ODNB	Oxford Dictionary of National Biography

List of contributors

Joanne W. Anderson is Reader in the History of Art at the University of Aberdeen. Her research focuses on the pre- and early modern art and visual culture of the European Alps, with specific interest in the circulation of visual types and travelling artisans in the context of parish and pilgrimage networks. She is author of *Moving with the Magdalen: Late Medieval Art and Devotion in the Alps* (Bloomsbury Visual Arts, 2019).

Céline Bonnotte-Hoover holds a PhD from the Department of French at the University of Toronto, Canada. Her studies focus on early modern travel literature and on questions of intercultural encounters and mediation. As such, her doctoral dissertation, titled "Le truchement: représentation d'une instance de médiation dans les récits de voyageurs français du XVIe au XVIIIe siècle", focuses on the analysis of the relationship between travellers and interpreters throughout the early modern period. She is currently a French Instructor at Carleton University, Ottawa.

Daniel Carey is Director of the Moore Institute and Professor of English at the National University of Ireland Galway. His work on travel writing includes edited volumes on *Asian Travel in the Renaissance*, *The Postcolonial Enlightenment*, and *Richard Hakluyt and Travel Writing in Early Modern Europe*. He is general editor, with Claire Jowitt, of the critical edition of Hakluyt's *Principal Navigations, Voyages, Traffiques, and Discoveries of the English Nation* (1598–1600). He is principal investigator for a project creating a database of early modern advice literature on travel (https://artoftravel.nuigalway.ie/).

Natalya Din-Kariuki is Assistant Professor in the Department of English and Comparative Literary Studies at the University of Warwick. Her research examines the literary and intellectual history of the sixteenth and seventeenth centuries, with a particular focus on travel writing, transnational and transcultural encounters, modes of cosmopolitanism, and rhetoric and poetics. She has held visiting fellowships at the

University of Leeds and the Folger Institute in Washington, DC. Previously, she was Lecturer in English at Worcester College, Oxford.

Robert John Clines is Assistant Professor of History and International Studies at Western Carolina University. His research has been funded by the Fondazione Lemmermann, a J. William Fulbright Scholarship, the Andrew W. Mellon Foundation, the Early Modern Conversions Project, and the American Academy in Rome. His book, *A Jewish Jesuit in the Eastern Mediterranean: Early Modern Conversion, Mission, and the Construction of Identity*, appeared with Cambridge University Press in 2019. His essays have appeared in *The Mediterranean Historical Review, Renaissance Studies, The Sixteenth Century Journal, The Journal of Early Modern History*, and *History Compass*.

Paul M. Dover is Professor of History at Kennesaw State University, Georgia, USA. He has authored numerous articles on the political and cultural history of late medieval and early modern Europe. He has written the historiography textbook *The Changing Face of the Past* (Cognella, 2014), is editor of *Secretaries and Statecraft in the Early Modern World* (Edinburgh University Press, 2016), and author of the forthcoming *The Information Revolution of Early Modern Europe*.

Gábor Gelléri is Lecturer in French at Aberystwyth University; he specializes on various aspects of early modern travel. His monograph *Philosophies du voyage: visiter l'Angleterre aux 17e-18e siècles* was published by the Voltaire Foundation in Oxford in 2016; his second monograph, *Lessons of Travel in Eighteenth-Century France: From Grand Tour to School Trips,* was published in 2020 by Boydell and Brewer. He is co-director of the Aberystwyth-Bangor Institute for Medieval and Early Modern Studies (IMEMS).

Marie-Christine Gomez-Géraud is Emeritus Professor of French Renaissance Literature at the Université Paris Nanterre. She is scientific co-director of the seminar series "Diptyique" on mystical discourse between the Middle Ages and early modernity. One of her main areas of interest is travel literature. She co-edited the *Navigations* of Nicolas de Nicolay and edited Jean Boucher's *Bouquet sacré*. Along with about a hundred articles on the topic, she is the author of two monographs: *Écrire le voyage au XVI*[e] *siècle* (PUF, 2000) and *Le Crépuscule du Grand voyage. Les récits de pèlerins à Jérusalem 1456–1612* (Champion, 1999). She understands travel as a form of intercultural meeting and is interested in the linguistic consciousness in its relationship to the notion of otherness. Her publication of the *Bible nouvellement translatée* by Sébastien Castellion was an opportunity to research the practice of translation during the Humanist era.

Eva Johanna Holmberg, FRHistS, is Academy Research Fellow and Faculty member of the Department of Philosophy, History, Culture and Arts Studies, University of Helsinki, where she is working on a five-year research project entitled Travel and Self-Description in Seventeenth-Century English Culture. She is the author of *Jews in the Early Modern English Imagination* (Ashgate, 2012). Her most recent article is "Writing the travelling self: travel and life-writing in Peter Mundy's (1597–1667) *Itinerarium Mundii*", published in *Renaissance Studies.*

David Nicol is Associate Professor of Theatre and Film Studies at the Fountain School of Performing Arts at Dalhousie University. He is the author of *Middleton and Rowley: Forms of Collaboration in the Jacobean Playhouse* (University of Toronto Press, 2012) and of numerous articles on Jacobean drama.

William T. Rossiter is Senior Lecturer in Medieval and Early Modern Literature at the University of East Anglia, having previously taught at the universities of Liverpool and Manchester. Publications include the monographs *Chaucer and Petrarch* (Boydell and Brewer, 2010) and *Wyatt Abroad: Tudor Diplomacy and the Translation of Power* (D. S. Brewer, 2014) and the edited collections *Literature and Ethics: from the Green Knight to the Dark Knight* (Boydell and Brewer, 2010) and *Authority and Diplomacy from Dante to Shakespeare* (Routledge, 2013). He is currently working on a biographical study of Pietro Aretino, which illustrates the intersections between early modern diplomacy and print culture.

Rachel Willie is Reader in Early Modern Literary Studies at Liverpool John Moores University, where her research broadly focuses upon seventeenth-century literary history and culture. Her monograph, *Staging the Revolution: Drama, Reinvention and History 1647–72* (Manchester University Press, 2015; paperback 2019) was shortlisted for the University English Early Career Book Prize 2016. With Kevin Killeen and Helen Smith, she has co-edited *The Oxford Handbook of the Bible in Early Modern England, c. 1530–1700* (Oxford University Press, 2015; paperback, 2018; winner of the Roland H. Bainton Prize in Reference Works, 2016). She is currently principal investigator on the AHRC-funded project, *Soundscapes in the Early Modern World.*

Introduction

Travel and conflict

Gábor Gelléri and Rachel Willie

A fragment of an Egyptian manuscript, set in the eleventh century BCE, documents the unfortunate travels of Wenamun. A priest of Amun, he is sent to Phoenicia to procure some cedar wood. Along the way, he is robbed and then seizes silver in turn; has disastrous diplomatic and mercantile encounters; and, after being detained for almost a year, sets sail back to Egypt. Unkind winds blow him off course, and he lands in Cyprus where, after nearly being murdered by angry locals, he gains the protection of the queen. The manuscript breaks off shortly after Wenamun finds a translator to assist him and he reaches a partial resolution to this situation.[1]

While there is growing acceptance that "The Report of Wenamun" is a literary tale rather than a report of an actual voyage, Egyptologists continue to speculate over what kind of text it is.[2] Part of the uncertainty may lie in the generic conventions of Ancient Egypt being different to those of contemporary Western Europe, meaning that it does not sit comfortably in any modern genre. This has led it to be loosely categorized as a "tale of travel" and interpreted as an allegory of Egypt's waning power.[3] But beyond this possible allegorical reading, we can read Wenamun's tale as a meditation on the perils of travel as experienced by the pilgrim, the trader, the diplomat and the (poor) linguist. His tribulations in the Levant illustrate that, even with careful planning and good connections, journeys could go awry. Wenamun experiences travails of the body and of the mind as well as financial hardship, along with disorientation from being a stranger in an unknown land. Regardless of whether the manuscript is an authentic account of an actual journey, this text, one of the earliest detailed narratives of a travel experience, is a narrative of travel and conflict deeply intertwined.

Travel and Conflict in the Early Modern World posits that conflict lies at the heart of the idea and practice of travel. This period witnesses an expansion of the known world, perhaps most strikingly illustrated by the (re)discovery of the Americas, and the establishment of trade routes across the transatlantic world. Spain, the leading Counter-Reformation power in the sixteenth century, grew wealthy from mining gold and silver in South America, and Western European countries profited through colonial expansion, trade, and piracy. Encounters with the other, whether for mercantile, religious, colonial, or diplomatic purposes, inevitably gave rise to the potential for conflict. Numerous new cultures were encountered for the first time by Europeans, and, as Marie-Christine Gomez-Géraud's contribution to this volume highlights, the discrepancy between the explorers' often ideologically driven expectations and the situations on the ground was rife with potential for linguistic, cultural, and armed clashes. As anxieties regarding Christian devotion, worship, and the individual's relationship with God intensified, missionaries travelled on evangelising missions or, as Robert Clines's chapter in this volume shows, on missions that sought to return to the fold isolated Christian communities believed to be in danger of apostasy.

This expansion of the known world also led to the remapping of space. Throughout the late medieval and early modern period, European cartographers gradually moved away from *Mappa Mundi* – the plotting of humanity's place and history within a Christian cosmos – and instead sought to plot the material land upon which people lived. Yet, as has been long recognised, even apparently neutral and innocuous mappings of space demonstrate how land is appropriated and politicised. According to Andrew Gordon and Bernhard Klein, "the cartographic image is revealed as more than a mere functional tool, or neutral scientific record, emerging instead as a crucial representational site of cultural and historic change".[4] Cosmography, cartography, and navigation were closely intertwined and aided travellers as they traversed the globe. As Surekha Davies highlights, illustrated Renaissance maps of the New World offered synthetic images for their various users regarding the physical features of the land and its population (real and mythical) and presented the map reader with possibilities for colonial expansion.[5] Regardless of the linguistic, cultural, and physical conflicts and tensions that arose from settling in the New World, the Americas were seen by Europeans as unclaimed territory and, as such, a land that offered both dangers and possibilities. Maps and tales of travel each presented readers with information about the world they inhabited and directed how they interpreted space. Both could also help the reader to avoid the dangers associated with travel. In 1559, William Cunningham made the following observation about cosmography:

> I had almost (throoough making ouer much hast) forgotten to resight the benefits we receiue of Cosmographie: in that she deliuereth vs from greate continuall trauailes. For in a pleasaunte house, or warme study, she sheweth vs the hole [*sic*] face of all th' Earthe, withal the corners of the same.[6]

In being able to see the entire (known) world through looking at a map or through reading a travelogue, the perils of travelling are both contained and neutralised by the map or the book; this is highlighted by the grateful entries in the *Album amicorum* of Abraham Ortelius, creator of one of the earliest great map collections, the *Theatrum orbis terrarum* (first published in 1570).[7] The page thus becomes a way to know the world while keeping out of danger. As Gábor Gelléri illustrates in this volume, travelling through the reading of travelogues could even be deemed to be the only appropriate, desirable, safe, and virtuous way to travel – particularly for women.

Despite its popularity and the essential role it played in the transmission of information, during the early modern period, travel literature was considered notoriously unreliable. This suspicion is already present in antiquity: the *Geography* of Strabo (c. 20 BCE) opined that "Everybody who tells the story of his own travels is a braggart".[8] As if to emphasise how even a translation can become a form of conflict, the term ἀλαζών ("braggart") was, in French dictionaries of the seventeenth and eighteenth centuries (including the *Encyclopédie*), systematically translated as "menteurs" (liars).[9] Intellectuals such as Voltaire avoided writing in the unstable and un-prestigious form that is the travelogue.[10] Reading travelogues for the gathering of information was similarly deemed dubious: John Locke debated with Bishop Stillingfleet on the reliability of this literature, although both had an extensive collection of travel books,[11] while Jean-Jacques Rousseau was vocal on the futility of reading them, which did not stop him from making an extensive use of them in various works.[12] Earlier writers such as Roger Ascham and Joseph Hall advised for foreign mobility to be eliminated or at least strictly controlled due to the conflicts, physical and moral dangers that (as Joanne Anderson's chapter in this volume attests) lurked on the road; but even travelogues presented perils. These texts could render the reader vulnerable by inspiring immoral thoughts, inculcating a fascination with the foreign, and possibly motivating a dangerous desire to travel. Thus, the Swiss officer turned Pietist preacher Béat de Muralt, himself a former traveller, suggested in his *Lettre sur les Voiages* ("Of Travelling", written c. 1700) that reading travelogues is pernicious, and the only acceptable form of travel remains using one's imagination to travel back in time.[13] Yet, as Rachel Willie's chapter in this volume demonstrates, even imaginary travel could have negative cognitive effects upon the reader. Travels towards imaginary spaces such as utopias are, by definition, rife with conflict: as Daniel Carey highlights in his chapter, "Narratives of utopia thrive by exposing home itself as the locus of conflict, and the alternative to home as a place of resolution and harmony, realizing an element of wish-fulfilment while censuring those held responsible for abridging it".[14] These alternative spaces, from semi-fictional pastures to locations that make no pretence of being "real", exist exactly because of the productive conflictual relationship that mobility, real or imagined, makes possible.

Reading travelogues relied upon the reader investing in the veracity of the narrative. Several scholars have proposed the notion of a "pact" of travelogues, similar to Philippe Lejeune's "autobiographical pact" where author and readers enter into a pact whereby each invests in the veracity of how the author presents their life story.[15] A travelogue works on the basis of a "double" pact: the reader is willing to suspend disbelief, to accept both that the text is the fruit of an experience of travel and that what it narrates is the truth of this travel. As Natalya Din-Kariuki illustrates in this volume, the "art of travel", to a large extent, overlaps with the art of rhetoric. The way in which accidents and encounters experienced and negotiated on the road were narrated and ordered could enable readers to suspend their credulity and enter into a pact of travel.

Authors could claim the possibility of the existence of the world they describe: in the proem to Book 2 *of The Faerie Queene* (1590), Edmund Spenser reminded his readers, after listing numerous newly discovered territories, "Why then should witlesse man so much misweene/That nothing is, but that which he hath seene?" (2, proem 3, 4–5). Ultimately, the only way for the reader to check the veracity of a travelogue is to retrace the steps of the writer. Regardless of whether an aspirant traveller believed descriptions of sea monsters circling far-off lands; tales of violent headless Blemmyes whose faces were on their stomachs; or even the outlandish warnings of elephants, hippopotamuses, and cannibals who lived in exotic climates, embarking upon a voyage of one's own posed its own difficulties. Aboard ship, malnutrition, piracy, and fierce storms were distinct possibilities, as was sailing off course and getting into serious trouble on long sea voyages. Many of the tribulations that impeded Wenamun continued to adversely affect the traveller in the early modern period.

Undertaking travel was dangerous and generated possible conflicts both with the natural world and with fellow travellers, people encountered on the journey, and at the destination. The stranger could be greeted with suspicion and, as Céline Bonnotte-Hoover illustrates in this volume, intercultural mediators such as interpreters were not always trusted by the people they assisted to communicate. In detailing Henry Blount's travels in the Levant, Eva Johanna Holmberg shows how the solitary traveller needed courage and the ability to dissimulate when confronted with complex situations; the travelogue narrating the trip could be used to document the traveller's skills and resourcefulness. Yet, for all its difficulties, travel was often perceived as a necessary part of a gentleman's education, and possibly the only genuine opportunity to change one's status. In a letter to Robert Cecil, Earl of Salisbury, dated 8 October 1611, the widowed mother of Thomas Cromwell (the future Earl of Ardglass), Frances Cromwell writes on behalf of her young son, whose "father left him so poore and of so smale meanes and habilitie, as he hathe in a manner little ells but the bare title of Nobilitie; And if his other wants be not supplied by his better education, he maie hold himself to be moste vnfortunat".[16] With the son left in a financially insecure position following the death of his father, his mother believes the means by which his fortune may be restored to prosperity is education, more specifically education gained through mobility. She begs the authorities to allow her son to travel as a way to acquire the knowledge he needs:

> My humble sute therefore ... is. That ... (in fauor of his calling to giue me the better hope of his future fortunes) to moue the ... moste Excellent ma[s]tere for his Licence to trauell into Fraunce & to remain there some three or fower yeres: By w[hi]ch tyme hauing attained the language, and some better vnderstanding of him selfe (as I hope he wille) he wille become more fit to doe his dutie & service ... and by his experience abroade in his more riper yeres maie (w[i]th Gods fauor) prove a profitable member in the Comon wealth.[17]

Travel is here presented as the ultimate pedagogical tool: through journeying to (and sojourning in) France, young Thomas Cromwell not only gains fluency in one of the essential European vernacular languages of the period, he also achieves a greater understanding of himself. Temporary displacement to a foreign land enriches the individual linguistically and emotionally because travellers acquire insights abroad about themselves (*inquisitio*) while also obtaining knowledge of the wider world (*conquisitio*).[18] These revelations further enrich the commonwealth to which the traveller returns: the lessons learned and the experiences gained on the road enable the traveller to become a useful citizen, a candidate for various positions where an active knowledge of foreign lands and mores is needed, and the information that travellers might have uncovered during their travels could also be put into use by their state administration. Travel thus has the appearance of being a force for the public good, but the reasons that led to Cromwell petitioning for her son to be allowed to go abroad demonstrate a more pressing need for travel. The anxiety to improve the young man's financial outlook demonstrates how the conflicts of fortune can be the catalyst to travel. Far from being solely the means to develop a good citizen, travel – a form of "orchestrated conflict", during which the young pupil's exposure to difficulties and otherness is carefully mediated and regulated through the lens of pedagogy – thus becomes the means to negate the anxieties and hardships caused by familial poverty.

Sending sons to France to learn French was a common elite practice in Britain during most of the seventeenth and eighteenth century. Fifty years after Frances Cromwell sought permission for her son to travel, another young man embarked upon a voyage to France. A series of letters from the tutor to Thomas, the youngest son of Richard Sackville, Fifth Earl of Dorset, report on Thomas's ill-fated travel. Following the similar, successful trips of some of his older brothers – the eldest of whom was the future poet and courtier Charles Sackville – Thomas was sent to France to learn French. From the moment of his departure, Thomas was melancholy about having to leave his mother. His tutor notes with some relief that Thomas suffered only a little from seasickness during the crossing of the Channel. The visit then seems to follow the familiar pattern of such instructive sojourns as the student and tutor proceed to pay a visit to the English ambassador to Paris, and then to settle in Blois. At this point, the pedagogical project begins to fall apart: identifying errors in the journey and its planning, the tutor reports that there are too many English visitors in Blois who prove a distraction and that Thomas Sackville was given an English page when they should have hired a French boy. A French servant could have assisted Thomas in his language learning, but the tutor concludes that they are too immersed in the English language and the society of the English living in Blois to adequately learn French. To rectify this error and get away from the English, they move on to Saumur, but the tutor reports with despair that Thomas continues to prefer playing to learning. This disappointing trip then takes a tragic turn when the young man falls ill and dies; the last letters in this set of correspondence refers to the embalming of Thomas's body and having it shipped back to England.[19]

During Thomas Sackville's trip, everything that could go wrong did go wrong: he experienced emotional and physical distress all the way to his final illness, and the pedagogical rationale for the trip – learning French – was poorly carried out. The exasperated tutor who accompanied Thomas Sackville on his visit to France found that conflict emerged between his desire to stimulate his charge's interest and curiosity, which were the goals of the costly and risky endeavour of studying abroad, and Thomas's disregard for his studies. One of the reasons why this visit proved so disastrous might be that Sackville was aged only thirteen when he was sent abroad; this was considerably younger than most of his peers. While some, such as the essayist Michel de Montaigne, advocated allowing children to travel this young, the prevailing opinion in England during the second half of the seventeenth century was that travel ought to be delayed until the traveller was in their late teens, if not older.[20]

While this is certainly an intriguing – albeit tragic – story, Thomas Sackville's demise should not be read as an invitation to consider "travel and conflict" only as "the tribulations of the unfortunate traveller". As with the fate of Sackville, the true or imagined misadventures of Wenamun with which we began this introduction represent more than a story of hapless travel: each of these narratives provide insights into attitudes to mobility. Wenamun is sent to purchase materials to build a ship, which will be used to transport the cult image of Amun: his first journey was intended to herald further voyages associated with religion. Sackville's travels were meant to prepare him for a future that may have involved taking up a public function – something that is explicitly present in the letter by Frances Cromwell discussed above.

Reducing the study of the relationship between travel and conflict to one of difficulty on the road therefore overlooks how conflict in travel is encountered as both a potentially productive cross-cultural confrontation and an opportunity for cross-fertilisation. Conflict is certainly a reality of travel, where a host of minor and major difficulties arise. As Paul Dover's contribution to this volume illustrates, everyday realities of travel (such as road conditions or the need for fresh horses) were essential not only in the development of travelling practices, but also in the emergence of the "new diplomacy" of which residency was a core tenet. The contributions to this volume by both Joanne Anderson and Eva Johanna Holmberg highlight the dangers represented by unwelcome encounters on the road. Travellers also faced more complex difficulties and needed to negotiate multifarious political, religious, and philosophical conundrums that present barriers to travel.

Conflict is also an inspiration behind, and an instigator of, mobility. The very idea of having young men travel is born out of a form of cultural and intellectual conflict that requires young men to learn skills that can only be gained on the road while at the same time asking travellers to not immerse themselves too deeply in the cultures that they encounter. It emerges from the perceived shortage of, and need to develop expertise in, language, statecraft, and administration that can only be gained through travel while at the same time seeing the world beyond the borders of the native country as lacking the refinement of home. As Sarah Goldsmith contends, by the eighteenth century, the risks taken, dangers encountered, and conflicts experienced while travelling lead the Grand Tour to be fundamental to the fashioning of elite masculine identities and asserting their dominance and right to hold power.[21] Travel thus forms the basis of an education designed to advance and transform the individual into someone who can be of public good, despite – or, indeed, possibly because of – the risks and dangers to which the traveller was unintentionally or intentionally exposed.

Constructive conflict?

The cases of Sackville and Cromwell, in different ways, point to how peace – albeit a tragic peace in Sackville's case – can be gained at the end of travel. Béat de Muralt's *Lettre sur les Voiages*, mentioned above, stated that the peace and rest at the end of the travel is the only worthy reason for all the tribulations and vanities of travel. However, as William Rossiter states in his essay in this volume, it would be reductive to consider travel as the end of conflict, and conflict the end of travel. As his discussion of Aretino demonstrates, sometimes "peace" could be found on the battlefield when the court is the site of conflict and even mortal danger. The battlefield (which can be the courtier's temporary *domus*) is not only a site of conflict: once the battle is fought, it may also serve as a space for conflict resolution, however brief the truce may be. Mobility could thus be inspired as much by the intention to prevent or solve conflict as it could be the method to generate it.

Yet, as the peace of the battlefield implies, the relationship between conflict and conflict resolution is not linear. In a slightly provocative article intended to eliminate the "bad name" given to conflict by the range of associations it has to armed conflict, to psychopathology, or to social disorders, Morton Deutsch defined conflict as a necessary and crucial element of human interaction. He sees it as the root of change, as a form of assessing oneself, and as a key feature in creating and demarcating individual and group identities. Conflict could be seen not only as the problem, but also as the sensation or action that heralds the solution. More generally, Deutsch contends, "Conflict prevents stagnation, it stimulates interest and curiosity".[22] In her chapter, Marie-Christine Gomez-Géraud highlights how conflict facilitates a better understanding of the world by being one of the ways in which the traveller experiences the corporeal reality of the world instead of an idealist fabrication. During one of the public discussions organised during the 1630s by the French intellectual Théophrase Renaudot and later published by his son, five opinions are expressed regarding the desirability of travel. Highlighting the ambivalent attitudes towards travel in seventeenth-century France, the opinions are mixed: two longer opinions are against travel, and three shorter ones are in favour, showing that no overall conclusion was reached. In one of the pro-travel opinions, travel appears as a synonym of "action", something he contends is always desirable to humans in comparison to inaction and passivity.[23] Likewise, conflict is neither inactive nor passive. Drawing on this association between travel, action, and conflict, we could easily replace the word "conflict" with "travel" in Deutsch's study, "Travel prevents stagnation, it stimulates interest and curiosity", and we could also switch the words in other initial statements in the same article, "travel is the root of personal and social change" and "travel is often part of the process of testing and assessing oneself". It is, indeed, for many early modern thinkers, the method par excellence to gain knowledge about oneself and the world and to instigate change both in the self and in the world. In extreme cases, the change could be, as in the case of Rousseau's Émile, reaching the state when he realises that he does not wish to change. In making this point, Normand Doiron argues, Rousseau initiated a major cultural conflict by endeavouring to dismantle a Humanist culture that was predicated upon progress-based thinking.[24]

Rousseau was in a minority among his contemporaries in expressing such scepticism. In the late eighteenth and early nineteenth centuries, intellectuals such as Georg Wilheim Freidrich Hegel suggested that the expansion and development of ideas over time is one form of the narrative of human progress; in this process, conflict is both a necessary and a positive force. In *The Phenomenology of Spirit* (1807), Hegel posits that, rather than focusing upon the negative aspects of conflict, in looking at the wider context, we see how, without war and conflict, individuals cease to work for the common good and instead become weak and lazy. Conflict thus lies as the heart of the Hegelian vision of human history and is a catalyst for progress and change.[25] Previously, the Scottish Enlightenment thinker, Adam Ferguson, in his *Essay on the History of Civil Society* (1767), rejected both the notion that human development was a narrative of progress and also Rousseau's belief that humanity was born into a state of innocence that preceded the formation of civil society. Instead, Ferguson argued that humans are essentially social beings, but he also believed that humanity's proclivity to conflict could be a source for good:

> We applaud, as proceeding from a national or party spirit, what we could not endure as the effect of a private dislike; and, amidst the competitions of rival states, think we have found, for the patriot and the warrior, in the practice of violence and stratagem, the most illustrious career of human virtue.[26]

For Ferguson, conflict can be identity forming, a virtue, and of social benefit if carried out for a collective interest. In each of the chapters that follow, we see that conflict and travel are interwoven and instigate change.

Travel and conflict, in their connectedness, enable the acquisition of knowledge, not only of the wider world but also of the relationship between the self and the world. In the preface to an interdisciplinary edited collection on the "dynamisms" of conflict, Bernard-Marie Garreau tackled the many faces of conflict.[27] Here, he invites us to study conflict not as a series of antagonistic events; instead, he asks us to focus on how it is represented and functions and the efforts taken to mediate it. Conflict can be a positive force as well as a negative: it can be essential in the formation and definition of individual or collective identity, a "triumphant result of opposing forces".[28] Enacted exile, as illustrated by David Nicol's analysis of Thomas Drue's *The Duchess of Suffolk*, imagines how the disruption of forced mobility reaffirms the self and also the nation. As this present volume attests, conflict can define not only interpersonal relationships, but also intercultural perceptions – all of which we encounter repeatedly in accounts of, and about, travel.

Travel and conflict in the early modern world

The essays in this volume are arranged thematically and grouped into four sections to illuminate the relationship between travel and conflict: part one, "language, translation, and assimilation"; part two, "travel, religion, and the violence of the road"; part three, "war, diplomacy, and dissimulation"; part four, "the art of travel and imaginary journeys". Part one is concerned with how language affected communication and shaped identity and alterity, along with the role of the interpreter as conduit for communication and miscommunication. Part two addresses how religious difference acts as a motive and a *topos* in accounts of travel and interactions between people of different faiths, and part three explores how war, diplomacy, and exile each presented their own challenges to the traveller. Finally, in part four, we turn to the problems of imaginary travel and question how reading travelogues and instructions on how to travel might conflict with the knowledge and experiences of the non-traveller.

In the opening essay, Marie-Christine Gomez-Géraud discusses the role played by language in the development of conflicts in the accounts of two early encounters with the Other during the expansion of Europe's geographical horizon, the diary of Columbus's first trip and Jacques Cartier's *Brief recit* (1545). The amount of manipulation required to reduce linguistic barriers suggests how important language can be in revealing alterity and as an element of conflict. Both travellers had to abandon their early optimism regarding languages. Columbus had to realize that neither the Hebrew translator he brought nor his own deep feeling of instinctively understanding the utterances of the Native Americans can help him to avoid conflict and that there was not one "fixed" language spoken by Native American tribes. The more pragmatic Cartier was well-prepared with interpreters but naively held Pentecostal dreams of intercultural harmony and peaceful expansion for the glory of God and King. Drawing from these two texts and from theoretical discourses regarding the conditions of communicating with "newly discovered nations", Gomez-Géraud argues that language is the foremost form of alterity; attempts to avoid this difficulty systematically lead to undesired moments of conflict.

Language, as the form and mediator of cultural contact, could thus reveal as well as deceive. Céline Bonnotte-Hoover attends to the question of language learning and the role played by interpreters during the early Jesuit missions in China through a study of the account of the mission of Matteo Ricci. Among the difficulties faced by the Jesuits in the establishment of their mission in the country, issues of language were the most difficult to overcome: regional idioms, lack of teachers, and radical linguistic differences to European languages all hindered communication, which, in turn, impacted the success of the conversion mission. The Chinese authorities condemned fraternising with outsiders, which made interpreting dangerous work, but the role of these intercultural mediators was crucial for a variety of reasons ranging from language learning and acquisition of cultural codes to official interpreted communication. Bonnotte-Hoover traces the role of several (unnamed) interpreters who often helped – and sometimes hindered – the settling of missionaries. As the missionaries were severely restricted and eventually forbidden from learning Chinese, it was through the interpreters that Ricci and his fellow Jesuits acquired linguistic and cultural knowledge that enabled them to communicate more directly with the Chinese population.

Following this discussion of how relationships were established and maintained between the Chinese and the Jesuits at a time in which China remained almost hermetically closed off to foreigners, Natalya Din-Kariuki concludes this section by examining *Sir Thomas Smithes voiage and entertainment in Rushia* (1605), an account of Thomas Smythe's embassy to Russia in 1604 and the significance of rhetorical disposition (the "placing" or "ordering" of material) in travel narratives. William Scott accompanied Smythe on his journey and sent an account of it in a letter to Robert Cecil. By placing Smythe's *voiage* in conversation with Scott's account and his English translation of "Le Premier Jour" and "Le Second Jour" from Du Bartas' *La Sepmaine* (1578), this chapter examines the use of "accident" in travel writing. The chapter considers how the account narrates events during the Time of Troubles, a period of civil unrest in Russia, illustrating how Smythe's text employs some of the same rhetorical strategies found in

Scott's writings, especially the use of *peripeteia*, tragic "accident" or "reversal". The chapter also discusses the importance of the virtue of prudence in both travel writing and in rhetoric: the prudent orator, like the prudent traveller, knows how to navigate accidents. In this narration of conflict, Smythe emerges as a model of prudence, managing to navigate the "strange accidents" of the civil war that interrupts his embassy.

Smythe's *voiage* thus employs rhetoric to fashion the traveller, illustrating how ambassadors occupied a liminal space between home and abroad where utterances are communicated to different audiences. The interpreters studied by Gomez-Géraud and Bonnotte-Hoover also sit at the threshold between cultures and languages, functioning as a vehicle for communication and miscommunication. As all the chapters in the first part demonstrate, inter-linguistic mediation was necessary for communication while travelling (whether physically in person, or metaphorically in text) and could be mediated by the exigencies of religion. Part two turns to how travel answered but also inevitably provoked conflict when it came to matters of faith and how faith helped to negotiate conflict on the road. Faith was needed when navigating dangerous terrain, weather, and bandits. Joanne Anderson examines the visual culture of the early modern road through exploring the *Viamala* ("the bad way"), one of the main routes used by people as they journeyed through the Rhaetian Alps in the fourteenth and fifteenth centuries. Crossing the Alps required planning, resources, and courage to cope with the often-treacherous terrain, pathways, and climate, but dangers could also take on human form: armies travelled the roads, and banditry was common. The Alps, therefore, were a bustling space, where violence could be spontaneous or coordinated; in this context, faith, expressed through art, guarded and fortified the *Viamala*. As Anderson illustrates, on the walls of churches, palaces, fortresses, and town gates, saints stand watch and guard against the threat posed by outlaws and strangers. In so doing, art mediates this culture of itinerant violence through projecting sacred archetypes of martial life. What emerges is that art is an active agent in the Alps, where Alpine visual culture interacts with the landscape of travel and conflict.

Visual culture highlights the conflicts presented to the traveller in the untamed landscape or in uncivilised lands. In late sixteenth-century Jesuit accounts of missions, Maronite Catholics in Lebanon are applauded for maintaining their faith despite being isolated from their European co-religionists, but these travelogues also lament Maronite acculturation with Islam. Robert Clines examines these cross-cultural conflicts that impeded attempts to restore Catholicism as the universal faith by focusing upon accounts by the Jesuit missionaries Girolamo Dandini and Diego de Salazar. Both laud the Maronite Catholics of Lebanon for appearing to be a remnant of the hermits of late antiquity, yet they are also troubled by how the Maronites deviated from orthodoxy and orthopraxy. Dandini and Salazar used these communities' environment and ritual practice as a means of demonstrating that Catholic renewal hinged on believers' active and outward participation in the return of the church to its Late Antique origins as well as their rejection of the cultural excesses of heresy and schism. These texts emphasise the material life of the Maronites as being both a relic of the Christian past and evidence of cultural deterioration due to cross-cultural exchange. Attempts to restore Catholicism as the universal faith, Clines contends, were thus informed by Europeans' travel experiences, conflicts with non-Catholics, and their attempts to grapple with the religious and cultural spectrum of the early modern world.

Christian practices in the Levant were both a source of comfort and of concern for early modern Jesuits: the dangers menacing those communities are mirrors of dangers perceived in post-Tridentine Europe. Inversely, Marian Catholicism proves to be the catalyst to travel in Thomas Drue's *The Duchess of Suffolk* (1624). Set in the 1550s, Drue's play dramatises the journey of the devout Protestant duchess, Catherine Willoughby, as she is pursued across Europe by Catholic persecutors, but the drama has been read as a political allegory on the plight of Elizabeth Stuart (Queen of Bohemia and daughter to King James VI of Scotland and I of England). Elizabeth and her family were exiled following the Habsburg invasion of Bohemia in 1620: this was one of the events that precipitated the Thirty Years War, which ravaged much of Europe. At the time of the play's performance, militant Protestant pamphleteers protested against James's pacific response to his daughter's plight and agitated for military participation in what they saw as an apocalyptic war against Catholicism. Challenging scholarship that assumes *The Duchess of Suffolk* concurs with these pamphlets, David Nicol argues that, although the play evokes sympathy for Elizabeth, it does not favour war and expresses a distaste for violence at every episode. By attending to the journeying play genre, Nicol reveals an underlying structure in which violence is questioned. Sitting at the threshold of radical militancy and loyalist pacifism, Drue thus utilises the trope of travel to reject violent conflict.

Following on from this balancing of militant Protestantism with pacifism, part three explores travelling for diplomacy, both for the state and the prince, to hone one's skills as a courtier and to enhance one's social status. Paul Dover examines fifteenth-century ambassadorial dispatches to show how, for all its disruptions, travel was always fundamental to diplomacy and transformed Italian diplomatic praxis. The key development in the history of Italian diplomacy in the fifteenth century was the emergence of residency in diplomatic representation. This had significant repercussions for the everyday experiences of the ambassador, which embraced both long-term residence and extraordinary embassies with specific, time-sensitive political tasks. Travel also engaged many recurring concerns of ambassadorial service: remuneration, lodging, the size and maintenance of a staff, courtly protocol and precedence, and the care of horses. It simultaneously compromised and enhanced the demands of information gathering. Complaints about the burdens of travel were legion, especially from republican patricians, who felt keenly the financial and social costs of extended absence from their metropole. The broadly shared expectation that fifteenth-century ambassadors would write and send dispatches nearly every day provides us with largely unmediated and nearly real-time accounts of travel experiences. Dover mines this vast and vivid epistolary record to discuss the history of mobility intertwined with that of diplomatic practices and epistolary culture.

Italian ambassadorial dispatches formed part of a wider political epistolary culture, and letter-writing and diplomacy are also interconnected in the actions of Pietro Aretino. William Rossiter attends to how Aretino married personal ambition and diplomacy with satire, to study the impact of travel and conflict on Aretino's writing. In 1525, Aretino survived an assassination attempt ordered by the papal datary in the aftermath of the Battle of Pavia – a battle which saw the Holy Roman Emperor, Charles V, take Francis I of France prisoner. These events each led to dislocation: Francis I was taken to the Certosa of Pavia and then to Madrid, where he was held until the following year; Aretino fled Rome for Venice, where he became a prominent figure of vernacular print culture. Immediately after the battle, Francis penned a public letter to his mother lamenting his misfortune, and another letter to Charles V, proposing clemency. This chapter argues that Aretino's own public letter to Francis (published 1538) was triggered by Francis's letter to his mother and is designed to further Aretino's ambitions as a non-official public diplomat. This form of political intervention is strikingly different from, but also complements, Aretino's viciously satirical "pasquinades" written in Rome, and both types of texts had a role in their author being repeatedly stabbed and left for dead. What emerges is that conflict begets travel, and both beget a reflective textual culture.

Aretino's flight was spurred on by a desire to preserve life, even though travel was often risky and life-threatening. Eva Johanna Holmberg shows how Henry Blount's travelogue exemplifies the way in which he utilised the potentially hostile space of the Levant to hone and display his resourcefulness. Early modern travel advice books usually gave some general advice on how to avoid conflict during travel; Holmberg questions whether travellers put these to practice and whether the same rules of comportment were applicable in every setting. Many cultural practices in the Ottoman Empire had the potential for creating conflict and misunderstandings; they were used by authors of travel literature as modes of self-fashioning and as indicators of their vulnerable status as Western Christians in "the Islamic world". In her chapter, Holmberg explores the writing of the self in Henry Blount's *A Voyage into the Levant* (1636), examining how Blount represents Turkish civilisation and the "customs of the country". Holmberg contends that the many conflicts and "travel dangers" were employed to advertise the travellers' wit and resourcefulness. The Ottoman Turks, their customs, and the dangerous situations presented by travel thus functioned as backdrops for a demonstration of Blount's skills: his representation of the Turk largely conforms to early modern European ideas about Ottoman ethnoreligious make-up and, through his ability to dissimulate, Blount is revealed as possessing the situational awareness useful for an intelligencer about town or the capacities required of an able courtier.

Blount's travels in the Levant equipped him with the skills required of a courtier: upon his return to England, he entered service at the court of Charles I and then received a knighthood in 1639. In 1611, Frances Cromwell hoped that her son may gain knowledge of himself through travel, and Blount shows that learning about the self and acquiring skills through ethnoreligious encounters could lead to preferment at court. Cromwell's interjection is an example of how women shaped and influenced travel, even as they were forced to remain mostly static; educational travel and travels of curiosity, the forms of mobility most readily associated with the notion of "travel", were essentially forbidden to them. Although women have been largely written out of the history of early modern travel, and perhaps especially from the *ars apodemica* ("art of travel") genre, part four on the art of travel and imaginary journeys opens with Gábor Gelléri's discussion of women's role in this genre. In the majority of cases, women appear in *ars apodemica* as individuals who are excluded from travels and occasionally as (serious or futile) topics of observation for the male traveller to whom instructions are directed. In this chapter, Gelléri examines the writing of three women – all established authors – who ventured into this genre: Margaret Cavendish, Julie de Giovane, and Mme de Genlis. In attending to their texts as well as cases where the private "instructions of a father to a son" (presented to sons before they embarked upon the Grand Tour) happen to be motherly instructions, we can see how women shaped the epistemology of travel. This chapter also addresses the only early modern text that, to our knowledge, is written directly for and about the female traveller: Mme de Genlis's educational novel *Adelaide and Theodore* (1781). Written as an answer to Rousseau's male-centred *Emile* (1762), an ideal travelling practice for children of both sexes is modelled. Collectively, these women writers show us how women engaged with travel as a scientific, educational, recreational, literary, and political activity.

By examining *ars apodemica* ("advice to a son" literature) and early pedagogical fiction by women, Gelléri shows how women, whose physical mobility was systematically perceived as problematic, engaged with travel as an intellectual thought process. Conversely, utopian fiction offers a means to reflect upon society through observations on an imaginary alien culture. Daniel Carey investigates travel as an enabling condition of utopia and considers the implications of the interaction between travel writing and utopian literature over a two-century period. He argues that utopian no-places enabled radical critiques of home nations, in sharp contrast to the anxiety to defend homelands which is often present in travel writing. From the inception of the utopian form in 1516 by Thomas More, travel has played an integral part in the structure and development of fictions, providing radically alternative political visions. Carey highlights that utopian fiction relies on a displacement of conflict: from the real or imaginary abroad, the place of conflict will be moved back to the *domus*. Whether consolatory or ironic, these visions present a different way of living; next to a dimension of wish fulfilment, they also carry within them their own implicit critique. No traveller can leave their own identity at home – and this ironic dimension of utopia highlights that, while humans might dream up worlds of perfection (u-topias as eu-topias), we are inherently unable to experience these utopian worlds.

Carey addresses how fantastical travel can be a site of conflict; in the final chapter of the volume, Rachel Willie explores how reading fantastical travel writing created conflict. Aphra Behn's 1687 play, *The Emperor of the Moon*, presents a protagonist driven insane from reading lunar travel narratives. This essay addresses the relationship between the transformative effects of reading, staged spectacle, imaginary journeys, and fabricated travel in this farce. The protagonist, Dr Baliardo, has his senses corrupted by reading and believes the moon houses a utopian society. The suitors to Baliardo's daughter and niece impersonate lunar nobility to win the doctor's consent to marry. This text connects reading, travelling, and deception with epistemology. Fabricated travel becomes a way to engage with natural philosophy as the ontological barriers between real and imagined space grow fractured and become a site of conflict. The play highlights the plurality and mutability of words and the senses: words deceive Baliardo into believing that there is a lunar society, and optical illusions trick him into believing he is conversing with noble extraterrestrials. Sensory and empirical knowledge are called to question as the material and the imagined converge. Behn presents utopian travel as a futile no-place that can only ever deceive. Armchair travel thus leads to conflict between what is imagined and what is real; this conflict can only be resolved by enacting the absurdities of the texts that are read.

What emerges from these essays is that the physical and material conflict of the road is occasionally emphasised in both actual and fabricated accounts of travel to enhance the value of the heroic traveller, but more often it is implicitly present in the very act of movement. When conflict is a reason for movement, the movement initiated or suffered could lead to a variety of consequences, from the tortured displacement of exile to fruitful cultural transfer. Finally, conflict could be exposed, and occasionally resolved, by the practices of real and imagined mobility.

Notes

1 A partial translation of the fragment can be found in James Henry Breasted, "The Report of Wenamon," *The American Journal of Semitic Languages and Literatures* 21 (1905): 100–09.

2 See Benjamin Sass, "Wenamun and his Levant—1075 BC or 925 BC?" *Ägypten und Levante/Egypt and the Levant* 12 (2002): 247–55.

3 Steve Vinson, "The Accent's on Evil: Ancient Egyptian 'Melodrama' and the Problem of Genre," *Journal of the American Research Center in Egypt* 41 (2004): 33–54 (45).

4 Andrew Gordon and Bernhard Klein, eds., *Literature, Mapping and the Politics of Space in Early Modern Britain* (Cambridge: Cambridge University Press, 2001), 1.
5 Surekha Davies, *Renaissance Ethnology and the Invention of the Human* (Cambridge: Cambridge University Press, 2016).
6 William Cunningham, *The Cosmographical Glasse* (London, 1559), A6r.
7 The manuscript is held today in Pembroke College, Cambridge; it features entries from 1574 to 1596, and was published by Jean Puraye: *Album amicorum Abraham Ortelius* (Amsterdam: A. L. van Gendt & co, 1969).
8 *The Geography of Strabo*, book 1, 2, 23, trans. Horace Leonard Jones. 8 vols. (London: Heinemann, 1917–1933) 1:110–11.
9 For a survey of these entries, including the fate of this Strabo quotation, see Gábor Gelléri, *Lessons of Travel in Eighteenth-Century France: From Grand Tour to Schol Trips* (Martlesham: Boydell and Brewer, 2020), 37–45.
10 On Voltaire's gradual moving away from travel writing through his various texts about England, see Gábor Gelléri, *Philosophies du voyage: visiter l'Angleterre aux 17e-18e siècles* (Oxford: Voltaire Foundation, 2016), chapter 5. Gelléri also cites other authors, such as abbé Morellet and Brissot de Warville, who explicitly avoided travelogues as not sufficiently serious genre.
11 See, for example, Daniel Carey, "Travel, Geography and the Problem of Belief: Locke as a Reader of Travel Literature," in *History and Nation*, ed. Julia Rudolph (Lewisburg: Bucknell University Press, 2006), 97–136.
12 As Georges Pire has shown, despite Rousseau's claim to have spent extensive time reading a vast array of travel literature with little benefits, he mostly knew this literature through the mediation of abbé Prévost's popular collection of travelogues, the *Histoire universelle des voyages* (published from 1746). See Georges Pire, "Jean-Jacques Rousseau et les relations de voyage," *Revue d'Histoire Littéraire de France* 56 (1956): 353–78.
13 The *Lettre sur les Voiages* was published with Muralt's *Lettres sur les Anglois et les François*, which saw numerous publications starting from 1725.
14 See p. 266 in this volume.
15 See Philippe Lejeune, *On Autobiography*, trans. Katherine Leary (Minneapolis: University of Minnesota Press, 1989). On the idea of a "reading pact" in travelogues, see for example Réal Ouellet, "Le statut du réel dans la relation de voyage," *Littératures classiques* 11 (1989): 257–72.
16 Frances Cromwell, Lady Cromwell. "Lady Frances Cromwell to Salisbury." October 8, 1611. MS Secretaries of State: State Papers Domestic, James I, 1603–1640. SP 14/66 f.118. Frances Rugge or Repps (d. before 1631) was the second wife of Edward Cromwell, later wife of Richard Wingfield of Powerscourt. The child for whom the request to travel is formulated in 1611 must be Thomas Cromwell, then aged 16–17, great-grandson and namesake of Henry VIII's Lord Chancellor.
17 *Ibid.*
18 On the dual presence of *conquisitioinquisitio* during travel, see Juliette Morice, *Le monde ou la bibliothèque: voyage et education à l'âge Classique* (Paris: Les Belles-Lettres, 2016), 77 sqq.
19 "Letter from W. Stevenson, tutor to Lord Dorset's youngest son Thomas and to Sir Charles Bickerstaffe's son" (1674–1675), Kent Archive & Library Centre, U269/C27.
20 The debate on the "ideal age" for a foreign travel experience was omnipresent in *ars apodemica* treatises on the art of travel. See Daniel Carey, "Advice on the Art of Travel", *The Cambridge History of Travel Writing*, ed. Nandini Das and

Tim Youngs (Cambridge: Cambridge University Press, 2018), 392–407 (397); more specifically on Montaigne's view on youthful trips in opposition to those of Rousseau who suggested travelling in the early twenties, see Juliette Morice, "Voyage et anthropologie dans l'*Emile* de Rousseau," *Revue de métaphysique et de morale* 77 (2013): 127–42 (138).

21 Sarah Goldsmith, Danger, Risk-taking and Masculinity on the British Grand Tour to the European Continent, c. 1730–1780. Unpublished PhD dissertation, University of York, 2015.

22 Morton Deutsch, "Conflict Resolution: the Basis for Social Control or Social Change?" in *Conflict management and problem solving: interpersonal to international perspectives*, ed. Dennis J. D. Sandole and Ingrid Sandole-Staroste (New York and London: New York University Press, 1987), 38–49 (38).

23 Eusèbe Renaudot ed., *Recueil général des questions traitées dans les conférences du bureau d'adresse, sur toutes sortes de matières, par les plus beaux esprits de ce temps* (Paris, 1655–1656). Dialogue 87, on the demonstration of the immortality of the soul and on travels, starts 2:408, the discussion on travels features 417–23. For a reading of this debate, see Gelléri, *Lessons of Travel*, 31–33.

24 Normand Doiron, *L'art de voyager: le déplacement à l'époque classique* (Sainte-Foy and Paris: Presses Universitaires de Laval & Klincksieck, 1995), 187–201.

25 See Georg Wilheim Freidrich Hegel, *The Phenomenology of Spirit*, ed. and trans. Terry Pinkard (Cambridge: Cambridge University Press, 2018).

26 Adam Ferguson, *An Essay on the History of Civil Society*, 5th ed. (London, 1782), 40.

27 Bernard-Marie Garreau ed., *Dynamiques du conflit. Actes du colloque du Centre de recherche en littératures, linguistique et civilisation CRELLIC, Lorient, 20, 21 et 22 novembre 2002* (Lorient: Université de Bretagne-Sud, 2003), Introduction.

28 *Ibid.*, 12.

Bibliography

Breasted, James Henry. "The Report of Wenamon." *The American Journal of Semitic Languages and Literatures 21* (1905): 100–09.

Carey, Daniel. "Travel, Geography and the Problem of Belief: Locke as a Reader of Travel Literature." In *History and Nation*, edited by Julia Rudolph, 97–136. Lewisburg: Bucknell University Press, 2006.

Carey, Daniel. "Advice on the Art of Travel." In *The Cambridge History of Travel Writing*, edited by Nandini Das and Tim Youngs, 392–407. Cambridge: Cambridge University Press, 2018.

Cromwell, Frances. "*Lady Frances Cromwell to Salisbury.*" October 8, 1611. MS Secretaries of State: State Papers Domestic James I, 1603–1640. SP 14/66 f.118.

Cunningham, William. *The Cosmographical Glasse*. London, 1559.

Davies, Surekha. *Renaissance Ethnology and the Invention of the Human*. Cambridge: Cambridge University Press, 2016.

Deutsch, Morton. "Conflict Resolution: The Basis for Social Control or Social Change?" In *Conflict Management and Problem Solving: Interpersonal to International Perspectives*, edited by Dennis J. D. Sandole and Ingrid Sandole-Staroste, 38–49. New York and London: New York University Press, 1987.

Doiron, Normand. *L'art de voyager: le déplacement à l'époque classique*. Sainte-Foy and Paris: Presses Universitaires de Laval & Klincksieck, 1995.

Ferguson, Adam. *An Essay on the History of Civil Society*, 5th ed. London, 1782.

Garreau, Bernard-Marie ed. *Dynamiques du conflit. Actes du colloque du Centre de recherche en littératures, linguistique et civilisation CRELLIC, Lorient, 20, 21 et 22 novembre 2002*. Lorient: Université de Bretagne-Sud, 2003.

Gelléri, Gábor. *Philosophies du voyage: visiter l'Angleterre aux 17e-18e siècles*. Oxford: Voltaire Foundation, 2016.

Gelléri, Gábor. *Lessons of Travel in Eighteenth-Century France: From Grand Tour to School Trips*. Martlesham: Boydell and Brewer, 2020.

Goldsmith, Sarah. *Danger, Risk-taking and Masculinity on the British Grand Tour to the European Continent, c. 1730–1780*. Unpublished PhD dissertation, University of York, 2015.

Gordon, Andrew and Bernhard Klein eds., *Literature, Mapping and the Politics of Space in Early Modern Britain*. Cambridge: Cambridge University Press, 2001.

Hegel, Georg Wilheim Freidrich. *The Phenomenology of Spirit*, edited and translated by Terry Pinkard. Cambridge: Cambridge University Press, 2018

Jones, Horace Leonard, ed. and trans. *The Geography of Strabo* , 8 vols.London: Heinemann & Harvard, 1917–1933.

Lejeune, Philippe. *On Autobiography*. Translated by Katherine, Leary. Minneapolis: University of Minnesota Press, 1989.

Morice, Juliette. "Voyage et anthropologie dans l'*Emile* de Rousseau". *Revue de métaphysique et de morale* 77 (2013): 127–42.

Morice, Juliette. *Le monde ou la bibliothèque: voyage et education à l'âge classique*. Paris: Les Belles-Lettres, 2016.

Muralt, Béat Louis *de, Lettres sur les Anglois et les François et sur les Voiages*. Geneva: Fabri et Barillot, 1725.

Ouellet, Réal. "Le statut du réel dans la relation de voyage". *Littératures classiques 11* (1989): 257–72.

Pire, Georges. "Jean-Jacques Rousseau et les relations de voyage". *Revue d'Histoire Littéraire de France*, *56* (1956): 353–78.

Puraye, Jean, ed. *Album amicorum Abraham Ortelius*. Amsterdam: A. L. van Gendt & co, 1969.

Renaudot, Eusèbe ed. *Recueil général des questions traitées dans les conférences du bureau d'adresse, sur toutes sortes de matières, par les plus beaux esprits de ce temps*. Paris, 1655–1656.

Sass, Benjamin. "Wenamun and his Levant—1075 BC or 925 BC?" *Ägypten und Levante/Egypt and the Levant* 12 (2002): 247–55.

Stevenson, W. *"Letter from W. Stevenson, tutor to Lord Dorset's youngest son Thomas and to Sir Charles Bickerstaffe's son"*, Kent Archive & Library Centre, U269/C27. 1674–1675.

Vinson, Steve. "The Accent's on Evil: Ancient Egyptian 'Melodrama' and the Problem of Genre," *Journal of the American Research Center in Egypt 41* (2004): 33–54.

Part I

Language, translation, and assimilation

1 Babel as a source of conflict

A case study of two discovery narratives

Marie-Christine Gomez-Géraud

Scholarship in the field of social attitudes in an intercultural context should encourage us to reread travel literature, and especially discovery narratives, with a new approach. French ethnologist Raymonde Carroll casts an interesting light on the phenomenon.[1] In her book, she grounds her analysis on an observation of daily interactions between French and American individuals, the better to highlight "zones of turbulence" around "implicit assumptions that govern our behaviour in spite of ourselves, whatever group we identify with and feel that we belong to".[2] One of the most interesting elements in Carroll's research is that her investigation deals with individuals whose cultures are very similar (Western culture) and whose linguistic communication is not hindered by obstacles of any sort. The notion she explores, "the invisible obvious" ("évidences invisibles" in the very perceptive French translation), covers all those unsuspected, implicit assumptions that can give rise to incomprehension, misunderstandings, and sometimes even conflicts.[3] In her interpretation, "It is possible that these situations [of cultural misunderstandings] were provoked by the strangeness of being in a foreign land, by a certain insecurity or ignorance due to the context".[4]

If even an encounter between French and American individuals at the end of the twentieth century is rife with the potential for cultural misunderstandings, thereby making these encounters an excellent research terrain to highlight these situations, we can easily imagine how many questions, conflicts and misunderstandings of all kinds must have arisen from the encounter between European explorers and the inhabitants of faraway lands at the dawn of the modern age. In these encounters, two worlds meet that could never have suspected that the other even existed. What can we know about these exchanges, beyond what the explorers themselves perceived and reported?

Travel literature, especially the kind produced by seamen eager to successfully complete the task assigned to them – extending the span of *terrae cognitae* on which they can plant the banner of their prince and oftentimes financial backer – can only be considered as a discourse whose first aim is to signify how Europeans represent the world and the way they relate to other cultures when they tell the tale of their first encounters with Native Americans.[5]

While the difficulty or downright impossibility of linguistic communication appears to be the deciding factor in how these relationships were established, the notion of "cultural misunderstanding" can perhaps help us give a more precise account of how language and what is beyond language interact, during the genesis of antagonistic processes that can lead to conflict. Revisiting two canonical "first encounter" narratives, one from the diary of Columbus's first journey, the other from Jacques Cartier's second Canada journey, will be our way of exploring this question.

Transparency: the most elementary form of conflict?

On Friday, 12 October 1492, Columbus and his crew catch a first glimpse of a luxuriant and pleasant land, and of the "gente desnuda" ("naked people") from the American continent.[6] It seems that this first contact with the Native Americans does not spark any doubt or questioning in the admiral. But Columbus is not discovering America: it is his dream he is discovering, a dream nurtured with long study of medieval sources, his gaze firmly trained on Behaim's world map. In his mind, he is setting sail for the Far East and the Garden of Eden; perhaps he thinks of himself as a "herald of the Apocalypse", entrusted with a divine mission.[7]

He is so sure of his objective that he has brought with him Luis de Torres, "que avía bivido con el Adelantado de Murcia y avía sido judío, y sabía diz que ebraico y caldeo y aun algo arávigo".[8] Torres will be the interpreter on the journey and, on these roads that should lead them not very far from earthly paradise, he will be able to put his knowledge of Hebrew to good use. Claude-Gilbert Dubois, considering the Renaissance period, points out how Hebrew enjoys a sort of "first-born right" among languages. In this view, the language spoken in the Garden of Eden must have been very close to Hebrew, and so Luis de Torres should be able to understand those who live so close to Eden.[9]

Columbus's project functions as an epistemological framework within which to make sense of the new. In the words of Francisco Morales Padrón, "this false, idealised and paradisiacal geography was Asia, for Colón, and he eagerly tried to convince the kings of Castile that the territories he had discovered were those Marco Polo was talking about".[10] Columbus is sailing towards the Orient, but he is sailing just as much towards a world that is already known, a self-evident world of foregone conclusions, whose meaning has already been surveyed, cross-examined and established by the great texts from antiquity and medieval science.[11]

From this point of view, the diary of the first journey is a constant surprise for a modern reader. Historiography has invited us to think about intercultural encounters in terms of "novelty", of "singularity", that is, in terms of otherness. But for the admiral, everything seems to be transparent and self-explanatory. He lands in the Indies; he talks about the famed land almost as if it was the garden next door. No matter; he knows what he wants, to annex the lands reached for the benefit of the Catholic kings, and his narrative subjects the whole reality he perceives to that programme. How he experiences language is very telling in this respect: Columbus does not speak the language of the Native Americans he meets on the Caribbean islands he reaches, but he *understands* what they say. Even complex messages, transmitted through signs, a code that is erroneously thought to be universal, are interpreted with no hesitation. Thus, on 13 October 1492, "Y por señas pude entender que, yendo al Sur o bolviendo la isla por el Sur, que estava allí un Rey que tenía grandes vases de oro y tenía muy mucho".[12]

In another exchange the next day, he reports that "entendiamos que nos preguntavan si éramos venido[s] del cielo".[13] The utterance itself, the words of the other party, matter less here than its interpretation ("entendíamos"/ "we understood"), which justifies months of uncertain and difficult sailing that has been bitterly contested by the crew.[14] In a troubling example of his rashness, Columbus hears and simultaneously translates the words of an old man and some other Native Americans, who call out to their tribesmen to join them nearer the boat, "vino uno viejo en el batel dentro, y otros a grandes bozes llamavan todos, hombres y mugeres: 'Venid a ver los hombres que vinieron del çielo, traedles de comer y de bever'".[15] In the manner of the omniscient novelist, the admiral literally and immediately takes up the words of the Other and ventriloquises them as direct speech. Columbus thus becomes the voice of the Native American, so to speak. The same appropriating attitude governs the interpretation he gives of the landscape, whose constituting elements immediately give rise to a translation in terms of commercial value. Having reached Punta de Mangle on the north coast of Cuba on 25 November, he converts into gold a river in which he sees stones glimmering; in the same way, amazingly tall trees are converted into ships, and other stones he sees on the beach soon become iron and silver.

It is a few more days before the admiral finally concedes that his knowledge is limited, as is the communication that was meant to flow in absolute transparency, "no sé la lengua, y la gente de estas tierras no me entienden, ni yo ni otro que yo tenga a ellos y estos indios que yo traigo, muchas vezes le entiendo una cosa por otra al contrario; ni fio mucho d'ellos, porque muchas vezes an probado a fugir", he confesses a little while later.[16] This is the first time he acknowledges his linguistic incompetence, made manifest in an addendum to the *Diary*. In the text, a connection is established between the failure to take possession – as evidenced in the attitude of Native Americans who eschew his authority and do not collaborate in his project – and the failure to communicate. To utter is not enough in the end: it will be necessary to go through language, to explain and to convince. Even if in a limited way, the admiral suddenly becomes aware of two things: the tangible realities of languages and the limits to his own power. At the same time that he becomes aware of alterity, Columbus catches a glimpse of the potential for a clash of wills; he has to renounce his fantasy of being all-powerful.

Even though his dream proves to be more powerful than the teachings of this experience, the same question crops up again on 3 December, on the north coast of Cuba. This failed encounter plays out in three steps. The setting is a true paradise, the earth so fertile that "que era gloria vella",[17] and his appraising gaze also rests on indigenous canoes as he notices their extremely clever design. In a village, the meeting with the local tribes is made easier at first by the presence of another Native American brought by Columbus. This marks the end of Act 1: suddenly the Native Americans are scared, perhaps the sign of brewing conflict, but that is not elucidated in the narrative. However, the *Diary* adds that Columbus manages to seize the crude weapons of the men, through a ruse, "varas [que tienen] al cabo d'ellas un palillo agudo tostado", by exchanging them against small gifts.[18] Is this what precipitates the last act of a meeting that might end in conflict? The admiral is already back in the rowboat when a Native American approaches:

> Hizo una grande plática que el Almirante no entendía, salvo que los otros indios de cuando en cuando alçavan las manos al çielo y davan una grande boz. Pensava el Almirante que los aseguravan y que les plazía de su venida, pero vido al indio que consigo traía demudarse la cara y amarillo como la çera, y temblaba mucho, diziendo por señas que el Almirante se fuese fuera del río, que los querían matar, y llegóse a un cristiano que tenía una ballesta armada y mostróla a los indios; y entendió el Almirante que le dezía que los matarían todos, porque aquella ballesta tirava lexos y matava; también tomo una espada y la sacó de la vaina mostrándosela, diziendo lo mismo. Lo cual oído por ellos, dieron todos a huir, quedando todavía temblando el dicho indio de cobardía y poco coraçón.[19]

Michel Mollat, in his analysis of this passage, emphasises the ambivalence of body language, which is just as difficult to translate as the language itself.[20] Tzvetan Todorov comments on the "failures" of this encounter, based on the linguistic and cultural inability to understand, but he also points out that "Columbus does not succeed in communications because he is not interested in them".[21] While the admiral, as usual, projects his own desires onto the attitudes and discourse of the Native Americans, this episode is also interesting for other reasons.[22] It makes the Columbine ethos, as constructed in the *Diary*, very clear: Columbus sees himself as sole master and decider, and intends to remain in this role even when death clearly threatens. As master of the signs, Columbus knows how to interpret the Native American's sudden pallor, and, master of his emotions, he remains unperturbed. There is an important new element here: for the first time, weapons are mentioned in the narrative, thus hinting at power struggles between the Native Americans and their conquerors. However, these weapons are diverted from their primary function (hurting, killing) and, at this stage, they are still objects to be exchanged, or the signifier of the power of the newly arrived conquerors; this is enough to prevent violence from erupting.

This constant misunderstanding is not a source of conflict yet: it would be possible to refer to a "missed connection" as much as a "missed conflict". Moreover, this scene remains an isolated episode in the *continuum* of discovery. After that, the same scenes of peaceful welcome aided by the trading of gifts continue in the same vein. After the *Santa Maria* runs aground on Hispaniola on Christmas Eve, the Native Americans help the sailors transport foodstuffs out of the hold. On the beach, bits of gold are traded against European artifacts again and again. This situation that we could call – adapting Roland Barthes's concept of "writing degree zero" – a "violence degree zero" comes to an abrupt end as soon as the *Carib*, "los de Caniba", are mentioned.[23] In his letter to Santangel, written after his return in February 1493, Columbus describes them as "muy ferozes los cualles comen carne umana" ("as very fierce and who eat human flesh").[24] He also mentions their weapons ("arcos y flechas", "bows and arrows") and the first project of a strategic alliance against the *Carib*, offered to one of the tribes he met. Over time, he becomes more adept at reading human groups and how they interact, and this gives rise to a perception of contexts in which violence could potentially develop:

> El Almirante le dixo [al jefe indio] por señas que los Reyes de Castilla mandarían destruir a los caribes y que a todos se los mandarían traer las manos atadas. Mandó el Almirante tirar una bombarda y una espingarda, y viendo el effecto que su fuerça hacían y lo que penetraban, quedó maravillado.[25]

More than the linguistic sign, which is often neglected, more than the symbolism of welcome and exchange rituals, the deafening report of the gun becomes the loudest voice and leaves the other "maravillado", speechless and amazed. At this stage, violence is only represented as a virtuality, a sheer demonstration of a power that is only biding its time. But the harm is done: by parading firearms, a stratagem often used by explorers in subsequent encounters to both stupefy and reassure indigenous tribes, those taking part in the discovery of the New World irremediably change the parameters of local conflicts by changing the traditional balance of existing relationships between the different human groups in the area.

The sailors continue to meet with Native Americans who have no previous experience of firearms, a fact that Columbus, in his dream quest for hegemony, always scrupulously notes. However, the first skirmish happens on 13 January, as Columbus and his crew start on their way back to Spain. It is immortalised in the name of the place where it happened: the Gulf of

Arrows.[26] It is the only time, during this first journey to the West Indies, that blood is drawn. It all starts on a beautiful beach, "hermosa playa", where an exchange with an unknown human group takes place. The Spanish explorers buy bows and arrows from the Native Americans and then persuade them to come aboard the caravel. For the first time, the Native Americans are described as "muy disforme en el acatadura",[27] as physically misshapen, their faces blackened with soot, their hair held up with a strange feather headdress.[28] This doesn't bode well, "juzgó el Almirante que debía de ser de los Caribes que comen los hombres".[29] The admiral enquires after gold as he is wont to do, but this time the frequent transparency of linguistic exchanges vanishes, for the Native American, "llamava al oro 'tuob' y no entendía por 'caona', como le llaman en la primera parte de la isla, ni por 'noçay', como lo nombravan en San Salvador y en las otras islas. Al alambre o a un oro baxo llaman en la Española 'tuob'".[30] What is this supposed *Carib* talking about? Of gold or of copper? And in what language? Suddenly Columbus reaches a new stage in his linguistic awareness, "entendía algunas palabras, y por ellas diz que saca otras cosas, y que los indios que consigo traía entendían más, puesto que fallava diferencia de lenguas por la gran distancia de las tierras".[31] He is forced to acknowledge that languages are several, an inconvenient reality he has long resisted because it puts a serious obstacle to his project of hegemony. He also admits that he doesn't know everything and that, on their own turf, the Native Americans know more than he does. The exchange then runs the familiar course of gifts of food and artifacts.

What follows, however, seems to be a repeat of 3 December. The Native American who followed them on board returns to the shore, having been ordered to bring back gold; meanwhile, Columbus's crew are trying to buy the weapons from the Native Americans, who feign to consent to it. But the script suddenly takes a new turn:

> tornaron con cuerdas en las manos para diz que atar los cristianos… arremetieron los cristianos a ellos, y dieron a un indio una gran cuchillada en las nalgas, y a otro por los pechos hirieron con una saetada; a lo cual visto que podían ganar poco, aunque no eran los cristianos sino siete y ellos cincuenta y tantos, dieron a huir que nos quedó ninguno, dexando uno aquí las flechas y otro allí los arcos.[32]

While the skirmish remains harmless, it signals the moment when the Native Americans rebel against these men who came from elsewhere and made them flee. It doesn't matter that they had very little to gain ("a lo cual visto que podian ganar poco"): for Columbus, the sky has darkened significantly. This realization that languages are a very real and impenetrable thicket; the dawning awareness of his ignorance; the clash of wills; the renunciation of his dream of being all-powerful, as signified by the Native Americans' flight; the unsuspected complexity of this newly discovered world: all these elements concur to darken the golden mirror in which the admiral had taken to admiring the world, superimposed with his own likeness. Conflict here becomes the experience that gives access to a new awareness of the world as world, and not as myth.

Impossible transparency: conflict through language

Some forty years later, Captain Cartier sets sail towards a world where he, too, expects to resolve conflict through myth. As I have discussed elsewhere, Cartier experiences a Babel of languages and clings to a Pentecostal dream.[33] In Hochelaga, a remote territory at the close of his journey, Jacques Cartier describes a world where, according to his description, concord and peace reign undisturbed among Native Americans who seem to have been deeply moved by the message of the Gospel that he read out to them, probably in Latin – not that it makes much of a difference to them. Just like Columbus's, Cartier's world is geocentric, conceptually unified, guaranteed by theology. It becomes even clearer in the *Relations* of 1534 and 1535: discovery and history progress hand in hand. In Cartier's eyes, the axis of history is Christian faith: following the course of the sun, it spreads its light from Orient to the extremity of Occident. In the Americas, this natural process finds its natural endpoint:

> Notre saincte foy a esté semee et plantee en la Terre saincte qui est en Asie à l'orient de notre Europpe et despuis par succession de temps apportee et divulguee jusques à nous et finablement en l'occidant de notre dicte Europpe à l'exemple dudict soleil portant sa clarte et chaleur de orient en occident comme dict est.[34]

Discovery thus slots into an order willed by God and guaranteed by the Christian prince whom Cartier serves. This is the backdrop against which the captain's whole mental universe unfolds. Any violence, any conflict will thus necessarily be interpreted like an obstacle put in God's way, a step backwards compared to History's salvation dynamics. That is why they will be immediately minimised in the narrative, in order to avoid jeopardizing the progress of the Discovery and also in order to ensure that King Francis I, who sponsored the expedition, will not withdraw his financial backing.

The 1534 narrative is enlightening in this respect. The islands of Canada are presented as rich and full of promise.[35] Cartier reiterates this impression in his prefatory letter dedicated to Francis I, which opens the *Brief recit*.[36] Relations with the Native Americans are either free of any friction, or the captain does not perceive it. Jacques Cartier takes note of fragments of rituals, like rubbing a guest's arms and chest in welcome, conversational phrases or a few words in indigenous languages.[37] While he is aware of the difficulties posed by the difference in languages when it comes to communication, he does not seem overly hindered, in his progress and in his work as an explorer, by the absence of a common language. The only tense moment happens during an encounter between the captain and the chief of the tribe who lives on a land where Cartier orders a tall cross planted, on the Gaspé peninsula, on 24 July 1534:

> Nous fit une grande harangue nous monstrant ladite croix et faisant le signe de la croix avec deux doydz et puis nous monstroit la terre tout alentour de nous comme s'il eust voullu dire que toute la terre estoit à luy et que nous ne devyons pas planter ladite croix sans son congé.[38]

The translation given by Cartier is quite plausible of course. However, we can note that the sailor gives utterance to his own project of exploration and possession, as if he had been understood, and possibly unmasked, by the chief. In order to answer the chief's concerns, he adopts a novel strategy: he coaxes the Native Americans into *approaching* so as to allay their fears. Only then does he give a reason for erecting the cross, "pour faire merche et ballise pour entrer dedans le hable".[39] One would like to be able to represent this signed conversation, which ends with the exchange of presents and the promise to return soon, bringing more manufactured goods. What is even more important is that Cartier takes away with him the chief's two sons, Taignoagny and Domagaya, who do not seem to object to leaving their tribe. They will become his interpreters during the next expedition. So much serenity is puzzling to witness: the cloud of incomprehension that floats over the narrative seems bathed in a diffuse light; as in the Colombus narrative, the captain's project seems stronger than any reality and seems to forestall any misunderstanding.

And still, conflicts will arise during the second journey. At first, the interpreters' presence does make contact easier, both with neighbouring tribes[40] and even within their own community.[41] In those instances, Cartier emphasises the interpreters' goodwill and the harmony that reigns among them all. But before long, the situation deteriorates, as soon as the project of advancing upstream towards Hochelaga is mentioned. Cartier notes that the interpreters are "tous changez de propoz et de couraige".[42] This is the beginning of mistrust, and, on 15 September:

> Taignoagny dict audit cappitaine que leur seigneur Donnaconna estoit marry dont ledit cappitaine et ses gens portoyent tant de bastons de guerre pour ce que de leur part n'en portoient nulz. A quoy respondit ledit cappitaine que pour sa marrisson ne laisseoit à les porter parce que c'estoit la coustume de France et qu'il le savoit bien.[43]

At the heart of the argument between Taignoagny and the captain are the weapons, symbols and instruments of power. But according to Cartier, these words are those of the interpreters only: by this he suggests that Taignoagny has failed in his duty to remain neutral, as dictated by his function as an interpreter. The next day, during a gathering that draws five hundred people and a dozen tribe chiefs, explorers and Native Americans are openly in conflict, "fut par Taignoagny dict audit cappitaine que le seigneur Donnacona estoit marry dont il allait à Hochelaga et que ledit seigneur ne voulloyt que luy qui parloit y allast avecques luy comme il avoit promis parce que la riviere ne valloit rien".[44] In the subsequent lengthy ritual, during which the chiefs give young children as presents to Cartier, it also transpires that the interpreters do not see eye to eye among themselves. The captain highlights contradictions in what is being related to him and deduces that the conflict is the result of Taignoagny's stratagems. Against this backdrop of mistrust, the firing of artillery is exploited by this same interpreter to evoke terror in the Native Americans:

> fit dire par interposees personnes que les compaignons du gallion qui estoient en la radde avoient tué deux de leurs gens de coups d'artilleryе dont tous se retirerent à grand haste ainsi que si les eussions voulu tuer. Ce que ne se trouva verité car durant ledit jour ne fut dudit gallion tyré artillerye.[45]

The narrative becomes very specific, in order to suggest that the interpreter has his own designs, designs that do not reflect the chiefs' words or desires. The character of the interpreter, like all go-betweens, can arouse suspicion.[46] However, in the general economy of the narrative, it also has its own function: the interpreter becomes responsible for all tensions, conflicts and misunderstandings that delay the exploration. Thanks to this device, Canada can remain the promised land that Cartier wants to present to the French king, a land with fertile soil and good-natured inhabitants.

Charles-André Julien explains the additional difficulties of the journey towards Hochelaga by laying the fault squarely at the feet of the interpreters, "It was no doubt their reports that inspired such aversion for this journey in Donnacona. He very probably feared that the French would put the diabolical power of their 'war batons' in service of a rival tribe".[47] Samuel Eliot Morison takes more seriously the idea that the interpreters did in fact relate the true wishes of Chief Donnacona, by taking into account certain realities that Jacques Cartier couldn't be aware of: strategic alliances between Native American ethnic groups.

> Their [Taignoagny's and Domagaya's] father did not wish Cartier to go to Hochelaga. The chief there, also a Huron, claimed dominion over Donnacona, who did not relish the relationship; but their respective positions would be reversed if the chief of Stadaconé [i.e. Donnacona *NDR*] could gain such powerful allies as the French, with their ships, firearms and iron tools.[48]

A curious incident the next day confirms Cartier's suspicions: he sees "great cunning" in it and increases the reader's perplexity.[49] The interpreters, we are told:

> firent habiller troys hommes en la façon de troys diables lesquelz avoient des cornes aussi longues que le bras et estoient vestuz de peaulx de chiens blancs et noyrs et avoyent le visage painct aussi noir que charbon et les firent mectre dedans une de leurs barques à nostre non sceu.[50]

During the customary encounter between Native Americans and sailors, on one of the boats appear three men dressed as devils ("apparessent estre trois diables"), "celui du meilleu [faisant] un merveilleux sermon en venant et passerent le long de noz navires avecques leur barque sans aucunement tourner leur veue vers nous".[51] After that comes a sermon "which lasted around half an hour".[52] Cartier interprets this stratagem as a spectacle designed to scare the sailors. It could be thought that he projects onto the Native Americans his own way of persuading others. Should we interpret this episode, as S.E. Morison does, as "Donnacona emptied his bag of tricks", by getting "three medicine men" to intervene?[53] Cartier might call this episode "the devils' comedy", but what meaning does it have for the Native Americans?[54] Is it a *bona fide* divination ritual? Is it a piece of magic designed to prevent the French from continuing on their way? Nothing would allow us to confirm or refute this. In the end, the interpreters are the ones to bring this puzzling spectacle to its denouement by the way in which they approach the sailors:

> Les mains joinctes et leurs chappeaulx soubz leurs coddes faisant une grande admyration. Et commança ledit Taignoagny à dire et proferer par la troys fois 'Jesus Jesus Jesus' levant les yeulx vers le ciel et puys Domagaya commança à dire 'Jesus Maria Jacques Cartier' regardant le ciel comme l'aultre. Et le cappitaine voyant leurs mynnes et serymonies leur commança à demander qu'il y avoit. Lesquelz respondirent qu'il y avoit piteuses nouvelles ... leur Dieu nommé Cudouagny avoit parlé à Hochelaga et que les trois hommes devant ditz estoient venuz de par luy leur anoncer les nouvelles qu'il y aurait tant de glasses et de neiges qu'ilz mourroient tous. Desquelles parolles nous prinsmes tous à rire et leur dire que leur Cudouagny n'estoit que ung sot et qu'il ne savoyt qu'il disoit et qu'ils le disent à ses messagiers et que Jesus les garderoit bien de froit s'ilz luy voulloient croyre. Lors lesdits Taignoagny et Domagaya demanderent au cappitaine s'il avoit parlé à Jesus et il leur respondit que ses prebstres y avoient parlé et qu'il feroit beau temps.[55]

The clash of conflicting wills intensifies as the function of the interpreter changes. He ceases to be a go-between in that instant. Instead of ensuring contact between Donnaconna and Cartier, Taignoagny and Domagaya explain the meaning of the presence and of the words of the devils and reveal a message with a supernatural origin. Cartier replies by speaking in the name of Jesus. The face-off between men from different horizons reveals the conflict between two transcendent wills and generally between two cultures that approach this encounter in terms of a power struggle. However difficult it can be to understand precisely the meaning of the scene as a whole, as a ritual or simply as a political ruse, it is to be noted that this clash of wills takes place on a transcendent plane and is imbued with a meaning that largely exceeds political considerations.

It is interesting to note that, at the end of the journey upstream, the stop at Hochelaga marks the triumph of a fantasy of cohesion that brushes off any linguistic obstacle. It is the high endpoint of a narrative in which messianic references abound, healing the "miserable savages" in preparation for the salvation to come.[56] All the sick people in the area are brought to Jacques Cartier, "couchant aupres dudit cappitaine pour les toucher tellement qu'il sembloyt que Dieu fust là descendu pour les guerir".[57] Conforming to what was at the time a customary rite in healing prayers, Cartier reads the Prologue to John's Gospel and the narrative of the Passion of Christ to a reverentially silent audience, before moving on to the exchange of small gifts. This scene is the ultimate vindication of Cartier against Taignoagny, of Jesus against Cudragny, and brooks no argument.[58]

This episode in Hochelaga is beyond linguistic communication and is a considerable displacement of the question of domination, which is at the root of the conflicts that mark the history of the Discovery. Flouting "the invisible obvious", that substratum of all discourse anchored in a given culture, the episode displaces the problem into the realm of myth, which imposes its own unequivocal reading. This is, of course, a fiction: the absence of interpreters does not make for easier exchanges: far from it! Conversely, once the interpreters disappear, Cartier can carry out unhindered his project of becoming Christ's curate in the Americas and the devoted servant of the prince, who is the ultimate commander of the expedition. Just like Columbus, Cartier's only aim in discovering the Americas is to increase the dominion of his prince, both terrestrial and celestial. Vested, in their own eyes, with a superior authority, can they even understand the reasons why conflicts sprang up? This insistence in reverting to myth, palpable in both discoverers, explains their attitudes towards indigenous languages. Todorov remarks on Columbus's "failure of attention to the other's language".[59] This is also very much the case for Jacques Cartier, even if we can find, at the end of his *Brief recit*, a somewhat developed glossary which contains a few puzzling entries as well. How can we explain the presence of "cinnamon" and "cloves"? This dream of the Orient knows no bounds, that it should bring forth spices in the middle of Canada? It is quite possible, as Michel Bideaux suggests in a footnote in his edition of the narrative, that Cartier's interpreters played the sailors' obsessions against them in order to get rid of them:

> Should we think that Taignoagny and Domagaya, discovering the existence of spices during their stay in France, could have given Cartier

> an approximate translation in their own language? As was the case with the riches of Saguenay and the fables of Donnacona, the greedy fascination that the French expressed for spices and where to find them made them the victims of a new mystification.[60]

According to the reasoning of Bideaux, the Native Americans might have understood the way the discoverers functioned, better than the discoverers themselves, blinded by certainty, ever understood the social systems that ruled the human groups they met. This would lead us to reconsider the part played by all agents in the history of the Discovery. This is an impossible task since the documents at our disposal only come from one of the parties. There is a very real danger, five centuries later, of scrutinizing the intentions of the Native Americans as they are subject to an external assault, by imposing analyses that would bear the seal of contemporary Western thought as regards the discovery of the Americas – it is impossible in such analyses to avoid a neocolonial bias.

Conclusion

In these early scenes of the Discovery, and in great contrast to what will happen later, what seems to dominate is less conflict than mutual surprise and incomprehension. This is, at least, the discourse that is elaborated upon in the narratives. By comparing Columbus and Jacques Cartier, we can piece together two conceptions of the importance of language in the process of discovery, in the emergence of conflicts or in the success of an encounter. Todorov points out that Columbus has no interest in people or in ethnography, nor in any of the nuances that this type of outlook can produce. In the eyes of Columbus, myth offers a ready-made solution, even governing the admiral's choice of interpreter. Cartier, on the other hand, a sailor and a practical man, does not brush off the question of the plurality of languages or that of the technical difficulties of communication, as evidenced by his endeavours to train interpreters and to establish glossaries of native languages. But he, too, resorts to myth as soon as conflict threatens his dream of a world governed by a unifying principle. In the end, the experience of conflict has no initiatory value for Cartier: on the contrary, it heralds a retreat back into the horizons of myth, back to a world of "invisibly obvious" assumptions, in which implicit cultural assumptions triumph without violence, precluding any rational perception of reality. The light that bathes the upstream encounter with the Native Americans is blinding and forbids seeing. The realities of language and of culture that rob the relation between societies that remain strangers to each other of any possible transparency would facilitate a growing, productive awareness of alterity. It is probably too early for that in the history of the discovery of new horizons, and the linguistic awareness of the different parties is insufficiently developed. One century after Cartier, Jean de Brébeuf, a missionary, among many other examples, will attempt to immerse himself in

Huron society, dedicating himself to learning the language, of which he will become an accomplished scholar.[61] His sole aim is to convert the First Peoples to Christianity, and he knows that such an enterprise cannot forgo a deep knowledge of native languages and cultures, of their myths and rites. These two case studies are significant in this regard. However, is it not an illusion to believe that only those misunderstandings generated by implicit cultural assumptions can give rise to lethal conflicts? This answer sounds like a last resort, grasping for the myth that transparency is in fact possible.

Notes

1 Raymonde Carroll, *Cultural Misunderstandings: The French-American Experience* (Chicago and London: University of Chicago Press, 1987).
2 Gilles Louÿs and Emmanuelle Sauvage, *De la singularité dans la communication interculturelle: approches transdisciplinaires* (Paris: L'Harmattan, 2014), 9.
3 "The invisible obvious" is the title of the French translation: *Evidences invisibles: Américains et Français au quotidien* (Paris: Seuil, 1987).
4 Carroll, 114.
5 Tzvetan Todorov's *La Conquête de l'Amérique. La question de l'autre* (Paris: Seuil, 1982); English translation *The Conquest of America: The Question of the Other* (New York: Harper & Row, 1984), was instrumental in theorizing these questions.
6 Cristóbal Colón, *Textos y documentos completos*, ed. Consuelo Varela (Madrid: Alianza editorial, 1982), 30. English translations are from Christopher Columbus, *The Journal of C. C. and documents relating to the voyages of John Cabot and Gaspar Corte Real*, ed. Clements R. Markham (London: Hakluyt Society, 1893), 37.
7 This phrase is inspired by Denis Crouzet's book, *Christophe Colomb, Héraut de l'Apocalypse* (Paris: Payot, 2006).
8 Colón, 50; "Luis de Torres, who had served in the household of the Adelantado of Murcia, and had been a Jew, knowing Hebrew, Chaldean, and even some Arabic" (Columbus, 66).
9 See Claude-Gilbert Dubois, *Mythe et langage au XVIe siècle* (Paris: Ducros, 1970), 67.
10 Francisco Morales Padrón, *Historia del Descubrimiento y Conquista de América* (Madrid: Editora Nacional, 1981), 114.
11 Morales Padrón, 217. There is a fine example of Columbus's extensive reading: when he throws around hypotheses about earthly paradise, Columbus resorts to quoting Strabo, Bede, Isidore of Seville, Saint Ambrose, the *Historia Scholastica* and Duns Scotus. See also p. 219: this time, it's the turn of Pliny, Nicolas de Lyre, Aristotle and St Augustine to be roped in. Edmond Buron describes at length the influence of Pierre d'Ailly by taking into account Columbus's marginalia (Buron, in his edition and translation of the *Ymago Mundi* (Paris: Maisonneuve et frères, 1930)). Nicolas Brémaud reinterprets the influence of these texts on the Columbian enterprise and gives an interesting summary, "In this delirium, in which Columbus 'felt imperiously called' [quote from Lacan] the discovery of the New World can be considered as the concrete realisation of the inner New World he built for himself, the world he constructed over time, steeped in the certainty that he was the Chosen one, sent and guided by God's hand" (Brémaud, "L'Élu et le Nom: le Nouveau Monde de Christophe Colomb" ("The Chosen and the Name: Christopher Columbus's New World"), *La Clinique lacanienne* 23 (2013): 133–48.

12 Colón, 32; "by signs I was able to make out that to the south, or going from the island to the south, there was a king who had great cups full of gold, and who possessed a great quantity" (Columbus, 39).

13 Colón, 33; "We understood that they asked us if we had come from heaven" (Columbus, 41).

14 Tzvetan Todorov dedicates a chapter to "Columbus as Interpreter" ("Colon herméneute" in the original), and he studies the different modes of linguistic misunderstanding that come to the fore during the first journey.

15 Colón, 33; "One old man came into the boat, and others cried out, in loud voices, to all the men and women, to come and see the men who had come from heaven, and to bring them to eat and drink" (Columbus, 41).

16 Colón, 67; "I do not know the language, and these people do not understand me nor any other in my company; while the Indians I have on board often misunderstand. Moreover I have not been able to see much of the natives, because they often take to flight" (Columbus, 85). The English translation here weakens the original phrase, which speaks volumes, "entiendo una cosa por otra al contrario", Columbus confesses that he understands the opposite of what is meant.

17 Colón, 70; "It was a glory to look upon them" (Columbus, 95).

18 Colón, 71; "wands, on the point of which a short piece of wood is fixed, hardened by fire" (Columbus, 95).

19 Colón, 71; "[he] made a long speech which the Admiral did not understand. At intervals the other Indians raised their hands to heaven, and shouted. The Admiral thought he was assuring him that he was pleased at his arrival; but he saw the Indian who came from the ship change the colour of his face, and turn as yellow as wax, trembling much, and letting the Admiral know by signs that he should leave the river, as they were going to kill him. He pointed to a cross-bow which one of the Spaniards had, and showed it to the Indians, and the Admiral let it be understood that they would all be slain, because that cross-bow carried far and killed people. He also took a sword and drew it out of the sheath, showing it to them, and saying the same, which, when they had heard, they all took to flight; while the Indian of the ship still trembled from cowardice" (Columbus, 95–96).

20 Michel Mollat, *Les Explorateurs du XIIIe au XVIe siècle* (Paris: J. C. Lattès, 1984), 197.

21 Todorov, 33.

22 This is what Todorov sees as "wishful thinking" (32).

23 Colón, 100.

24 *Ibid.*, 145.

25 *Ibid.*, 145; "The Admiral said, by signs, that the Sovereigns of Castille would order the Caribs to be destroyed, and that all should be taken with their heads tied together. He ordered a lombard and a hand-gun to be fired off, and seeing the effect caused by its force and what the shots penetrated, the king was astonished" (Columbus, 137). We must point out a probable mistake in the English translation: the Spanish text mentions "manos" ("hands") and the English translation speaks of "heads".

26 Colón, 118; Columbus, 164.

27 Colón, 114; "very wanting in reverence" (Columbus, 159).

28 This detail is mentioned several times in this episode. Is it because the headdress is new to him, or is it because Columbus sees in it a trace of animality that becomes relevant when he mentions the *Caniba*? Frank Lestringant develops this theme by showing how Columbus "invented the Cannibal", and how, in the Diary, "the *Cynocephalus* comes before the *Cannibal*" (Lestringant, *Le Cannibale. Grandeur*

et décadence (Paris: Perrin, 1994), 44–45). See also p. 46 for a note on the etymology of the term, "word roots become blurred, and accompany the osmosis of images". Colombus associates *caniba* to *canis*, and then to "Great Khan", "evoking the growing threat of dog-headed anthopophagi".

29 Colón 115; "The Admiral supposed that he belonged to the Caribs, who eat men" (Columbus, 159).

30 Colón, 115; "He called gold *tuob*, and did not understand *caona*, as they call it in the first part of the island that was visited, nor *nozay*, the name in San Salvador and the other islands. Copper is called *tuob* in Española" (Columbus, 159).

31 Colón, 115; "he understood a few words, and the Indians who were on board comprehended more, there being a difference in the languages owing to the great distance between the various islands" (Columbus, 160).

32 Colón, 115–16; "They were running back to pick up their bows and arrows where they had laid them aside, and took cords in their hands to bind the boat's crew...the Spaniards attacked the Indians, and gave one a stab with a knife in the buttocks, wounding another in the breast with an arrow. Seeing that they could gain little, although the Christians were only seven and they numbered over fifty, they fled, so that none were left, throwing bows and arrows away" (Columbus, 161).

33 See Marie-Christine Gomez-Géraud, "Portrait de deux truchements: Taignoagny et Dom Agaya", catalogue of the exhibition *La Renaissance et le Nouveau Monde* (Québec: Musée du Québec, 1984), 52–54 and "La figure de l'interprète dans les récits de voyage français au XVIe siècle", in *Voyager à la Renaissance*, actes du XXVe colloque d'Études Humanistes de Tours. Edited by Jean Céard and Jean-Claude Margolin (Paris: Maisonneuve et Larose, 1987), 319–35.

34 Jacques Cartier, *Relations*, ed. Michel Bideaux, (Montréal: Presses de l'Université de Montréal, 1984), 127. "our most holy faith was sown and planted in the Holy Land, which is in Asia to the east of our Europe, and afterwards by succession of time, it has been carried and proclaimed to us, and at length to the West of our Europe, just like the sun, carrying his light and its heat from east to west, as already said forth" (Henry Percival Biggar, *The Voyages of Jacques Cartier, published from the originals with translations, notes and appendices* (Ottawa: F. A Ackland, 1924), 89). Notes 5 to 7 in the edition by Michel Bideaux are in teresting in order to understand Cartier's point of view: his cosmology remains totally traditional and geocentric (344). Cartier's *Brief recit* is published in 1545, a year after Copernicus's *De revolutionum orbium celestium*, as M. Bideaux remarks (344, n. 7); Copernicus describes the heliocentric system, but his findings "had little impact on geography". For a brief outline of the growing acceptance of heliocentrism, see Rachel Willie's chapter in this current volume.

35 See Réal Ouellet, *La Relation de voyage en Amérique (XVIe-XVIe siècles)* (Paris: Hermann, 2015), 26–27. He comments on Cartier's perspective on landscape and the modes of orientation of the text towards "the myth of Paradise, marked by harmony and natural plenty".

36 See Cartier, "la bonté et fertilité d'icelle [terre] la innumerable quantité des peuples y habitans la bonté et paisibleté d'iceulx et pareillement la fécondité du grand fleuxe qui decourt et arrouse le [pays]" (128). "fertility and richness, of the immense number of peoples living there, of their kindness and peacefulness, and likewise of the richness of the grand river" (Biggar, 90).

37 Cartier, 110; "nous disant en leur langage *napou tou daman asurtat* et autres paroles que nous n'entendions". "saying to us in their language: *napou tou daman asurtat* and other words, we did not understand" (Biggar, 50). On speaking a few words of indigenous language, see Cartier, 115; "these are almost all names of foods" (Biggar, 63).

38 Cartier, 116; "made us a long harangue, making the sign of the cross with two of his fingers; and then he pointed to the land all around about, as if he wished to say that all this region belonged to him, and that we ought not to have set up this cross without his permission" (Biggar: 65).
39 Cartier: 117; "the cross had been set up to serve as a land-mark and guide-post on coming into the harbour" (Biggar, 67).
40 As they see boats approaching not far from Tadoussac, "en une grande peur et crainte" but "ils peurent entendre l'un de noz sauvaiges qui se nomma et fict sa cognoissance et les fict venyr seurement à bort" (Cartier, 135); "in such great fear and trembling that one of our Indians who gave his name and told who he was and made them come alongside in all confidence" (Biggar, 115).
41 See Cartier, 137–38.
42 Cartier, 139; "altogether changed in their attitude and goodwill" (Biggar, 128).
43 Cartier, 141; "Taignoagny began to make a speech and to say to the Captain, that Chief Donnacona was vexed that the Captain and his people carried so many weapons when they on their side carried none. To this the Captain replied that for all Donnacona's grief, he would not cease to carry them since such was the custom in France and Taignoagny well knew" (Biggar, 129).
44 Cartier, 142; "Taignoagny told the Captain that Chief Donnacona was annoyed because he [Cartier] intended to go to Hochelaga, and was most unwilling that Taignoagny should accompany him, as he had promised to do; for the river was not worth exploring" (Biggar, 131).
45 Cartier, 143–144; "Taignoagny sent word by others, that the sailors on board the bark (*Emérillon*), that lay out in the roadstead, had fired their cannon and had killed two Indians whereupon all scurried off in such haste that one would have thought we had wished to destroy them. The report proved false; for no shot was fired that day from the bark" (Biggar, 135).
46 For a discussion of how the interpreter's privileged status as an intermediary between stranger and the community to which they travel arouses suspicion, see Céline Bonnotte-Hoover's chapter in this current volume.
47 Charles-André Julien, *Les Voyages de découverte et les premiers établissements* (Brionne: Gérard Monfort, 1979), 126.
48 Samuel Eliot Morison, *The European Discovery of America* (New York and Oxford: Oxford University Press, 1971), 407.
49 Cartier, 144-46; Biggar, 136–40.
50 Cartier, 144; "They dressed up three Indians as devils, arraying them in black and white dog-skins, with horns as long as one's arm and their faces coloured black as coal, and unknown to us put them into a canoe" (Biggar, 136).
51 Cartier, 144; "the one in the middle made a wonderful harangue, but they passed by our ships without once turning their faces towards us" (Biggar, 137).
52 Biggar, 137.
53 Morison, 407.
54 Paul Zumthor reminds us that "In the hierarchy that descends from God and the Pope, down to the poorest, [the savage] clings to a rung right under them, just before demons". This element, belonging to an imaginary world still very much active in Cartier's time, might have influenced how the sailor tried to interpret a ceremonial that he was incapable of interpreting. See Zumthor, *La Mesure du monde. Représentations de l'espace au Moyen Age* ("The Measure of the World. Representing space in the Middle Ages") (Paris: Seuil, 1992), 280.
55 Cartier, 145; "with their hands joined, and their caps under their arms, pretending to be much astonished. And Taignoagny began to speak and repeated three times '*Jesus*', '*Jesus*', *Jesus*', lifting their eyes towards heaven. Then Dom Agaya called out '*Jesus*', '*Maria*', '*Jacques Cartier*' looking up to heaven as the

other had done. The Captain seeing their grimaces and gesticulations, began to ask them, what was the matter ... They replied that there was bad news ... their god, Cudouagny by name, had made an announcement at Hochelaga, and that the three above-mentioned Indians had come in his name to tell them the tidings, which were that there would be so much ice and snow that all would perish. At this we all began to laugh and to tell them that their god Cudouagny was a mere fool who did not know what he was saying; and that they should tell his messengers as much; and that Jesus would keep them safe from the cold if they would trust in him. Thereupon Taignoagny and his companion asked the Captain if he had spoken to Jesus; and he replied that his priests had done so and that there would be fine weather" (Biggar, 138–39).

56 On October 2, the sailors reach Hochelaga and are welcomed by dances. The narrative mentions that the Indians bring victuals and bread, "faict de groz mil qu'ilz gectoient dedans noz barques en sorte qu'il sembloit qu'il tombast de l'air" (Cartier, 150); "they brought us quantities of fish, and of their bread which is made of Indian corn, throwing so much of it into our long-boats that it seemed to rain bread" (Biggar, 150). He alludes here to manna in the desert (Ex 16). Even more significant is the moment when Cartier lands on one of the islands on the river, on the way to Hochelaga "l'un d'iceulx hommes print le cappitaine entre ses bras et le porta à terre aussi liegerement comme si c'eust esté ung enffant de cinq ans tant estoit cest hommes fort et grand" (148); "one of those Indians took the Captain in his arms and carried him on shore as easily as if he had been a six-years old child, so strong and big was the Indian" (Biggar, 147). The allusion to the legend of Saint Christopher carrying the child Christ across the river is plain to see. For a discussion of how the image of Saint Christopher guided travellers through the Alps, see Joanne Anderson's chapter in this volume.

57 Cartier, 154–55; "laid out near the Captain, in order that he might lay his hands upon them, so that one would have thought Christ had come down to earth to heal them" (Biggar, 165).

58 Cudragny (or Cudouagny) was presented by Cartier as the God of the populations he met along the St Lawrence River. As Michel Mollat writes in his summary of Cartier, Cudragny is a deity "governing meteorological phenomena, the survival [in the afterlife] that represents a permanent movement between the stars and abundant fields, and the power of the sorcerers" (*Les Explorateurs du XIIIe au XVIIe siècle* (Paris: Lattès, 1984), 228–29).

59 Todorov, 31.

60 Cartier: 409, n. 661.

61 See Marie-Christine Gomez-Géraud, "Entre savoir expérimental et science livresque: le voyageur face aux langues du Canada", in *Les Représentations de la Nouvelle-France et de l'Amérique du Nord*, ed. Sophie Linon-Chipon, Raymonde Litalien and Hélène Richard (Paris: Éditions du CTHS, 2013), 147–60.

Bibliography

d'Ailly, Pierre. *Ymago Mundi*. Edited and translated by Edmond Buron. 3 vols. Paris: Maisonneuve et frères, 1930.

Brémaud, Nicolas, "L'Élu et le Nom: le Nouveau Monde de Christophe Colomb." *La Clinique lacanienne* 23 (2013): 133–48.

Carroll, Raymonde. *Cultural Misunderstandings: The French-American Experience*. Chicago and London: University of Chicago Press, 1987.

Cartier, Jacques. *The Voyages of Jacques Cartier*. Translated by Henry Percival Biggar. Ottawa: F. A. Ackland, 1924.

Cartier, Jacques. *Relations*. Edited by Michel Bideaux. Montréal: Presses de l'Université de Montréal, 1984.

Colón, Cristóbal. *Textos y documentos completos*. Edited by Consuelo Varela. Madrid: Alianza editorial, 1982.

Columbus, Christopher. *The Journal of C. C. and Documents Relating to the Voyages of John Cabot and Gaspar Corte Real*. Translated by Clements R. Markham. London: Hakluyt Society, 1893.

Crouzet, Denis. *Christophe Colomb: Héraut de l'Apocalypse*. Paris: Payot, 2006.

Dubois, Claude-Gilbert. *Mythe et langage au XVIe siècle*. Paris: Ducros, 1970.

Gomez-Géraud, Marie-Christine. "Portrait de deux truchements: Taignoagny et Dom Agaya." *Catalogue of the Exhibition La Renaissance et le Nouveau Monde*. Québec: Musée du Québec, 1984.

Gomez-Géraud, Marie-Christine. "La figure de l'interprète dans les récits de voyage français au XVIe siècle." In *Voyager à la Renaissance, actes du XXVe colloque d'Études Humanistes de Tours*. Edited by Jean Céard and Jean-Claude Margolin, 319–35. Paris: Maisonneuve et Larose, 1987.

Gomez-Géraud, Marie-Christine. "Entre savoir expérimental et science livresque: le voyageur face aux langues du Canada." In *Les Représentations de la Nouvelle-France et de l'Amérique du Nord*. Edited by Sophie Linon-Chipon, Raymonde Litalien and Hélène Richard, 147–60. Paris: Éditions du CTHS, 2013.

Julien, Charles-André. *Les Voyages de découverte et les premiers établissements*. Brionne: Gérard Monfort, 1979.

Lestringant, Frank. *Le Cannibale. Grandeur et décadence*. Paris: Perrin, 1994.

Louÿs, Gilles and Emmanuelle Sauvage. *De la singularité dans la communication interculturellle: approches transdisciplinaires*. Paris: L'Harmattan, 2014.

Mollat, Michel. *Les Explorateurs du XIIIe au XVIe siècle*. Paris: J. C. Lattès, 1984.

Morales Padrón, Francisco. *Historia del Descubrimiento y Conquista de América*. Madrid: Editora Nacional, 1981.

Morison, Samuel Eliot. *The European Discovery of America*. New York and Oxford: Oxford University Press, 1971.

Ouellet, Réal. *La Relation de voyage en Amérique (XVIe-XVIe siècles)*. Paris: Hermann, 2015.

Todorov, Tzvetan. *The Conquest of America: The Question of the Other*. Translated by Richard Howard. New York: Harper and Row, 1984.

Zumthor, Paul, *La Mesure du monde. Représentations de l'espace au Moyen Age*. Paris, Seuil, 1992.

2 Language, mediation, conflict and power in early modern China

The roles of the interpreter in Matteo Ricci's *Journals*

Céline Bonnotte-Hoover

In 1582, Matteo Ricci, a member of the Society of Jesus, travelled to Macau with the intention of spreading the Christian word in China. As one of the first missionaries to enter the country and to establish a mission on the mainland, his life is intrinsically linked with the success of the Chinese missions. His *Journals* are a precious testimony of the struggles faced by the Jesuits, of the conflicts that arose, and of the subsequent success of their Chinese endeavours. The published *Journals* are, however, not Ricci's original writings. In 1612, the Jesuit Nicolas Trigault, who had travelled to China in 1610 and reached Canton (Guangzhou) after Ricci's death, was entrusted by the Society with the promotion of the Chinese missions. Trigault set about translating Ricci's writings from their original vernacular Italian into Latin[1] and not only improved on Ricci's style but also "insert [ed] material either where there was an editorial need or where Ricci had left lacunae".[2] When analyzing it, we have to consider the text's origins, and in particular the fact that its publication had the primary objective of promoting missionary interests. The *Journals* provide a report of Jesuit activities and in so doing aim to increase the merit of missionary work in the eyes of the public and to emphasise the difficulties that had to be overcome. At the centre of our present reflection are two of the challenges that the Jesuits faced in China: the complexity of the local language and the political situation within the country.

As they recount the efforts of the Jesuits to settle in the country, Ricci's *Journals* lend themselves perfectly to an analysis of language, mediation, conflict, and power in early modern China. The process of learning the Chinese language was slow, and, as a result, the Jesuits recruited interpreters and teachers to assist them in communicating and learning. This solution to the linguistic challenge, however, triggered a political one: an inhospitable if not hostile attitude in a country that openly rejected and threatened foreigners as well as those who associated with them. This hostility was exacerbated in the case of the Jesuits as they came and learned local languages in order to disseminate Christian ideas. Michaela Wolf's statement that "Western expansion into non-Western societies always entails seizing power of the respective language or languages" is therefore crucial when considering the Jesuits.[3] Knowledge of the local language was a tool of empowerment that allowed foreigners to travel independently, to communicate with whomever they chose, and therefore to discuss any topic without supervision. As a result, this knowledge was seen as a threat by some of the authorities of a country that had historically been closed off to outsiders, and especially so when this knowledge was in the hands of men as determined as the Jesuits. The interpreters recruited by the Jesuits found themselves at the heart of this conflict, entangled in a web of multiple allegiances, to their country and to the foreigners they served. Their position was complex as they were expected to perform in such a way as to help the missionaries who hired them but were also targeted by authorities who did not approve of the foreigners' progress.

This chapter thus aims to provide an analysis of the relationship between language and power as it was experienced by the Jesuits, from their first attempt to set foot on mainland China until their position was firmly established. It first focuses on the strategic objective of language learning before exploring the influence of the Chinese political situation on the process of language learning; on the availability of interpreters; and, as a result, on the status of the Jesuits in the country. Finally, it analyzes a few key scenes portraying the Jesuits and their interpreters, both good and bad, in order to study their interaction when conflicts subside and the influence of politics is removed.

The study of language as the key to succeed in China

The development of vernacular languages during the early modern period led to a higher awareness of the variety of languages present in Europe and in the world. Though travellers showed an increasing interest in learning some of the local languages they encountered, no one matched the Jesuits in their systematic approach to contact and language. Most documents of the period testify to the central role that language learning played in the establishment of Jesuit missions as it was used as a tool in the project of conversion. Jesuits were trained early to learn foreign languages and were selected for the missions based on their dedication to the cause and their spiritual, physical, and linguistic abilities.[4] Their process in China was no different than in their other settlements: mastering the local language was at the heart of the Jesuit project.

The first Jesuits, who paved the way for the success of Matteo Ricci, arrived in China at a time when no other missionary organization had been able to establish itself on the mainland and when most Europeans were restricted to the Macau peninsula. The Ming dynasty's isolationist policies led to the closing of the Chinese borders. Macau was rented to the Portuguese who had settled on it, and it became the centre of Portuguese, and thus European, trading in the area. Alessandro Valignano, appointed Visitor to the East, believed that if the Jesuits were to learn the Chinese language, they would obtain opportunities thus far unprecedented, including entering the country: as a result, he instituted a strict policy of language training. Despite the fact that China was closed off to foreigners, resulting in the Jesuits having only limited contact with the local population, Valignano's policy was put into place in 1579 and led to the success he had envisioned. The importance of mastering the language was ingrained in the missionaries even before their arrival in Macau. The learning of Chinese started abroad, at the college in Goa, but was soon centralised to Macau's Saint Paul's College, making it a hub for all missionaries destined to work in both China and Japan.

Although a command of the language was essential to the missions, the task of learning it was a challenging one and the complexity of the Chinese language is often mentioned:

> I would venture to say that no other language is as difficult for a foreigner to learn as the Chinese. By the grace of God, however, and by unremitting labor, the members of our Society...have learned to speak their language (28).

This topos of "hardship" has a dual role in the *Journals*: though it stems from a truly complex situation, it also glorifies those who endured it. As the publication of the text included a clearly promotional component, it is difficult to determine where the line can be drawn between the desire to depict the reality of the Jesuit experience in China and the desire to promote the work done in that country by the missionaries. There is, however, no doubt that the language represented a great challenge, a fact to which Michele Ruggieri, Ricci's predecessor and the first to follow Valignano's new policy, attested. The *Journals* identify several reasons for this challenge, the first being the variety of dialects that existed in China. Ruggieri was still in India when he began learning the Chinese language and was told to focus on "the curial dialect of the language, as it was called, the particular dialect in universal use throughout the kingdom" (131). This would help the Jesuits as they aimed to integrate themselves into Chinese society and, more specifically, within the community of the literati. However, once in China, Ruggieri was confronted with the fact that an important portion of the population did not speak this "curial dialect", but rather other provincial varieties, and that a "universal" language did not exist in China.[5] This linguistic diversity, although comparable to that of several European countries during the same period, was not anticipated, and represented the first obstacle to the attainment of the Jesuits' objectives. Furthermore, while Portuguese was common in the area dominated by merchants affiliated with Portugal, it was of limited use among the Chinese and unable to serve as a lingua franca. This complex linguistic state of affairs could thus only lead to difficult communications.

Beyond the sheer complexity of the language, further difficulties consisted in the lack of available teachers and in the fact that interpreters were not always present to facilitate the interactions when a teacher was found. The *Journals* point out the compounding difficulties that students of Chinese faced:

> If the acquisition of knowledge were measured solely by diligence of application, one could probably make considerable progress in this particular study in a shorter space of time. This, however, was not the case in the present instance, because the Chinese language, being the most difficult and most intricate ever heard of or read about, was made doubly difficult by the lack of teachers who could teach it. (131–32)

Though the Jesuits were experienced language learners, their progress was slowed by the internal complexities of the language and by external factors. Consequently, the study of the Chinese language was presented as a necessary hardship by the Jesuits who dedicated themselves to it.

The challenges of Ruggieri's apprenticeship, and that of other Jesuits after him, did not put an end to Valignano's policy. Language learning continued despite these obstacles and was proof of the Jesuits' determination to settle in China. Over time, they were able to improve their learning environment: they found teachers and recruited interpreters who contributed to their progress.[6] Although Ruggieri never became proficient in the language, he became skilled in Chinese protocols and used this knowledge to the advantage of the Society. When the Jesuits were attempting to obtain permission to establish a residence on the mainland, Ruggieri was invited to meet with the viceroy of Guangzhou (the Governor-General of the province of Guangdong),[7] Chen Rui, whom he impressed with the knowledge he had acquired. Unable to speak without a go-between, Ruggieri was accompanied during this encounter by a third party who remained nameless.[8] His attempt to learn the local language was seen in a positive light by Chen Rui, and the meeting, successfully mediated by the interpreter, resulted in Ruggieri being invited to reside on the mainland.[9] The good grace of the viceroy towards the Jesuits is here presented as the direct result of Ruggieri's dedication not only to learning the Chinese language but also to familiarising himself with Chinese culture. This attitude, in line with the policy of accommodation pursued by Jesuits throughout the world, "please[d]" the viceroy, who was won over by them. This was a major achievement for the mission as the protection of its members by a powerful man would allow them to begin what they had come to accomplish.

On a broader scale, the rewards of language learning can be seen in the trajectory of Matteo Ricci. Talented in languages and highly educated, he was able to bring European scientific knowledge to the Chinese in fields such as astronomy and cartography; his command of the Chinese language allowed him to translate European scientific texts and discuss them with the local elite.[10] His reputation grew as time passed, despite the low number of converts to Christianity, and "despite his missionary zeal, Ricci freely admitted that among the reasons for his popularity, his religious teachings came last".[11] Indeed, he fostered a reputation of himself as a scholar, and his accomplishments in China were largely credited to his ability to communicate, which allowed him to make important friends.[12] Linguistic competence allowed the Jesuits the opportunity to communicate with a range of people, to make powerful allies, and to gain protectors, benefits that offset the many difficulties associated with the language learning process.

The politics of interpretation

As can already be seen in the encounter between Ruggieri and viceroy Chen Rui, the beginnings of the Jesuit settlement in China were fundamentally linked to the country's political situation.[13] As a result, they were full of setbacks that reflected the ambivalent attitude of the Chinese towards foreigners. Indeed, George Minamiki explains that "[t]he Chinese ... looked upon their own country as ... the centre of culture and civilisation, yet they were shut in upon themselves and entrenched in their superior tradition".[14] This meant that before foreigners could be respected or accepted into society, they had to convince the Chinese that they were not "barbarians",[15] which the Jesuits did in part through the use of European science.

The Jesuits were unlike other travellers; as a result, the discussion of their language learning needs to be resituated in the more global context of Jesuit objectives and needs to take into consideration their impact on early modern China. Far from being simple visitors, the missionaries aimed not only to settle in the country permanently but also to convert its population. Their presence was thus perceived in a different light than that of passing merchants and travellers as it upset the order of a Chinese society that was "tolerant as long as a new way did not interfere with or overturn the basic structure of their own society".[16] Despite their policy of accommodation, Jesuit missionaries brought change and modified the social fabric of the country, altering the relation of the Chinese people to religion and cultural practices by introducing Christian elements, which often led to a conflict of values. As a result, they were met with suspicion, as was anyone who offered them assistance – and, particularly, their interpreters.

An edict posted at the gates of the city of Ansam (Xiangshan) by Liu Jiezhai, the new viceroy of the Guangdong and Guangxi provinces, showed that the relationship between missionaries, their interpreters, and the local authorities was delicate and could even turn dangerous:

> Serious complaints have been made that various crimes and abuses of the law have been perpetrated at Macau, which are attributed to the Chinese interpreters employed by the foreigners. These interpreters are soliciting the strangers and teaching them the ways of our people. Most serious of all, we are well informed that they have persuaded certain priests from abroad to learn the Chinese language and study Chinese

> letters, and that now these priests are demanding a residence at the Capital, in order to build a church and a private home. This we declare to be injurious to the realm, to which no benefit can accrue by admitting foreigners. The said interpreters will be put to a cruel death if they do not immediately desist from the practices mentioned (143).

By targeting interpreters, the edict reveals the instability of the position of mediator, even within their own community. The viceroy's accusations were twofold: the interpreters had reportedly encouraged foreigners to break the law as well as convinced the said foreigners to learn the Chinese language and to become familiar with the country's culture. Claiming to have been "well informed" about this behaviour – although the reader never finds out who does the informing – the viceroy was nonetheless wrong on the second count. As we will later see in the case of an interpreter described as a "trickster", there was merit to the accusation that interpreters enabled illegal behaviour. However, the second accusation overstated the initiative they showed. While they were a key factor in contributing to the progress of their students in learning the Chinese language and, by consequence, to the overall progress of the missions, interpreters were far from being the driving motivation behind Jesuit language learning.

The viceroy's attack stemmed from anxiety to protect the Chinese empire. The reasons listed to justify his decision reveal that he had understood the intentions of the missionaries and had anticipated the consequences of the Jesuit's settlement in his country. Deeming their plans and their overall presence as "injurious to the realm", he rejected the foreigners and established himself as the defender of the traditional Chinese way of life ("the ways of our people") while adding to the xenophobia that appears frequently throughout Ricci's *Journals*. In an attempt to put a stop to a situation that worried him as it risked producing more linguistically and culturally competent (and therefore uncontrollable) Jesuits, he targeted precisely the support without which missionaries would be incapable of communicating. The threat to the interpreters' lives was not to be taken lightly: the viceroy's declaration was not a warning but rather a guarantee of a swift ("immediately") and irreversible ("death") punishment that would not proceed through humane methods ("cruel"). The conflict between the viceroy and the Jesuits was set, and interpreters, already in the middle of communication, now also found themselves at the heart of it.

The criticism of the viceroy, despite being partly ill informed, singled out the very qualities that made interpreters both necessary and desirable to travellers: their resourcefulness, their willingness to bend or break the law for their masters, and their motivation to teach them their language. The criticism confirms, as a matter of fact, that the interpreters in the service of Ricci and Ruggieri were indeed capable and useful while at the same time reminding us that the position of mediator was not always easy to hold: the Chinese interpreters risked their lives if they continued to assist them.

However, since the Chinese political situation was in a period of fluctuation (particularly where outsiders were concerned), the Jesuits benefited from a change of policy and, a short time later, received a notice from the same viceroy authorising the construction of a house and a church and likely lifting the threat made against the interpreters.

As mediators, interpreters were at the centre of all interactions and were responsible for ensuring their smooth proceeding. Even as the status of the Jesuits improved in China, the interpreters, acting as their representatives, were still subject to the risks associated with contact with outsiders. The *Journals* recount one such occasion where a silent interpreter was blamed for the Jesuit's behaviour during an official audience. After being expelled from their house by the viceroy Liu Jiezhai, the Jesuits were offered compensation of 60 gold pieces, which they declined. Offended by this refusal, the viceroy summoned Ricci to appear before him. He arrived accompanied by a man who "was brought along as a companion, and for show, because the Father needed no interpreter" (219). Although the presence of the man remains entirely unexplained in the text, the comment following his mention provides valuable information: Ricci did not need assistance during the encounter and therefore likely spoke a certain amount of Chinese.[17] His companion, rendered unnecessary by Ricci's sufficient linguistic competence, was therefore a silent witness to the Jesuit's conversation with the viceroy. That Ricci could communicate on his own and that he used his knowledge to hold his own against a person of such authority confirms the power brought about by mastering the local language.

Ricci's obstinacy in refusing the 60 crowns and his statement that he did not need financial help from the man responsible for his expulsion caused the situation to deteriorate very quickly. In an unexpected turn of events, Ricci's companion was blamed for the words of the missionary:

> [The viceroy] turned toward the interpreter and continued, "The cause of all this trouble is this villain here who told this other man what to do", and in his excitement he ordered his court officers to bring out an iron chain and fasten it around the interpreter's neck. At these words, the poor man, in consternation, tried to remove the blame from himself by declaring his innocence and asserting that Father Ricci, being hurt by the loss of his house, had become stubborn in this matter of the money. Ricci agreed to this, freeing the interpreter from all fault and taking the full blame upon himself. (219)

The viceroy's accusation took the other men by surprise and came as an echo of the earlier edict in which he had targeted the Jesuits' interpreters. The charge remained unchanged: the interpreter, described as wholly bad (a "villain"), was blamed for influencing Ricci's behaviour and encouraging him, if not to break the law, then to go against the wishes of the viceroy. Responsible for translating their masters' words, interpreters were held accountable for the content of the message and could find themselves in vulnerable positions. Here, the vulnerability of the man is amplified by his description as a silent observer: his mere presence, and the viceroy's assumptions as to his behaviour, lead to an accusation that is implicitly condemned in the *Journals*. The decision to punish the interpreter is presented as arbitrary, resulting from pure emotion (here, "excitement") and occurring with a rapidity which did not anchor it in reason. The accusation thus almost becomes a performative act as the statement leads immediately to the preparation of the punishment. The innocence of the companion is contrasted with Ricci's admission of his own "guilt", namely through the use of the parallelism between "freeing from all fault" and "taking the full blame". It took Ricci's intervention and his appeal to the assurance of friendship that the viceroy had previously given him to spare his unlucky companion. His intervention also led to the revocation of the expulsion order: Ricci was permitted to choose, albeit within a strictly demarcated area, where a new mission would stand.

While this particular incident ended happily, it demonstrates how risky any association with foreigners could be in China. A whim of the authorities was enough to endanger the life of the mediator, proving the instability of the position. Indeed, interpreters were convenient scapegoats when the stakes were high: this is a recurring theme in the travel literature of the early modern period since many travellers were quick to disavow their interpreters if it served their interest.[18] There were, however, some travellers, including Ricci, who did protect their interpreters even at their own risk, showing a faithfulness to these collaborators and an awareness of their value.

Sometimes, rather than originating from the authorities, the threats came from the interpreters themselves as they occasionally decided to work against their employers. This could happen in several ways; through lies, bribery, and even treason.[19] The theme of the treacherous interpreter is another *topos* of travel literature and is summed up in the popular phrase *traduttore, traditore,* which inextricably links translation and betrayal.[20] Throughout the early modern period, travellers reported various sufferings at the hand of their interpreters, and the Jesuits themselves were not spared. Once they established their residence in Shaozhou (Xaucea),

> [i]t was decided also not to lodge the interpreters in the house because they were not trustworthy. They had a habit of putting a wrong interpretation upon what they saw, and then reporting everything that was done in the house to the whole outside world. Experience had taught that they generally turned out to be trouble makers. (229)

This behaviour of the interpreters, detrimental to the Jesuits' objectives, ties in with one of the central questions concerning their qualification: can they be trusted? In their position as go-betweens, they had access not only to people and information but to sensitive insider knowledge and thus to power.[21] This placed them in a unique position, and they were often perceived as problematic figures by the very same travellers who relied on them. As to the trustworthiness of the Chinese interpreters in question, the answer is given directly in the *Journals*: the Jesuits were well aware of the fact that their interpreters were "not trustworthy", misinterpreted what they observed, and subsequently reported their observations to outsiders. This time the accusations originate from the Jesuits and are presented as based on proof (since "experience had taught" them), which offers a noteworthy contrast to the way the viceroy's earlier allegations were described.

It is impossible to know what specific trouble was caused as no details are provided, yet the terms "habit" and "generally" indicate that the Jesuits suffered from their interpreters' behaviour on multiple occasions. That interpreters spied on the Jesuits was linked to the same concerns already apparent in the edict issued by the viceroy Liu Jiezhai. Indeed, as Dejanirah Couto reminds us,

> the service and the defense of a city which was part of a xenophobic Empire provided, nevertheless, another dimension to this espionage: it did not serve disparate strategies individual or collective, instead it represented the interests and the political action of a city-state in a cohesive manner.[22]

While Couto writes specifically on Macau, the same applied to the remainder of the country that, for a time, only tolerated missionary presence. By revealing what they witnessed in the residence, the interpreters, turned spies, deprived the foreigners of privacy and allowed the outside to look in. This reversal of private and public spheres in turn meant that the authorities could take action. The interpreters now became partners of the same officials that had previously targeted them and joined the rank of the Jesuit's antagonists. Though their access to the house, along with their access to the missionaries' discussions, was limited when their duplicity was discovered, these untrustworthy interpreters were not easily dismissed from employment. In fact, the Jesuits still required their assistance in order to communicate and to learn the language, and, unable to do without them, they were forced to tolerate their behaviour for the sake of their mission.

As the three examples discussed above show, the political situation in China, embodied by the authorities, and more specifically by the viceroy of the Guangdong and Guangxi provinces, had a serious impact, not only on the Jesuits, but also on the interpreters and companions who occupied a key role in their progress. Targeting them put a strain on their relationship with the missionaries and served as a deterrent. Controlling the interpreters was a source of power in itself since it allowed them to influence the behaviour of the Jesuits who needed them. Politics and language, in particular communication and language learning, were intrinsically entangled in the fight between the viceroy and the Jesuits over the control of the missionaries' ability to communicate.

The relationships between Jesuits and their interpreters

When issues of power are removed and the political situation allows the Jesuits more freedom, their relationship with their interpreters is generally portrayed as good and the latter overall reveal themselves useful to the former. Throughout their first period of settlement in China, the Jesuits were assisted by these men who facilitated communication; sometimes saved their lives; and, finally, trained them enough for their own presence to become unnecessary. Very little information is given about these interpreters in the *Journals* as it was common for early modern travellers, and writers, to erase any trace of their interpreters. Indeed, travellers often aimed to portray themselves as the heroes of their travel relations and therefore either concealed the presence of people who assisted them or only mentioned them in passing.[23] As the *Journals* mostly follow this second trend, I will dedicate the remainder of this chapter to a selection of examples from Ricci's *Journals* that show how interpreters and Jesuits interacted when the authorities did not prevent them from working together. This analysis will further our understanding of what these mediators did for the Jesuit missionaries in China.

Alida Metcalf's three-part typology of mediators proves enlightening in the analysis of the interpreters who will appear below. Basing her reflection on Stephen Greenblatt's concept of "go-between", she defines "physical go-betweens" as the people who created a physical link between countries and peoples, "transactional go-betweens" as those who "facilitated social interactions", and "representational go-betweens" as the ones who "created writings, drawings, maps, oral traditions".[24] These terms will assist us in better defining the roles of some of the interpreters employed by the Jesuits and will help clarify how these men were successful.

The episode of the "trickster", appearing in the early stages of the Jesuit efforts to enter China, is a testimony to the resourcefulness demonstrated by some interpreters, an essential quality highly valued by missionaries and travellers alike. In 1582, upon realizing that a legal return to the mainland was impossible, Ricci and Ruggieri attempted to travel inland to Guangzhou by breaking the rules. They relied completely on the talents of their interpreter who negotiated their passage by boat. His role therefore encompasses both the "physical" and "transactional" categories defined by Metcalf. Indeed, the *Journals* report how the "young man... somewhat of a trickster, showed the pilot the letters of the former viceroy and won him over to his confidence" (141). The ploy was risky: the viceroy, Chen Rui, had been dismissed and, strictly speaking, his letters held no power. A discovery of the lie would have brought about severe sanctions as travel without authorization was forbidden and the Jesuits had been expressly ordered to return to Macau. As the key actor in the Jesuit's attempt to travel inland, the interpreter himself risked punishment.[25]

The passage demonstrates some of the qualities that were required of interpreters. They needed to be familiar with the customs and laws of the country in which they operated and, since they were expected to be able to handle a variety of social situations, resourcefulness was essential. In this particular scene, the interpreter's "trick" relied on his ability to manipulate the objects of symbolic power that were the viceroy's letters and to persuade the captain: we can imagine his confidence showing the no longer valid letters and convincing the captain of his trustworthiness.[26] Nonetheless, despite the interpreter's initial success, other factors influenced the outcome of the interaction. Due to the pressures of other passengers who saw "foreign devils"[27] amongst them and refused to travel with them, the captain forced the missionaries to disembark. The incident shows that although mediators acquired power by virtue of their position, they were unable to control every aspect of interactions (in this case the xenophobia of the population) and the "trickster" failed in his endeavour through no fault of his own.

This "trickster" was not the only competent interpreter who assisted the Jesuits, and the *Journals* attest to how several others came to their aid. The adventure of the fathers Emanuele and Rodrigues, who travel within China without permission in order to recover the body of a deceased missionary, is another instance in which an interpreter uses his skills to benefit the Jesuits. This particular adventure takes place in a context of fear. Having almost been caught by the unexpected inspection of the boat they were travelling on, the two Jesuits call on a Portuguese merchant to rescue them: "[t]his man, with considerable risk to himself, sent over his interpreter to bring them across to his boat" (557). Although the merchant risked punishment if his involvement was discovered, it was his interpreter who assumed the physical risks of the rescue mission. In addition to his role as mediator, the man served as a physical and "faithful guide" (559) until the Jesuits were safely returned home. The interpreter first operated on a physical level, using his body as the tool in the rescue of the two Jesuits, but his contributions also encompassed language and made him a transactional go-between. Having "joined [a] fisherman and bargained with him" (558), he was responsible for securing a passage back to Macau aboard a small fishing boat. The combination of his two roles made him a particularly effective mediator, and he successfully saved both Jesuits in his care. However, as useful and competent as he was, it is pertinent to note that, once again, no information other than the nationality of his employer was provided about him.

Despite success in their recruitment of some interpreters, a series of partially incompetent, and thus unhelpful, men confirmed the fallibility of third parties. The many degrees of linguistic competence of interpreters often led to criticism of mediators in general[28] but also had the effect of motivating travellers, and especially missionaries, to learn the local languages. As a result, when interpreters' shortcomings were revealed, the Jesuits took over despite results that clearly reflected the challenges and limitations they faced themselves:

> [F]requent mention was made of the Christian religion, but without much effect, because one of the Fathers, knowing nothing of the Chinese language could only listen in silence, and the other, who could only struggle with it and spoke in a faltering and hesitant manner, had only an inexperienced old man as an interpreter. (178–79)

The fact that the missionaries travelled with such an "inexperienced" interpreter shows how difficult finding qualified interpreters could be.[29] Yet again, no information is given as to the identity of the man, but this raises more questions as to how, being so unskilled, he obtained the role of interpreter. The man, expected to behave as a transactional go-between, is unable to facilitate the interactions sought by the missionaries, and the repetition of the restrictive formula "could only" reflects the frustration of the two Jesuits (here Michele Ruggieri and António de Almeida). Experience and practice could reduce the difficulties though did not erase them: the transition from being forced to remain silent to struggling when speaking accurately reflects the complexity of the learning process and emphasises the time needed to master the language.

The Jesuits never wavered from their original goal of mastering the Chinese language themselves, an accomplishment that would rid them of mediators and bring them closer to their religious objectives:

> At times some of the more learned Chinese, who were interested in the religious customs of the Christian world, requested that we discuss the whole question more openly and freely and also the question of Chinese idols. These discussions were carried on through interpreters and at times without them, but the latter method was rather awkward because of a lack of knowledge of the language, which forced the foreigners to say what they could, rather than what they wanted to say. (156)

As the Jesuits became known in China, interest grew in the new religion they brought with them: the "request" of the Chinese literati shows a curiosity and a will to discuss "more openly and freely", which encouraged the missionaries and contrasted with the hostility of the authorities around the same period. With time, some conversations even took place without interpreters, proof that the missionaries were improving their skills and starting to communicate on their own. The latter are made explicit in the text by the opposition between "what they could, rather than what they wanted", which points to the complexities of the language while subtly reminding the reader of the Jesuits' usual eloquence when dealing with matters of faith.[30] The topics of conversation also contributed to the challenge of communication, religious matters being the most difficult to discuss since they required the translation of Christian concepts into the Chinese language.[31] Finding acceptable religious equivalents proved a complex task for interpreters, travellers, and missionaries worldwide. In China, it led to the controversy about religious terminology (*la querelle des termes*) and contributed to the Chinese rites controversy; both severely impacted the missionaries' position in the country.[32]

This complexity encouraged the Jesuits to redouble their efforts to master the language and to communicate clearly the concepts of their faith. In so doing, they transformed their relationship with interpreters, first by using them as teachers of the target language, then by communicating increasingly without them, and finally by replacing them. Indeed, in Ricci's *Journals,* interpreters appear very sparsely and usually solely in relation to missionaries who had recently arrived in the country and who were therefore in greater need of assistance. After years of learning and practice, Ricci's progress in the Chinese language and culture enabled him to communicate without go-betweens and to write several books in Chinese, including a *Treatise on Friendship*, which confirmed his status as an intellectual amongst Chinese scholars. His skills allowed him to assist other missionaries in the country while his reputation and connections contributed to securing the Jesuit position in China.

Conclusion

The difficulties of the language and the complexity of the linguistic situation in China presented one of the first and utmost obstacles for foreigners who attempted to enter the country. The radical difference between the Chinese language and the European languages known to the missionaries, the sheer number of regional idioms and their dialectal varieties, and the lack of "teachers" who could facilitate the learning process created a recurring theme in the *Journals*: the Jesuit missionaries' struggle to overcome obstacles of communication. Furthermore, the distinct political situation of early modern China offered a unique framework for the Jesuits who wished to take up residence in the country and to promote Christian beliefs and values. The Jesuits not only encountered a strongly hierarchised society but found themselves at the weaker end of the power structure and facing conflicts they had not expected. Rejected by a country that closed itself off to outsiders, they were initially utterly dependent on others to communicate.

Although they tried to dispense with interpreters by learning the Chinese language themselves, the particular nature of their interactions quickly made it clear that mediators would continue to be an essential and vital part in their communication with Chinese society and with the authorities. The recruitment of third parties who acted either as interpreters or language teachers – or both – cannot, however, be separated from questions of power; the very attempt to remedy the Jesuits' predicament created further complications of a socio-political nature. Consequently, Ricci's *Journals* reveal a hierarchy between the various actors of communication that remains stable for as long as the Jesuits are deprived of steady and powerful allies. Indeed, the interpreters the missionaries employed in order to remediate the imbalance of power in their favour were, like their employers, themselves subject to socio-political constraints. Their knowledge of the Chinese language and culture did not protect them from the high stakes of the situations in which they found themselves involved. This state of affairs further deprived the missionaries of the control they sought over their own destiny. Already subject to the authority of Chinese officials whose behaviour was not always consistent – as was the case of the viceroy who expelled the Jesuits but revoked the order almost immediately – the Jesuits were also dependent on interpreters who were themselves at the mercy of the very same authorities and whose position at the heart of this conflict meant that their very lives were threatened.

This crucial, formative period of the establishment of the Jesuit mission in China is portrayed well in Ricci's *Journals*. To this day, they provide the reader with a rich and comprehensive portrait of the challenges faced by the Jesuits in their attempts to negotiate extreme cultural differences, to learn the local idioms, and to integrate in a society that was far from welcoming of the Western intrusion. The role of communication and therefore language acquisition is at the foundation of this interface, and the description of its facilitators (i.e. interpreters and language teachers) is a key part of any story of intercultural contact. As our analysis of the *Journals* makes clear, the anonymity and general short drift given to these mediators should in no way be understood as minimizing the roles they played.

Notes

This research was made possible by the financial support of the Social Sciences and Humanities Research Council of Canada.

1 Nicolas Trigault (and Matteo Ricci), *De Christiana expeditione apvd Sinas svscepta ab Societate Jesv/ex P. Matthæi Ricij eiusdem Societatis com[m]entarijs. Libri V ad S.D.N. Pavlvm V in quibus Sinensis Regni mores, leges atq. instituta & novæ illius Ecclesiæ difficillima primordia accurate & summa fide describuntur/auctore P. Nicolao Trigavtio Belga, ex eadem Societate* (Augsburg, 1615).

2 David E. Mungello, *Curious Land: Jesuit Accommodation and the Origins of Sinology* (Stuttgart: Steiner, 1985), 46. Trigault's work was then later translated into English by another Jesuit. Where there are only page numbers after quotations, they refer to this English edition: Nicolas Trigault (and Matteo Ricci). *China in the Sixteenth Century: the Journals of Matthew Ricci, 1583–1610*, trans. Louis J. Gallagher (New York: Random House, 1953).

3 Michaela Wolf, "Translation as a Process of Power: Aspects of Cultural Anthropology in Translation," in *Translation as Intercultural Communication*,

ed. Mary Snell-Hornby, Zuzanah Jettmarová and Klaus Kaindl (Amsterdam and Philadelphia: John Benjamins, 1997), 123–33 (125).

4 On Jesuit language learning practices see, among others, François de Dainville, *L'Éducation des jésuites (XVIe-XVIIIe siècles)* (Paris: Éditions de Minuit (collection 'Sens commun'), 1978); Luce Giard, ed., *Les jésuites à la Renaissance. Système éducatif et production du savoir* (Paris: Presses Universitaires de France, 1995); and Philippe Rocher, *Le goût de l'excellence: quatre siècles d'éducation jésuite en France.* (Paris, Beauchesne, 2011).

5 "The Chinese who had become Christians at Macau ... and those who visited the place from inland for commercial purposes, were all inexperienced in the use of the curial dialect and of Portuguese as well. The provincial traders all understood the curial idiom but spoke it very poorly, as they were accustomed to conversing in their own dialect" (132).

6 The text indicates that "[w]ith the help of interpreters, [Ruggieri] gave much time to Chinese books" (134).

7 The central government delegated power to these men who governed over one or more provinces of the territory.

8 "Father Ruggieri...though speaking through an interpreter ... reminded the Governor that he had already taken up the study of the Chinese language and literature. This seemed to please him" (137).

9 Although Ruggieri remained on the mainland for some time, his invitation was revoked by the viceroy, Chen Rui, who was dismissed from his position. Accused amongst other things of attaining power through bribes and "fearing critics would use his patronage of foreigners as one more accusation once he left office", Chen Rui chose to cut his ties to the Jesuits who were forced to return to Macau. See R. Po-Chia Hsia, *A Jesuit in the Forbidden City: Matteo Ricci, 1552–1610* (Oxford: Oxford University Press, 2010), 77.

10 See among others: Paul Dreyfus, *Mattèo Ricci: l'homme qui voulait convertir la Chine* (Cahors: Éditions du Jubilé, 2004); Michel Masson, ed., *Matteo Ricci, un jésuite en Chine: les savoirs en partage au XVIIe siècle* (Paris: Édition Facultés jésuites de Paris, 2009).

11 Lia Xia, "Cultural and Historical Perspectives on Translation in China," in *Translating China*, ed. Luo Xuanmin and He Yuanjian (Bristol, Buffalo and Toronto: Multilingual Matters, 2009), 13–38 (25).

12 C. Shelke and M. Demichele, eds., *Matteo Ricci in China: Inculturation through Friendship and Faith.* (Rome: Gregorian & Biblical Press, 2010); Yiu Liu, *Harmonious Disagreement: Matteo Ricci and his Closest Chinese Friends* (New York: Peter Lang Publishing, 2015).

13 See also note 9.

14 George Minamiki, *The Chinese Rites Controversy: From Its Beginning to Modern Times* (Chicago: Loyola University Press, 1985), 1.

15 Ana Carolina Hosne, *The Jesuit Missions to China and Peru, 1570–1610: Expectations and Appraisals of Expansionism* (London and New York: Routledge, 2013), 83.

16 Minamiki, 1–2.

17 Ricci's competence allowed him to speak for himself, but the text indicates that his knowledge of the language was still imperfect: "[a]t first, the Viceroy did not fully understand what Ricci had said, but a Captain standing at his side, bent his knee and made clear to him just what the Father did say" (220). The statement in the *Journals* seems to have amplified Ricci's ability at the time. As the text was an object of missionary publicity, this amplification was likely a result of the desire either of Ricci (as the original writer) or of Trigault (as the translator and editor) to point out the progress made in China.

18 See (for the Far East) Geneviève Bouchon, *Inde découverte, Inde retrouvée 1498–1630, Études d'histoire indo-portugaise* (Lisbonne: Fundação Calouste Gulbenkian, Paris: Commission nationale pour les commémorations des découvertes portugaises, 1999), 308; (for North America) Normand Doiron, "Le mensonge de l'interprète: idéologie et utopie dans le *Quatriesme Voyage de Champlain* (1613)," *Littératures* 3 (1989): 65–75 (73).

19 Doiron, *ibid.*

20 Dejanirah Couto gave this *topos* as title of her article on the ambiguous role of two Portuguese interpreters in India ("The Role of Interpreters, or Linguas, in the Portuguese Empire during the 16th Century," *Electronic Journal of Portuguese History* 1(2003): 1–9).

21 Michael Cronin, "The Empire Talks Back: Orality, Heteronomy and the Cultural Turn in Interpreting Studies," in *The Interpreting Studies Reader*, ed. Franz Pöchhacker and Miriam Shlesinger (London: Routledge, 2002), 387–97; Maria Tymoczko and Edwin Gentzler, eds., *Translation and Power* (Amherst: University of Massachusetts Press, 2002).

22 Couto, 8.

23 Marie-Christine Pioffet, *La tentation de l'épopée dans les* Relations *des jésuites*. (Sillery: Septentrion, 1997); Isabelle Morlin, "Récits de voyages marchands dans la seconde moitié du XVIIe siècle: portrait du négociant en héros," *Études littéraires* 40 (2009): 13–29; Slyvie Requemora-Gros, *Voguer vers la modernité: le voyage à travers les genres au XVIIe siècle* (Paris: Presses de l'Université Paris-Sorbonne, 2012).

24 Alida C. Metcalf, *Go-betweens and the Colonization of Brazil, 1500–1600* (Austin: University of Texas Press, 2005), 8–12. See also Stephen Greenblatt, *Marvelous Possessions: The Wonder of the New World* (Chicago: University of Chicago Press, 1991).

25 The fear of punishment could also deter the interpreter from getting involved with travellers. One such example appears in the *Journals*, during the summary of the expedition of Ricci's predecessor Francis Xavier who attempted to enter China in the year 1552 and found a volunteer interpreter to assist him in his attempt. This man is described only as someone possessing the required skills to act as a functional interpreter. He was "not only skilled in his own language but versed in its literature" and "offered himself gratis to undertake the proposed expedition" (124). However, although his offer had been accepted, the realization of the dangers surrounding Xavier's plan to illegally enter the country frightened the interpreter and he "changed his mind and lost all interest in the affair" (125). Indeed, the laws of sixteenth-century China forbade any foreigners from entering the country without authorization and promised severe punishments for anyone who dared break the law, as well as for any complicit parties. These risks therefore deprived Xavier of a competent interpreter, and he decided to carry out his plan on his own – though he died before he could set out for the mainland.

26 The *Journals* do not make clear if the news of the viceroy's dismissal had reached the public but the Jesuits, aware of the situation, describe as a "trick" what might otherwise have simply been a normal interaction.

27 Hsia, 78–79.

28 Although criticism is frequent and praise is rare, the thorough analysis of a corpus of French travellers to North and South America and to South Eastern Asia and the Far East, completed for my PhD dissertation, suggests that most interpreters did behave in a satisfactory manner. The complaints about them would then be more likely motivated by concerns stemming from the preconceived ideas of the time than by real proof and were often linked to the desire

of travellers to portray themselves as heroes overcoming the obstacles of travel (which included dealing with interpreters).

29 The abilities of persons hired as interpreters were sometimes very questionable. Ricci's predecessor, Francis Xavier, arrived in Asia accompanied by Antonio, "a former pupil at the school in Goa" (124). The mention of Goa indicates he would have been educated by the Jesuits as the Society had opened schools in 1545 and a college in 1557. We know he was of Chinese descent but little else other than that he "had already rendered himself practically useless by forgetting his native language, for lack of practice" (124). Francis Xavier therefore found himself in a rather comical, although highly problematic, situation: relying on a man who was utterly unable to help him. While quickly dismissed, the obstacle represented by the gaps in Antonio's knowledge is noteworthy as it raises the question of how a person destined to serve as interpreter, and called explicitly by that title, would have been able, and allowed, to forget one of the languages required to perform his task effectively.

30 Margaret J. Leahey discusses this topic while supporting her analysis with excerpts from missionaries in New France. See "'Comment peut un muet prescher l'évangile?' Jesuit Missionaries and the Native Languages of New France," *French Historical Studies* 19 (1995): 105–31.

31 Sangkeun Kim, *Strange Names of God: The Missionary Translation of the Divine Name and the Chinese Responses to Matteo Ricci's "Shangti" in late Ming China, 1583-1644* (New York: Peter Lang, 2004).

32 Jacques Gernet, *China and the Christian Impact: A Conflict of Cultures*, trans. Janet Lloyd (Cambridge: Cambridge University Press, 1985); Minamiki; Mungello.

Bibliography

Bouchon, Geneviève. *Inde découverte, Inde retrouvée 1498-1630, Études d'histoire indo-portugaise.* Lisbon and Paris: Fundação Calouste Gulbenkian and Commission nationale pour les commémorations des découvertes portugaises, 1999.

Couto, Dejanirah. "The Role of Interpreters, or Linguas, in the Portuguese Empire during the 16th Century." *Electronic Journal of Portuguese History 1* (2003), 1–9.

Couto, Dejanirah. "'*Traduttore, traditore*' Vicissitudes linguistiques d'une ambassade portugaise dans les deltas du Gange et de l'Irraouaddy (1521)." In *Les langues de la négociation. Approches historiennes.* Edited by Dejanirah Couto and Stéphane Péquignot, 163–90. Rennes: PURennes 2017.

Cronin, Michael. "The Empire Talks Back: Orality, Heteronomy and the Cultural Turn in Interpreting Studies." In *The Interpreting Studies Reader*. Edited by Franz Pöchhacker and Miriam Shlesinger, 387–97. London: Routledge, 2002.

Dainville, François. *L'Éducation des jésuites: XVIe-XVIIIe siècles*. Paris: Éditions de Minuit 1978.

Doiron, Normand. "Le mensonge de l'interprète: idéologie et utopie dans le *Quatriesme Voyage de Champlain* (1613)." *Littératures 3* (1989), 65–75.

Dreyfus, Paul. *Mattèo Ricci: l'homme qui voulait convertir la Chine.* France: Éditions du Jubilé 2004.

Gernet, Jacques. *China and the Christian Impact: A Conflict of Cultures*. Translated by Janet Lloyd. Cambridge and New York: Cambridge University Press; Paris: Éditions de la Maison des sciences de l'homme, 1985.

Giard, Luce. *Les jésuites à la Renaissance. Système éducatif et production du savoir*. Paris: Presses universitaires de France, 1995.

Gomez-Géraud, Marie-Christine. "La figure de l'interprète dans les récits de voyage français au XVIe siècle." In *Voyager à la Renaissance, actes du XXVe colloque d'Études Humanistes de Tours*. Edited by Jean Céard, Marie Madeleine Fontaine and Jean-Claude Margolin, 319–35. Paris: Maisonneuve et Larose, 1987.

Greenblatt, Stephen. *Marvellous Possessions: The Wonder of the New World*. Chicago: University of Chicago Press, 1991.

Hosne, Ana Carolina. *The Jesuit Missions to China and Peru, 1570–1610. Expectations and Appraisals of Expansionism*. London and New York: Routledge, 2013.

Hsia, R. *Po-Chia A Jesuit in the Forbidden City: Matteo Ricci, 1552–1610*. Oxford: Oxford University Press, 2010.

Kim, Sangkeun. *Strange Names of God: the Missionary Translation of the Divine Name and the Chinese Responses to Matteo Ricci's "Shangti" in late Ming China, 1583–1644*. New York: Peter Lang, 2004.

Leahey, Margaret. J. "'Comment peut un muet prescher l'évangile?' Jesuit Missionaries and the Native Languages of New France." *French Historical Studies 19* (1995), 105–31.

Liu, Yiu. *Harmonious Disagreement: Matteo Ricci and His Closest Chinese Friends*. New York: Peter Lang, 2015.

Masson, Michel, ed. *Matteo Ricci, un jésuite en Chine: les savoirs en partage au XVIIe siècle*. Paris: Édition Facultés jésuites de Paris 2009.

Metcalf, Alida C. *Go-betweens and the Colonization of Brazil, 1500–1600*. Austin: University of Texas Press, 2005.

Minamiki, George. *The Chinese Rites Controversy: From its Beginning to Modern Times*. Chicago: Loyola University Press, 1985.

Morlin, Isabelle. "Récits de voyages marchands dans la seconde moitié du XVIIe siècle: portrait du négociant en héros." *Études littéraires 40* (2009), 13–29.

Mungello, David E. *Curious Land: Jesuit Accommodation and the Origins of Sinology*. Stuttgart: Steiner 1985.

Mungello, David E., ed. *The Chinese Rites Controversy: Its History and Meaning*. Nettetal: Steyler Verlag 1994.

Pioffet, Marie-Christine. *La tentation de l'épopée dans les Relations des jésuites*. Sillery: Septentrion 1997.

Requemora-Gros, Slyvie. *Voguer vers la modernité: le voyage à travers les genres au XVIIe siècle*. Paris: Presses de l'Université Paris-Sorbonne 2012.

Rocher, Philippe. *Le goût de l'excellence: quatre siècles d'éducation jésuite en France*. Paris: Beauchesne 2011.

Shelke, Christopher and Mariella Demichele, eds. *Matteo Ricci in China: Inculturation through Friendship and Faith*. Rome: Gregorian & Biblical Press 2010.

Trigault, Nicolas (and Matteo Ricci). *De Christiana expeditione apvd Sinas svscepta ab Societate Jesv/ex P. Matthæi Ricij eiusdem Societatis com[m]entarijs. Libri V ad S.D.N. Pavlvm V in quibus Sinensis Regni mores, leges atq. instituta & novæ illius Ecclesiæ difficillima primordia accurate & summa fide describuntur/auctore P.* Nicolao Trigavtio Belga, ex eadem Societate. Augsburg, 1615.

Trigault, Nicolas and Matteo Ricci. *China in the Sixteenth Century: The Journals of Matthew Ricci, 1583–1610*. Translated by Louis J. Gallagher. New York: Random House, 1953.

Tymoczko, Maria and Edwin Gentzler, eds. *Translation and Power*. Amherst: University of Massachusetts Press, 2002.

Witek, John W. and Michel Reis, eds. *Religion and Culture. An International Symposium Commemorating the Fourth Centenary of the University College of St Paul*. Macau: Instituto Cultural de Macau and Ricci Institute for Chinese-Western Cultural History, University of San Francisco 1999.

Wolf, Michaela. "Translation as a Process of Power: Aspects of Cultural Anthropology in Translation." In *Translation as Intercultural Communication*. Edited by Mary Snell-Hornby, Zuzanah Jettmarová and Klaus Kaindl, 123–33. Amsterdam and Philadelphia: John Benjamins 1996.

Xia, Lia. Cultural and Historical Perspectives on Translation in China." In *Translating China*. Edited by Luo Xuanmin and He Yuanjian, 13–38. Bristol, Buffalo and Toronto: Multilingual Matters 2009.

3 "Strange accidents"

Navigating conflict in *Sir Thomas Smithes voiage and entertainment in Rushia* (1605)

Natalya Din-Kariuki

In 1605, *Sir Thomas Smithes voiage and entertainment in Rushia*, an account of the 1604–1605 embassy of Sir Thomas Smythe (1558–1625) to Tsar Boris Godunov, was published anonymously.[1] A prefatory note "To the Reader" explains the origins of the account:

> Reader, the discourses of this voyage (at the comming home of the Gentleman that was chiefe in it and his company into England) affoorded such pleasure to the hearers, by reason the accidents were strange and Novell, that many way-laid the Newes, and were gladde to make any booty of it to delight themselves, by which means, that which of it selfe being knit together was beautifull, could not chuse but shew wilde, beeing so torne in peeces. So that the itching fingers of gain laid hold upon it, and had like to have sent it into the world lame, and dismembred. Some that picke up the crums of such feasts, had scrapt togither many percels of this Rushian commoditie, so that their heads being gotten with child of a Bastard, there was no remedy but they must be delivered in *Paules Church-yard*. But I taking the truth from the mouths of divers gentlemen that went in the Journey, and having some good notes bestowed upon me in writing, wrought them into this body, because neither thou shouldst be abused with false reports, nor the Voyage receive slaunder ... Read and like, for much is in it worthy observation.[2]

This note establishes a contrast between two different ways of presenting the "discourses" of Smythe's embassy. One focusses on a selection of "accidents", incidents or events, particularly those that happen unexpectedly, while the other offers a more comprehensive account, situating the aforementioned accidents within it. In the author's view, the latter produces work that is not only of a higher aesthetic quality ("beautifull", not "wilde"), but also more credible ("truth", not "false"). It is also more substantial, supplying the reader with "feasts" rather than "crums"; the related claim that the book contains material for "worthy observation" and not just pleasure recalls the Horatian sense of the poet's duty to profit as well as delight. Finally, the author argues that an account of this kind makes a more meaningful contribution to public life. Unlike the self-interested persons who obtained the "Newes" through disreputable means, he uses material from the "mouths" and "notes" of the travellers themselves. The author thus presents himself as a heroic figure who has "wrought" the disparate fragments of the voyage into a unified whole and, in so doing, benefitted the reputations of Smythe and his company, as well as the minds of English readers.

The author's reference to the defective text as "torne in peeces" recalls the way in which humanist editors often describe their role as a reunification of scattered parts, framing this reunification as a civic duty.[3] This language carries a particular salience in the context of travel, in which the act of editing travellers' accounts is conceived as a form of travel or travail itself. Richard Hakluyt, for example, states that his project is to "gather ... and as it were to incorporate into one body the torne and scattered limmes" of English voyages; these "labours", Hakluyt says, have been carried out "for the common-wealths sake", as well as the reader's.[4] Writing along similar lines, another editor going by the name "Theophilus Lavender" introduced a travel account by explaining that he had "gathered the matter" of several letters in order to "make one body of them", dedicating his "paines and labour" to the "publike good".[5] Like his contemporaries, the author of *Smithes voiage* emphasizes the importance of the proper arrangement of textual material in order to legitimize, and gain credit for, his labour.

While a concern for the reintegration of disparate material is something that *Smithes voiage* has in common with other published travel accounts, it also advances further, more unusual generic claims by referencing recent developments in Russian politics. The title page explains that the relation of Smythe's voyage is combined "With the tragicall ends of two EMPERORS, and one Empress, within one Moneth during his [Smythe's] being there: And the miraculous preservation of the now raigning EMPEROR, esteemed dead for 18 yeares". The assassinations of Boris Godunov; his son, Feodor II; and his wife, Maria Skuratova-Belskaya, followed by the arrival of a pretender to the throne who claimed to be the rightful heir, Dmitry Ivanovich, took place during a period of Russian history that has since been called the Time of Troubles (1598–1613) and that is often characterized as a civil war.[6] The language of the title page foregrounds the narrative shapes, or plots, of these events, presenting them as tragic "ends" followed by a "miraculous preservation" reminiscent of a romance.

Narrative shape is also a concern of the account proper, in which the author describes Russia's political conflict in terms of poetic craft. For instance, in a passage on the "tragedies" that befell the Godunovs, he calls for "some excellent pen-man" to "lively, naturally, or indeed poetically delyneare [*sic*] or enumerate these occurrents". As examples of excellent poets, he names "*Sydney* Prince of *Poesie*", "Lord *Salustius*" (the Huguenot poet

Guillaume de Salluste Du Bartas), "the noble *Foulk-Grevill*" (Fulke Greville), and "our Lawreat worthy *Benjamen*" (Jonson); he cannot match up to them, he says, but "*Hic labor, Hoc opus est*".[7] The Latin tag is doubly allusive: "*Hoc opus, hic labor est*" is a phrase from Virgil's *Aeneid*, which Sidney uses in his *Defence of Poesy* to describe the task of "moving" people to "well-doing".[8] By borrowing the phrase, the author implies that there is a parallel between the "work" of poetry and that of travel writing. Further, he suggests that this work has to do with the proper handling of "occurrents" (incidents or events), a term that reinforces the emphasis on accidents in the book's prefatory note.

Smithes voiage points to the inextricability of travel and conflict in the early modern period in four ways. First, by highlighting the political accidents that occurred during or shortly after Smythe's embassy, it calls attention to the frequency with which travellers, particularly diplomats, had to navigate conflict during their time abroad. Second, by demonstrating that travel writers thought carefully about their choices of style and form, particularly their presentation of accidents, it reveals that early modern travel writing is a site of aesthetic conflict. Third, by claiming that Smythe's reputation, as well as the integrity of the reader, is at stake, it shows that the navigation and narration of accidents involve ethical conflicts, which could have implications for the wider public. Finally, it is inflected by Smythe's experience of conflict in a different context. In 1601, he was suspected of involvement in the Essex rebellion and imprisoned in the Tower of London. After serving a short sentence and paying a substantial fine, Smythe was in favour again from 1603, when James VI and I appointed him to a number of prestigious positions. Smythe's appointment as ambassador to Russia was part of this upward trajectory. But the earlier encounter with scandal, however brief, left its mark on *Smithes voiage*. Analogizing the experiences of imprisonment and sea travel, the author claims that the extreme weather Smythe encountered on his journey to Russia meant that it was as if he "were again in the *Tower of London*, as he is in the ship: now tost, then becaulmed: in feare and hope within one moment". Although this is the only direct reference to the Essex rebellion in *Smithes voiage*, the analogy captures the relationships of travel, conflict, and accidents that preoccupy the account as a whole.

In order to extend our understanding of these interrelated issues, this chapter examines the treatment of accidents in *Smithes voiage* through its connection to William Scott (*c.* 1571–*c.* 1617), the author of a poetic treatise, *The Model of Poesy* (*c.* 1599), and of a partial English translation of Du Bartas's *La Sepmaine* (1578). Scott accompanied Smythe on his trip to Russia and later gave an account of it in a letter to Robert Cecil, Earl of Salisbury and secretary of state to James VI and I.[9] The central role of questions of plot, the ordering of accidents and occurrents, in Scott's writings resonates with the focus on plot in *Smithes voiage*, suggesting that Scott had a role in the composition of *Smithes voiage*, or, alternatively, that it was written for him.[10] By reading *Smithes voiage* alongside Scott's extant body of work, the chapter demonstrates that analogies with literary form, such as those used in the paratexts of *Smithes voiage* and underscored by the author's conception of himself as a labouring poet, are integral to travel writing: poetics supplied travellers with the narrative shapes with which to think about, and describe, their experiences.

This chapter opens by considering the virtue of prudence in humanist thought and travel advice, and its relationship to plot, to show that the account's characterization of Smythe as a man of "plot" and "provision", able to cope with sudden turns of events, is informed by these bodies of thought. Another key concept in the early modern understanding of plot is "disposition", the part of rhetoric that deals with the arrangement of material (what early English rhetorician Thomas Wilson calls the "settelyng or orderyng of thynges"[11]). I shall explore this by examining the association of plot and travel in Scott's translation of Du Bartas, in the *Model*, and in his letter to Cecil, placing particular emphasis on Scott's account of the narrative device of accident, also termed "reversal", or, in the transliterated Greek, *peripeteia*. Finally, I explore how its precepts outlined in Scott's writings play out in *Smithes voiage*.

The emphasis on plot in *Smithes voiage* is surprising, even paradoxical. We might assume that in a travel narrative, disposition is automatic: that its *fabula* and *sjuzhet* are necessarily simultaneous, and story-telling consecutive; we might assume, moreover, that the accidents of travel can only be unforeseen. But this chapter reveals the formal artfulness, the careful deployment of accident, and the poetic prudence that informs the composition of *Smithes voiage* and the theoretical thinking behind it. It shows that *peripeteia* is an apt device for the writing of travel, providing a tool with which to navigate uncertainty and unfamiliarity, mishap and error. In doing so, it establishes a connection between the matter of travel writing, its accidents and occurrents, and the way in which this matter is arranged.

Prudential plotting: the example of Sir Thomas Smythe

By the time he travelled to Russia, Smythe had already held the roles of trade commissioner; member of the Company of Merchant Adventurers; and governor of the Muscovy, Levant, and East India companies.[12] These years of experience in trade form the backdrop of his appointment as ambassador to the tsar. Smythe's mission was to acquire trading privileges, building on the work carried out by other English ambassadors and agents in Russia such as John Merrick. The opening of *Smithes voiage* explains this context, setting Smythe's embassy within the longer history of Anglo-Russian contact. It also comments on the importance of trade to international relations more generally, claiming that "Traffike" is what "tyes kingdoms together in mutuall Amitie", joining "handes" and "soules" together in "harmony" and "*Peace*".[13] As an introduction to Smythe's embassy, this depiction of trade is somewhat misleading. In practice, the negotiations in Russia involved navigating a set of interlocking conflicts. The clearest illustration of this is the instructions Smythe received from Robert Cecil before his departure, which provide guidelines for how to respond to questions about England's peace with Spain and how to reassure the tsar that England would not join forces with the Ottoman Turks to wage war on Russia.[14] As the instructions suggest, discussions about Anglo-Russian trade were bound up with broader geopolitical concerns, including the possibility of military conflict. This meant that Smythe had to be prepared to address these concerns, if the situation required it, in order to persuade the tsar to engage meaningfully with his requests.

Conflict, or the anticipation of it, took other, more local forms, too. *Smithes voiage* explains that Smythe delivered two speeches to his merchants at the beginning of the embassy, the first before they boarded their ships in Gravesend and the second before they departed Archangel, encouraging them to "love" and "delight" in each other, to "expresse [their] loves in helping and cherishing", and to show "love and curtesie" to the strangers they encountered, in keeping with the "message of peace and Amity" they brought to Russia.[15] As Alison Games writes, these speeches, which were designed to "preempt conflict", are an example of the efforts that English ambassadors, governors, and consuls made to maintain decorum within their communities abroad.[16] In *Smithes voiage*, such preempting is typical of Smythe, who is defined by his ability to look ahead in ways that align him with the virtue of prudence, as it is understood in humanist moral philosophy. One episode, in which Smythe invited all of the merchants on board his ship for a feast, demonstrates this with particular vividness:

> [T]hey found him like a good hous-keeper at home, and like a bountifull sir, command them the best cheare. In a word, the Ambas. own and ordinarie praise shall ease mee of more labor in proving it, al must confesse necessarily, which was this: That al the provisions whatsoever (yea his purse likewise) was every ones & no one, no not himselfe for all.[17]

The comparison of Smythe to "a good hous-keeper" resonates with the language and ideas of moral philosophical writing on prudence. Thomas Elyot's *Boke Named the Governour*, for example, defines prudence as "the captayne or maystres of vertues", which, like "the house wife ... doth investigate & prepare places apt and convenient, where other vertues shall execute theyr powers or offices".[18] Further, the description of Smythe's hospitality recalls the close etymological and conceptual links between "prudence", "providence", and "provisions", terms that connote foresight and preparation.[19] Elyot's description of prudence draws on these links, by turning repeatedly to metaphors of vision. Moreover, Elyot's view that providence or foresight is essential for the governance of the "publike weale" is shared by *Smithes voiage*, in which the sea is imagined as a "miserable commonwealth to live in", where "mens eies must ever be open to al dangers", and Smythe's success attributed to his "for-knowledge". Smythe's character, as described in *Smithes voiage*, thus resembles that of Elyot's ideal governor.

This characterization of Smythe is also in sympathy with the precepts of travel advice literature, which aimed to discipline the traveller's vision.[20] Like the humanist moral philosophy to which it was closely related, apodemic writing emphasized the importance of foresight, encouraging the inexperienced traveller to, in Fynes Moryson's words, "forecast with himself his owne purposes and ends".[21] Dealing with this subject, the Flemish scholar Justus Lipsius compares travel to throwing a javelin at a specific target: "Like those who, when throwing a javelin, do not do so in vain, but set their sights on the mark at which they aim: so too should you determine in advance what you are searching for in travel".[22] In John Stradling's occasionally rather loose translation of Lipsius, this is transformed into an image of archery, declaring that travellers must "choose themselves out some mark ... whereat they aime and shoote".[23] Sidney employs the same metaphor, advising his brother Robert to keep "imprinted in [his] mind the scope and marke you meane, by your paines, to shoot at" during his travels in France.[24] By the same token, writers like Elyot praise Ulysses, upheld as an ideal traveller in this period, for "wonderfull prudence and fortitude ... in his passage from Troye".[25] Smythe, the seventeenth-century ambassador undertaking a "passage" through Russia, is thus part of a lengthy genealogy that extends back to antiquity.

The prudence of Smythe's diplomacy, the quality that enables him to preempt and respond to conflict, has an important rhetorical dimension. While the aforementioned instructions from Cecil discipline his speech, by stipulating what, and how, he should "say" and "declare" to the tsar, his leadership of the merchants takes the form of oratory, enabling him to deliver the urgings and exhortations that are so crucial to the embassy's success. Further, *Smithes voiage* refers frequently to Smythe's eloquence, claiming that he speaks "very wiselie" and is so persuasive that his words act on listeners "like a *Spell*".[26] The emphasis placed on the prudence of Smythe's speech is apt because prudence plays a key role in the rhetorical tradition. It figures especially prominently in accounts of rhetorical disposition, which is defined as the domain of the "prudent man".[27]

The relationship between travel and rhetoric is reciprocal: just as the rhetorical tradition shapes travel, travel shapes the rhetorical tradition. A number of rhetoricians turn to metaphors of travel to explain the principles of disposition. Quintilian, who says that "prudent planning" is essential for the orator, explains that speaking without disposition is like "drifting without a pilot, incoherent, repetitive, incomplete, wandering in the dark, as it were, in unknown places, with no fixed beginning or end, and guided by chance rather than design".[28] Echoing Quintilian, Wilson explains that an oration "hath little force with it, and dothe smally profite, which is utterde without all order"; an orator who cannot dispose will "wander" and not know "how to goe", until he is left "gropyng in the darke".[29] These accounts of disposition, which parallel the language and assumptions of apodemic writing, illustrate that the orator, like the traveller, needed to exercise prudence in order to determine his desired destination as well as the route required to reach it.

This aspect of the rhetorical tradition thus informs the representation of Smythe in *Smithes voiage*. Indeed, Smythe himself specifically links speech and prudence, urging his company to "be wise and very circumspect what wee speake" in order to avoid conflict with the others they encounter.[30] Smythe's call for "circumspect" speech resonates with the emphasis on sight and foresight in rhetoric because "circumspect" is from the Latin *circumspicere*, "to look around". As we have seen, such foresight is specifically associated with the act of disposition. It is no coincidence, then, that Elyot's preferred verb for the activities of the "circumspecte governour" is "dispose". In its account of Smythe's embassy, then, *Smithes voiage* demonstrates the inevitability of conflict in travel in the early modern world, but it also reveals that this conflict could be mitigated, if not avoided entirely, by employing strategies drawn from moral philosophy, travel advice, and rhetoric, all of which prioritize prudence.

The task of prudent disposition in *Smithes voiage* is not exclusive to Smythe, however: it is also required of the narrator. Like Smythe, he prioritizes disposition and aims to "provide" for his reader. Exercising foresight, he often looks ahead to the parts of Smythe's voyage that are still to come, in order to decide which to relate immediately and which to "defer".[31] As we have seen already, he conceives of his writing as matter for consumption, selecting and ordering material with the reader's "feeding" in mind. His concern with disposition is particularly clear in a passage on the Northern Seas:

> Now being far entred the Northren [*sic*] Seas, I could here feed you: yea, make you a feast royall with little cost: and after that disport you excellent strangely, with Water-Tylts and Barryers: bring in the *Levyathan* with his *Oystershel-armed shield* … The gallant Sea-fight betwixte the huge *Whale, the Sword-fish, and the Thresher:* Two against one, Rapyer and dagger against your two-hand-sworde: Or I might speake verye much to little purpose of the *STURGEON*, whose Rowes makes the *Italians* delightfull *Caviare*, of which this Nation makes the most and best of any. The *Porpus*, *Belluga, Sharke, Seale, Grampas, Salmon*, &c…were it not unkindlye and unnaturall to feede you, with those creatures, that rather are satisfied with men, whereby it would prove a harde disgestion [*sic*]. Besides, I might discourse the Witchcraft of the Seas (as I may tearme it) when one Ship shall be

> becalmed, and Divers have excellent gales or undergo the description of the Capes, Sandes, Shelles, Islands, Rockes, Whirlepooles, &c.[32]

This passage is concerned with, and exemplifies, rhetorical *copia*, here defined as speaking "very much to little purpose", a mismatch of *res* (matter) and *verba* (words). The use of the term "purpose" reinforces the concern with the virtue of foresight expressed elsewhere in the account, while making its connection to travel more explicit: related to the term "propose", which is itself influenced by the Latin *propositus*, "put or set forth" and the Old French *poser*, "to place", one of the senses of "purpose" is "to be bound for", used in reference to a place or with an adverb of direction. Fittingly, one of the examples of this sense given in the *OED* comes from an early modern travel account, William Lithgow's *Totall Discourse*, in which Lithgow "purposed for Constantinople".[33]

In *Smithes voiage*, "purpose" is also associated with composition. The reference to ships that "undergo...description" explicitly connects the acts of writing and travelling; specifically, it suggests that a "description" is inherently meandering, taking routes around "Capes", "Islands", "Whirlepooles", and other geographical features, lacking a clearly defined destination. Further, the author's reference to the "Witchcraft of the Seas" is possibly an allusion to Odysseus. The passage as a whole shares the *Odyssey*'s sense of a perilous and convoluted sea journey and of the traveller's need to resist the siren temptation of such a narrative.

Yet in the course of passing over it, the author confirms that the thought of a divagating narrative is, indeed, tempting. This temptation, and the need to resist it, is explicitly related to digestion: the author considers telling the reader about the sea creatures that can be found there, but decides against it; it would be "unkindlye and unnaturall to feede" the reader this "feast royall" and "disport" them, as it would "prove a harde disgestion [*sic*]".[34] This language is apt, because the originary meaning of digestion is disposition, derived from the Latin *digerere*, to separate, divide, distribute, or dispose. In keeping with the dispositional emphasis of the account, "disport" implies a connection between "harde disgestion" and the passage through a place: to "disport" is to amuse or entertain, but it also connotes a physical movement, a leading, turning, or carrying away from more serious matters.[35] Finally, this passage is replete with conflict. It describes a sea at war with itself, as its creatures take up arms to do battle, participating, variously, in "Water-Tylts and Barryers" and a "Sea-fight". Comic as they are, these aquatic jousts and fights are analogous to the conflict that takes place on land, offering yet another example of the close relationships between travel, conflict, prudence, and disposition.

The author thus implies that the task of "feeding" the reader is bound up with the purposeful selection and arrangement of material. Further, he suggests connections between the ethical ideal of prudence, rhetorical disposition, and questions of narrative or plot. As we have seen, he presents himself as a prudential figure, able to foresee, and thus navigate, any accidents that may occur in the course of his composition. He conceives of his material in the language of poiesis and places particular emphasis on its narrative shaping, the various ways in which its events can be ordered, as well as the implications that this ordering can have. By describing the process of writing as a sea journey, the author implicitly parallels his own actions with those of Smythe: both are concerned with providing for their respective constituencies, and both try to maximize the profitability of their resources while observing, but steering clear of, accident and conflict.

In these respects, *Smithes voiage* resonates with the formal and ethical concerns of sixteenth-century prose fiction, which, as Lorna Hutson and Nandini Das have shown, also conceives of narrative in spatial or cartographic terms and implies a connection between the navigation of this narrative space – by the reader and the protagonists – and changes in fortune. These works were designed to be read in an active and pragmatic fashion, to be emplotted; readers were expected to use narratives as a mental map on which they could place material so that it could be invented, or found, in future.[36] *Smithes voiage*, a travel account, exploits the spatial connotations of narrative and also reinforces the link between narrative and fortune. But it differs from prose fiction in foregrounding the prudence and craft of the *author* rather than that of the reader. Consequently, it is more closely engaged with the *dispositio* of the text, and the prudential disposition of the author, as well as rhetorical precepts about the handling of narrative, particularly the relationship between accidents and the narrative as a whole, than with the reader's invention. In order to explain the significance of rhetorical disposition in *Smithes voiage*, the next section of this chapter turns to the writings of William Scott, in which the narrative device of accident, and concerns about travel and conflict, are central.

William Scott's theories of disposition

Scott's writings offer essential context for the study of *Smithes voiage* in particular and of issues of travel and conflict in general. As mentioned earlier, Scott participated in Smythe's embassy and possibly played a role in the authorship of *Smithes voiage*. His translation of Du Bartas and his treatise, the *Model of Poesy*, enrich and develop ideas of prudence, plot, disposition, and accident in ways resonant with both *Smithes voiage* and the rhetorical tradition with which *Smithes voiage* is engaged. The *Model* is especially significant, in that it contains the most detailed account of *peripeteia*, or plot "reversal", in the vernacular literature of the sixteenth century. Peripeteia is used in the description of conflict in *Smithes voiage* as well as in Scott's letter to Robert Cecil, which offers a condensed summary of the "accidents" which occurred during the embassy. The biographical and intellectual connections between Smythe and Scott mean that they should be read together, understood as companions in text as they were in travel.

Early modern travel advice suggests that Du Bartas was considered a traveller's poet. James Howell, for example, advises English travellers to read French poets, "specially Du Bartas", during their time abroad.[37] Fittingly, Scott's translation of Du Bartas anticipates the themes of travel, plotting, and entertainment that recur in *Smithes voiage*, as well as in the *Model*. For example, both Du Bartas and Scott describe writing as a kind of travel. In an "Invocation" to God ("Le poëte invoq Dieu") at the start of the "First Day", Du Bartas asks God to "refine" his spirit and "enrich" his work, in order that he may "[t]he diverse, and choyce wares of this wordles [*sic*] shopp display".[38] Scott extends and intensifies the "shop" metaphor in the following interpolation:

> And graunte this forraine Prize, (High God) I, weakling, maye
> Unto our coastes transporte, the readye, though rough waye;
> Yf fraught to heavie bee, errors pathes diverse wyde,
> Thy spiritt Gale, and Pilot, my barke dryve and guyde.[39]

Scott's lines literalize the etymology of "translation" as "moving" across.[40] Like a merchant trader, he asks for God's guidance in safely transporting Du Bartas's work, "this forraine Prize", to English shores over the rough seas and paths of error. Something similar happens in Scott's translation of the "Second Day", where the last two lines of Du Bartas's invocation read as follows:

> Thy holy Phare direct me, in my voyage right,
> That I may reach my Inne before my day loose light[41]

"Phare" is the French word for lighthouse, which derives from the Greek φάρος (*pharos*). Once again, Scott builds on Du Bartas's nautical imagery in an interpolation:

> And as my slender barck [*sic*] is forced in the mayne,
> From harbour quiet safe, the heavens grace me susteyne,
> As thou hast steerd the travailes of the former daye
> Thy spiritt be the Pilott, still to guide my waye[42]

By asking God to be the "Pilott" and "Phare" for their poetic voyages, "plots" of imagination, Du Bartas and Scott anticipate a sense of "plot" (in addition to the existing senses of "chart" or "map") that would emerge later in the century: that of a "sea-card", used by sailors to determine whether or not they were on course.[43]

The first sentence of Scott's *Model* draws on a similar set of images to those in his translation of *La Sepmaine*:

> Because all doctrine is but the orderly leading of the mind to the knowledge of something convenient and possible for us to know which before we knew not, and the way wherein we are to be led by this clew of discipline is the space between ignorance and science, we are therefore to make this passage by certain degrees and steps, of necessity taking that which is nearest our understanding and which giveth light to that that ensueth, still proceeding from things known to things unknown, until we have gone through all the mazy paths that might trouble or stop the voyage of our mind in discovery of those riches she naturally covets.[44]

Scott imagines learning as a voyage, and knowledge as the riches acquired through this voyaging, not unlike the "forraine Prize" he wishes to transport across the seas in his work as a translator. The figurative voyage is compared to wandering through the "mazy paths" of a labyrinth, while the *Model* is presented as a "clew" that will help the reader undertake this "passage". In classical mythology, a "clew", a ball of thread, provides the means by which Theseus finds his way through the labyrinth, thanks to the ingenuity of Ariadne, the daughter of King Minos.[45] For Scott, the image of the clew helps to introduce a commitment to logical procedure, the "orderly", that characterizes the *Model* as a whole.[46] As he subsequently explains, the quality that will help writers to proceed in such an "orderly" fashion is prudence, the "inseparable companion of the poet's".[47] By drawing a connection between travel, disposition, and prudence, the *Model* resonates with the language and ideas of humanist moral philosophy as well as with the rhetorical tradition.

Voyages are rarely as orderly as the opening of the *Model* suggests, however: in the "mazy paths" of a labyrinth, there is a high chance that there will *not* be a clew and that the traveller will get lost. First theorized in Aristotle's *Poetics*, peripeteia is a surprising or unexpected turn of events. It often, but not always, appeared alongside *anagnorisis* or recognition, which typically came after, and resolved, the ambiguities and uncertainties created by reversals.[48] As the example of *Smithes voiage* suggests, travel writing is centrally concerned with how to respond to, and narrate, such reversals, or "accidents".

Further sections of the *Model* help to explain why peripeteia is so well suited to the demands of accounting for one's travels. Following Julius Caesar Scaliger, Scott claims that there are four "virtues" of poetry.[49] His discussion of the third virtue, sweetness, is most pertinent. One of the best ways to achieve the quality of "sweetness", according to Scott, is the narration of "strange accidentes". He explains that when the development of a narrative goes against the reader's expectations, they will respond either with "gladness" to a "more glad and welcome" turn of events, as in comedy, or with "astonishing admiration" if events are "more sad and unworthy", as in tragedy.[50] To elicit these responses, the poet should deploy "smooth and cleanly alterations" that work by "cozening the expectation". Scott expands on this point by giving an example of peripeteia, which he calls an "indirect compassing of matters". He refers to an episode in Sidney's *New Arcadia*, in which Musidorus, who had lost his friend Pyrocles "upon the Sea", unexpectedly finds him again when they recognize each other in battle.[51] Thus in Sidney's romance, as in *Smithes voiage* and the *Model*, peripeteia is associated with travel. As these passages from the *Model* suggest, peripeteia is central to Scott's poetics.[52] Paradoxically, it is related to his emphasis on proceeding "orderly". Peripeteia, which means, literally, a sudden change, enables the writer to produce certain responses in the audience and also provides an instrument with which to navigate or "compass" the matter at hand.[53]

Scott draws on these ideas in his letter to Cecil, which discusses the handling of narrative in relation to the "strange accidents" that occurred during Smythe's embassy.[54] As in the *Model*, Scott emphasizes the significance of plot "turns" and recognizes that it is Cecil's wish to hear immediately about the "mayne or Cardenall accident" that occasions the letter's composition, the "death of *Borise*" and the rise of his challenger, "*Dmetree I Ivanowich*".[55] He gives a brief sketch of these events but then ends the letter by emphasizing that there are many aspects of the Russian situation that remain unexplained, particularly the sudden appearance of the man claiming to be Dmitri Ivanovich, heir to the throne. These knots in the narrative cannot be resolved or, in Scott's words, "will hardly be aunswered". But the lack of resolution has a purpose: Scott uses it to remind his correspondent that the letter is a work in progress ("slight drauftes and lineaments") that needs to be revised, completed, and integrated into a fuller narrative and combines this claim with a modesty *topos*, suggesting that Cecil is better suited than he at dealing with such "slight" provisional writing:

> [O]f all this and much more I am reddy to give account, and this is all I Can suddenly saye referring to yowr Lordships highe Judgement, that Can make more out of theis slight drauftes and lineamentes of so great an affaire, then I Can; drawe it to parfecter forme.[56]

Here, the quality of "Judgement" (often used as a synonym of prudence) is attributed to the reader rather than the traveller and writer. Scott offers Cecil the basic structure of the story and its main "accident" but not a complete resolution, expecting him to use his judgement to "drawe it to parfecter forme". This is typical: as mentioned earlier, early modern readers, particularly those involved in or preparing for a public role, were expected to read prudently and actively, to make sense of a text by "plotting" their way through it. Although Scott's letter to Cecil and *Smithes voiage* emerge from the same embassy and focus on the same set of conflicts, they differ in terms of emplotment. In the *Voyage*, the reader receives the events emplotted, while in the letter to Cecil, these events are given to Cecil to emplot himself: the letter thus offers up raw material that awaits Cecil's dispositional reading. However, the texts share the sense that there are two ways of presenting Smythe's travels, as a selection of "accidents" or as a "knitted together" narrative, and that the latter is preferable. With this contrast of narrative styles in mind, the final section of this chapter returns to *Smithes voiage*, to focus specifically on its description of the accidents that befell the Godunovs.

Accidents and occurrents in *Smithes voiage*

Halfway through *Smithes voiage*, the narration of the accidents named in the account's title ("the tragicall ends of two EMPERORS, and one Empress" and "the miraculous preservation of the now raigning EMPEROR") begin. The narrator explains that during a stay in "*Vollagde*" (Vologda), the ambassador received news of the tsar's "suddaine and untimely death". As befitting his previously demonstrated prudence, Smythe responded with "mature deliberation".[57] However, as the events unfold, Smythe begins to fade into the background, as the narrator draws extended comparisons with poiesis. One particular example stands out. Describing the death of Boris's son, Feodor II, he imagines that Feodor had anticipated his death, knowing

> that his fathers Empire and Government, was but as the *Poeticall Furie in a Stage-action*, compleat yet with horrid and wofull Tragedies: a first, but no second to any *Hamlet*; and that now *Revenge*, just *Revenge* was comming with his Sworde drawne against him, his royall Mother, and dearest Sister, to fill up those Murdering Sceanes; the *Embryon* whereof was long since Modeld, yea digested (but unlawfully and too-too vively) by his dead selfe-murdering Father.[58]

The narrator's choice of language is compelling. "*Poeticall Furie*" recalls the *furor poeticus*, a classical theory of poetic inspiration that was picked up and reinvigorated by neo-Platonists such as Ficino, Du Bellay, and Ronsard. Poetic fury had a secondary, less specific set of connotations that included creative madness, passion, and possession, which, as "*Stage-action*" suggests, found particular purchase in early modern drama.[59] There may also be an implicit pun on "fury" as the tragic revenging force, as dramatized in Aeschylus's *Eumenides.*[60] Further evidence of the author's familiarity with contemporary literature – seen earlier in the references to Sidney, Du Bartas, Greville, and Jonson – are his fleeting reference to *Hamlet* and his personification of Revenge, a figure that occasionally appeared as a character in early modern plays as well as in woodcuts and tapestries.[61]

The conceptual structures of poetics are present, too, in the phrase "the *Embryon* whereof was long since Modeld, yea digested". This vivid, albeit unsettling image of embryonic cannibalism refers to an earlier murder of a young person, a murder that Boris Godunov had arranged in an attempt to gain power nearly fifteen years previously. His victim was the nine-year-old Dmitri Ivanovich, then the legitimate heir to the throne. Ironically, Feodor II, Boris's son, later died a similar death, strangled by agents of a political opponent; in this sense, Boris "Modeld" his son's death. The narrator's claim is more than a simple appreciation of historical irony, however. By presenting Boris as a playwright, designer of the "*Embryon*" of "Murdering Sceanes", he invokes the analogy of poetic creation and reproduction that is commonplace in early modern writing.[62] This sense of the early or preliminary, a work *in potentia*, is reinforced by "Modeld". Obviously, "Modeld" recalls the title of Scott's *Model*. Borrowed from French and Italian, the term "model" is new to English at this time, possessing several interrelated senses, including the representation of a structure, such as the designs for a building, a small-scale representation of a person or thing, or an exemplary object of imitation. It could also denote scale, measure, or proportion, as in the Latin *modulus*, a unit of length used in architecture.[63] A further sense of "model" is "mould", and thus womb: this passage therefore combines the architectural and biological senses of "Modeld" in making a case for the unnaturalness of Godunov's actions.

Smithes voiage uses "model" in reference to tragedy. The unusual verbal form "Modeld" shifts emphasis from play to playwright, such that "modelling" is presented as a fundamental component of the writer's craft as well as, given the context of the metaphor, of statecraft. "Digested" continues the writing theme, recalling Hamlet's memory of "an excellent play, well digested in the scenes".[64] As I explained above, digestion, in its original sense, is disposition. Hamlet thus suggests that the play was "excellent" because it was well disposed. In *Smithes voiage*, the language of digestion suggests that Boris drew on his already "digested" experience of writing scenes to murder in order to write several more. The more literal, alimentary sense of the word is at work here, too, reinforcing the text's portrayal of his hunger for power, an appetite that is constantly consuming but never satisfied, which "digests" young male competitors in their early, embryonic stages. By stating that "the *Embryon* whereof was long since Modeld, yea digested", the narrator thus places the blame for Feodor's murder squarely on his father, who outlined the "Sceanes" that Revenge would go on to "fill". As this passage makes clear, in the early modern period, drama provided a rich conceptual framework with which to understand political crises. Metaphors of dramatic plot could be used to delineate plots of a material sort.

The best illustration of this poetical-political joining is the text's reference to Boris's "modelling" as "too-too vively". "Vively", a term that means means "vivid" or "lively", often featured in the context of aesthetics. For example, the Prologue in Ben Jonson's *The Magnetic Lady, or Humours*

Reconciled speaks out against the use of defamation in plays, instead preferring depictions that were "vively":

> It is the solmne [*sic*] vice of interpretation, that deformes the figure of many a faire *Scene*, by drawing it awry; and indeed is the civill murder of most good *Playes*: If I see a thing vively presented on the *Stage*, that the Glasse of custome (which is *Comedy*) is so held up to me, by the Poet, as I can therein view the daily examples of mens lives, and images of Truth, in their manners, so drawne for my delight, or profit, as I may (either way) use them.[65]

Jonson's "Glasse ... so held up to me" draws on the commonplace notion that imaginative works could serve a didactic purpose, display (or *model*) exemplary ways of being, and act as prompts for self-reflection. In both classical and early modern poetics, vividness is a valuable quality. But according to *Smithes voiage,* Boris's plotting takes vividness too far, exceeding the boundaries of both life and art. These lines from Jonson are instructive in another respect, in that they share the travel account's concern with the structure of dramatic scenes: libel "deformes the figure / of many a faire *Scene*, by drawing it awry". "Figure" is used in a sense cognate with "model": form, shape, or outline.[66] Meanwhile, "awry" signals a failure of modelling, the result of which is a scene that is uneven or crooked, and (by implication) both formally and ethically improper.[67] The persistence of dramatic language in *Smithes voiage*, and the ideas underpinning its account of how to describe "accidents" and how to "knit" them together, thus corresponds with the language and precepts of contemporary discourses of poesy.

Later in *Smithes voiage*, the writer makes yet another comparison of Russia's political problems to the structure of tragedy, explaining that

> as it happens for the more part with such unusuall and rare accidents (as the deposing of Princes and States-men) Tragicall events fall one in the necke of another so necessarily here: for now whilst the good were in consulting for the well ordering of this yet perplexed state, the evill were plotting more mischeefe by new stratagems and pollicies: for the verie soule of Envie was crased in this body, that the twins of Tirannie and Crueltie, strove for prioritie of byrth.[68]

The description recalls the etymology of accident as a "falling" of events.[69] Again, the writer characterizes prudence, the "well ordering" of the state, as a quality of the "good" and associates its opposite, the "plotting" of "mischeefe", with "evill". His reference to the state as "perplexed" recalls Scott's conception of peripeteia as a clew: related to the Latin *plecto*, "I plait, weave, braid", or "I twist, bend, turn", perplexed connotes a situation, or a material object, that is intertwined, tangled, or confused.[70]

Peripeteia is, in a word, complicated: a term that ultimately comes from the same root as intertwined, tangled, and confused.[71]

By comparing the accidents of politics to those of the dramatic stage and suggesting that the Russian state is tangled, in need of proper plotting, the writer reinforces the association of prudence, disposition, and narrative that we have repeatedly encountered. Further, his image of the "twins of Tirannie and Crueltie" recalls the earlier reference to the "*Embryon*" that had been "Modeld" by Boris Godunov, as well as the preface's comparison of defective texts, in which events are "torne in peeces", to a "child of a Bastard". Like the images of the "*Embryon*" and "child of a Bastard", the "twins of Tirannie and Crueltie" associate particular dispositions of plot with illegitimacy and monstrous birth. By contrasting "good" and "evill" kinds of plotting, this passage, like the account's other descriptions of the "strange accidents" in Russia, shows the prudence of Smythe's plotting in sharper relief: unlike most the other political figures described in *Smithes voiage*, Smythe does not act out of self-interest but in order to provide for others.

After describing these accidents, the writer briefly records Smythe's departure from Russia, and his safe arrival "uppon the sweet shores of our owne native Countrey *England*".[72] But *Smithes voiage* does not end there: in keeping with his insistence that accidents must be "knit together", their limbs "wrought" into a body, the writer includes a further section that explains how the situation in Russia developed after the conclusion of the English embassy. Specifically, it tells the reader that the real heir to the throne was found in Poland and that this "Prince of *Rushia*" subsequently returned to take his rightful place on the throne. In this section, as before, the author conceives of his narrative as a journey. Addressing the reader, he says "*Our Voyage* is ended, yet must I now intreate your immaginations [*sic*] to be carried backe againe into *Rushia*, where you shall receive so much and such perfect Newes, as will make up the precedent matter into a pleasing History".[73] "Perfect", from the Latin *perfectus*, stands for something that is fully realized, finished, or complete.[74] The author thus suggests that the "perfect Newes" provided in this section will illuminate retrospectively the "precedent matter" and give the reader a fuller understanding of the events that have been described. It is a providential structure, which justifies confusion by pointing to the clarifying end that is in view. Further, he suggests that this perfection of narrative will change the generic identity of the material in question, transforming it from "Tragicall ends" to a "pleasing History". Tragic peripeteia is enthralling: as Scott explains in the *Model*, it provokes "astonishing admiration", or wonder. But the author of *Smithes voiage* suggests that wonder is not enough. For him, as we have seen, travel writing that exclusively focusses on wondrous accidents does a disservice to its reader as well as to the voyage it describes.

The journey described in *Smithes voiage* is set as much in the realm of rhetoric and poetics as in Russia, revealing the productive traffic of language and ideas between travel writing and humanist learning in this period. Its engagements with contemporary drama, and the author's prefatory remarks about competing accounts of Smythe's embassy, suggest that travel writing could itself function as a site of literary criticism, testing out, and theorizing different modes of narration and description. In particular, it illustrates that travel writers thought carefully about where to "place" their material. As we have seen, this "placing" does not necessarily have much to do with geographical place but with the "places" of narrative and the ways in which writers and readers proceed through them. *Smithes voiage* also testifies to the connections between travel writing and the conceptual frameworks of humanist moral philosophy, which shaped the *ars apodemica*, and thus the practice of travel, in tangible ways. Further, this account demonstrates that travellers, like governors and military captains, were held to particular ethical standards. They needed to know how to deal with the contingencies of conflict, to find ample provision for their constituencies, even when resources were limited, and, at all times, to be prudent in their disposition.

Notes

1 For Smythe's biography, see Basil Morgan, "Smythe [Smith], Sir Thomas (*c.* 1558–1625)", "Smythe [Smith], Sir Thomas (c. 1558–1625), merchant." *Oxford Dictionary of National Biography*, https://www.oxforddnb.com/view/10.1093/ref:odnb/9780198614128.001.0001/odnb-9780198614128-e-25908.

2 *Sir Thomas Smithes voiage and entertainment in Rushia* (London, 1605), A2r. Hereafter "*Smithes voiage*".

3 The conception of textual fragments as scattered limbs is informed by the Horatian metaphor of "disjecta membra". See Horace, *Satires I* in *Satires, Epistles, and Ars Poetica*, ed. and trans. H. Rushton Fairclough (London: Heinemann, 1926), 4.62.

4 Richard Hakluyt, *The Principal Navigations, Voyages, Traffiques and Discoveries of the English Nation.* 3 vols (London: printed by George Bishop, Ralph Newberie and Robert Barker, 1598–1600) 1: xxxix. See Nandini Das, "Richard Hakluyt's Two Indias: Textual Sparagmos and Editorial Practice", in *Richard Hakluyt and Travel Writing in Early Modern Europe*, ed. Daniel Carey and Claire Jowitt (Farnham: Ashgate, 2012), 119–28.

5 William Biddulph, *The Travels of certaine Englishmen*... (London, 1609), Ar-v. "Theophilus Lavender" appears to be a persona invented by Biddulph. See Gerald M. Maclean, *The Rise of Oriental Travel: English Visitors to the Ottoman Empire, 1580–1720* (Basingstoke: Palgrave Macmillan, 2004), 55.

6 See Chester S.L. Dunning, *Russia's First Civil War: The Time of Troubles and the Founding of the Romanov Dynasty* (University Park: Penn State University Press, 2001), esp. 60–238.

7 *Smithes voiage*, K1v. Literally, "this is the work, this is the task".

8 Virgil, *Aeneid*, trans. H. Rushton Fairclough rev. G.P. Goold. 2 vols (Cambridge, MA: Harvard University Press, 1999–2000), 1:6.129. Philip Sidney, *Miscellaneous Prose of Sir Philip Sidney*, ed. Katherine Duncan-Jones and Jan van Dorsten (Oxford: Oxford University Press, 1973), 91.

9 For Scott's biography, see William Scott, *The Model of Poesy*, ed. Gavin Alexander (Cambridge: Cambridge University Press, 2013), xix–xxviii. Hereafter "*Model*". Scott's letter to Cecil is in TNA SP 91/1, 203-4). For a transcription of Scott's letter to Cecil, see *Model*, 249–53. My quotations from Scott's letter are taken from the Alexander transcription.
10 *Model*, xxii–xxv, xxiv.
11 Thomas Wilson, *The Art of Rhetorique*, 2nd ed. (London, 1553), fol. 3v.
12 See Morgan on Smythe in the *ODNB*.
13 *Smithes voiage*, Br.
14 Smythe's instructions from Cecil are in TNA SP 91/1/196. There is a reference to these instructions in *Smithes voiage*, Bv. For a discussion of this aspect of Smythe's embassy, see Tracey A. Sowerby and Jan Hennings eds., *Practices of Diplomacy in the Early Modern World c. 1410–1800* (London: Routledge, 2017), 1.
15 *Smithes voiage*, B2r, Dv.
16 Alison Games, *The Web of Empire: English Cosmopolitans in an Age of Expansion, 1560–1660* (Oxford: Oxford University Press, 2008), 170.
17 *Smithes voiage*, C1r–C2r.
18 Sir Thomas Elyot, *The Boke Named the Governour*, ed. H.H. Croft. 2 vols (London: C. Kegan Paul & Co, 1880), 1:240.
19 *OED*, s.v. "prudence", *n.*, senses 1–3; "providence", *n.*, senses 1–4; "provision", *n.*, senses 1b, 3, and 4a.
20 See Joan-Pau Rubiés, "Instructions for Travellers: Teaching the Eye to see", *History and Anthropology* 9 (1996): 139–90. See also Gábor Gelléri's chapter in this current volume.
21 Fynes Moryson, *An Itinerary Written by Fynes Moryson Gent.* (London, 1617), 3.1. 2. 9.
22 Justus Lipsius, *Epistolarum selectarum centuria prima* (Antwerp, 1586), Epistola XXII, 54: 'Ut enim ij qui iaculum mittunt, non in vanum dirigunt, sed scopum præscpeculantur ad quem collineent: sic tu præfini, quid peregrinatione istac quæras.' My translation.
23 Justus Lipsius, *A Direction for Travailers*, trans. John Stradling (London, 1592), A4r.
24 Robert Devereux, Philip Sidney, and William Davison, *Profitable Instructions Describing what Speciall Observations are to be Taken by Travellers in All Nations, States and Countries* (London, 1633), 77.
25 Elyot (1880), 60.
26 *Smithes voiage*, E2r-v.
27 See, e.g., Richard Sherry, *A Treatise of the Figures of Grammer and Rhetorike*, 2nd ed. (London, 1555), iii.
28 Quintilian, *The Orator's Education*, ed. and trans. Donald A. Russell. 5 vols. (Cambridge, MA: Harvard University Press, 2001), 3:7.1.3. On the relationship of rhetoric and prudence, see Victoria Kahn, *Rhetoric, Prudence, and Skepticism in the Renaissance* (Itahca: Cornell University Press, 1985).
29 Wilson (1553), fol. 83v.
30 *Smithes voiage*, sig. D1v. Prudent speech is a central theme in apodemic texts. Thomas Neale, for example, warns that "babling" and "ridiculous language" will make the traveller seem "imprudent" and "improvident". See Thomas Neale, *A Treatise of Direction* (London, 1643), 36–37, 41.
31 *Ibid.*, E4v, H2r.
32 *Ibid.*, C2r-v.
33 *OED*, s.v. "purpose", *v.*, sense 7; "purpose", *n.* and "propose", *v.* The *OED* cites the 1632 edition of Lithgow's *Totall Discourse*, but the phrase also appears

in his earlier account, *A Most Delectable and True Discourse* (London, 1616), 35.

34 *Smithes voiage*, C2v.

35 *OED*, s.v. "disport", *v.*

36 See Lorna Hutson, "Fortunate Travelers: Reading for the Plot in Sixteenth-Century England", *Representations* 41 (1993): 83–103; Nandini Das, "Romance Re-charted: The 'Ground-Plots' of Sidney's *Arcadia*", *The Yearbook of English Studies* 41 (2011): 51–67. Sidney (1973), 103. For a seminal study of pragmatic reading, see Anthony Grafton and Lisa Jardine, "Studied for Action: How Gabriel Harvey Read His Livy", *Past and Present* 129 (1990): 30–78.

37 James Howell, *Instructions for Forreine Travell* (London, 1642), 45.

38 Peter Auger, "William Scott's Translation from Du Bartas' *Sepmaine* [with text]", *English Literary Renaissance* 47 (2017): 21–72, 33. All quotations from Scott's translation are taken from this edition.

39 *Ibid.*, 33.

40 *OED*, s.v. "translation", *n.*

41 *Ibid.*, 53.

42 *Ibid.*, 53.

43 See *OED*, s.v. "plot", *n.* sense II. 3b.

44 *Model*, 5.

45 See *OED*, s.v. "clew", *n.*, esp. sense 3a.

46 See Michael Hetherington, "'An Instrument of Reason': William Scott's Logical Poetics", *The Review of English Studies* 67 (2016): 448–67.

47 *Model*, 30.

48 Aristotle, *Poetics*, ed. and trans. Stephen Halliwell (Cambridge, MA: Harvard University Press, 1995), 1452a.

49 *Model*, 33.

50 *Ibid.*, 39.

51 See Philip Sidney, *The Countess of Pembroke's Arcadia* (*the new Arcadia*) ed. Victor Skretkowicz (Oxford: Clarendon Press, 1987), 38.

52 For a further reference to peripeteia, see *Model*, 40.

53 *OED*, s.v. "peripeteia", *n.*

54 *Model*, 249.

55 *Ibid.*, 250.

56 *Ibid.*, 252.

57 *Smithes voiage*, H1v.

58 *Ibid.*, Kr.

59 See Duncan Salkeld, *Madness and Drama in the Age of Shakespeare* (Manchester: Manchester University Press, 1993), 26.

60 *OED*, s.v. "fury", sense 5a.

61 For references to this *Hamlet* allusion see, e.g., Tom Matheson, "Hamlet's Last Words", *Shakespeare Survey* 48 (1996): 113–22, 114–15; Margreta de Grazia, *"Hamlet" Without Hamlet* (Cambridge: Cambridge University Press, 2007), 45–8. On *Hamlet*'s engagement with diplomatic *relazioni* and other modes of foreign intelligence, see András Kiséry, *Hamlet's Moment: Drama and Political Knowledge in Early Modern England* (Oxford: Oxford University Press, 2016), 89–133. For an example of an early modern play which has a Revenge character, see Thomas Kyd, *The Spanish Tragedy*, 3rd ed., ed. J.R. Mulryne (London: New Mermaids, 2009).

62 For example, Sidney's Astrophil exclaims that he is "great with child to speak, and helpless in my throes", his pains eased only when he heeds his muse's call to write. See Philip Sidney, *Astrophil and Stella*, in *The Poems of Sir Philip Sidney*, ed. William A. Ringler (Oxford: Oxford University Press, 2012), 165.

63 *OED*, s.v. *model*, n. and adj., especially 1a, 1b, 2a, 4a, 4b, 9a, and 10.
64 William Shakespeare, *Hamlet*, ed. Harold Jenkins (London: The Arden Shakespeare, 1982), 2.2.436.
65 Ben Jonson, *The Magnetic Lady, or Humours Reconciled*, ed. Helen Ostovich (1640–41 folio text) in *The Cambridge Edition of the Works of Ben Jonson Online*, eds. Martin Butler, David Bevington, Karen Britland, Ian Donaldson, David L. Gants and Eugene Giddens, 2.7.27–31.
66 *OED*, s.v. "figure", *n.* 1a, 3a.
67 *OED*, s.v. "awry", *adv.*, *adj.*, and *v.*
68 *Smithes voiage*, L2v.
69 *OED*, s.v. "accident", *n.* II. 5a, 7.
70 *OED*, s.v. "perplexed", *adj.* and *n.*
71 *OED*, s.v. "complicated", *adj.*
72 *Smithes voiage*, M1v.
73 *Ibid.*, M1v.
74 *OED*, s.v. "perfect", *adj.*, *n.*, and *adv.*

Bibliography

Anon, *Sir Thomas Smithes Voiage and Entertainment in Rushia*. London, 1605.

Aristotle, *Poetics*. Edited and translated by Stephen Halliwell. Cambridge, MA: Harvard Univeristy Press, 1995.

Auger, Peter, "William Scott's Translation from Du Bartas' *Sepmaine* [with text]", *English Literary Renaissance* 47 (2017): 21–72.

Biddulph, William, *The Travels of Certaine Englishmen*. London, 1609.

Das, Nandini "Richard Hakluyt's Two Indias: Textual Sparagmos and Editorial Practice", In *Richard Hakluyt and Travel Writing in Early Modern Europe*. Edited by Daniel Carey and Claire Jowitt, 119–28. Farnham: Ashgate, 2012.

Das, Nandini, "Romance Re-charted: The 'Ground-Plots' of Sidney's *Arcadia*" in "Travel and Prose Fiction in Early Modern England", *Yearbook of English Studies* 41 (2011): 51–67.

Devereux, Robert, Sidney, Philip and Davison, William, *Profitable Instructions Describing what Speciall Observations are to be Taken by Travellers in All Nations, States and Countries. London*, 1633.

Dunning, Chester S.L., *Russia's First Civil War: The Time of Troubles and the Founding of the Romanov Dynasty*. University Park: Penn State University Press, 2001.

Elyot Sir Thomas, *The Boke Named the Governour*. 2 vols. London: C. Kegan Paul & Co, 1880.

Games, Alison, *The Web of Empire: English Cosmopolitans in an Age of Expansion, 1560–1660*. Oxford: Oxford University Press, 2008.

Grafton, Anthony and Lisa Jardine, "Studied for Action: How Gabriel Harvey Read His Livy", *Past and Present 129* (1990): 30–78.

Hakluyt, Richard, *The Principal Navigations, Voyages, Traffiques and Discoveries of the English Nation*. 8 vols. London, 1598–1600.

Hetherington, Michael, "'An Instrument of Reason': William Scott's Logical Poetics", *The Review of English Studies* 67 (2016): 448–67.

Horace, *Satires, Epistles, and Ars Poetica*. Edited and translated by H. Rushton Fairclough. London: Heinemann, 1926.

Howell, James, *Instructions for Forreine Travell*. London, 1642.

Hutson, Lorna, "Fortunate Travelers: Reading for the Plot in Sixteenth-Century England", *Representations 41* (1993): 83–103.

Jonson, Ben, *The Cambridge Edition of the Works of Ben Jonson Online*. Edited by Martin Butler, David Bevington, Karen Britland, Ian Donaldson, David L. Gants and Eugene Giddens. Cambridge: Cambridge University Press. 2014. http://universitypublishingonline.org/cambridge/benjonson.

Kahn, Victoria, *Rhetoric, Prudence, and Skepticism in the Renaissance*. Ithaca: Corbnell University Press, 1985.

Kiséry, András, *Hamlet's Moment: Drama and Political Knowledge in Early Modern England*. Oxford: Oxford University Press, 2016.

Kyd, Thomas, *The Spanish Tragedy*, 3rd ed. Edited by J.R. Mulryne. London: New Mermaids, 2009.

Lipsius, Justus, *Epistolarum Selectarum Centuria Prima (Antwerp, 1586)*; English translation *A Direction for Travailers*. Translated by John Stradling. London, 1592.

Lithgow, William, *A Most Delectable and True Discourse. London,* 1616.

Lithgow, William, *Totall Discourse*. London, 1632.

Maclean, Gerald M., *The Rise of Oriental Travel: English Visitors to the Ottoman Empire, 1580–1720*. Basingstoke: Palgrave Macmillan, 2004.

Matheson, Tom, "Hamlet's Last Words", *Shakespeare Survey 48* (1996): 113–22.

Morgan, Basil, "Smythe [Smith], Sir Thomas (c. 1558–1625), merchant", *Oxford Dictionary of National Biography*. Oxford: Oxford University Press, 2004. https://www.oxforddnb.com/view/10.1093/ref:odnb/9780198614128.001.0001/odnb-9780198614128-e-25908.

Moryson, Fynes, *An Itinerary Written by Fynes Moryson Gent*. London, 1617.

Neale, Thomas, *A Treatise of Direction*. London, 1643.

Purchas, Samuel, *Purchas His Pilgrimes*. 4 vols. London, 1625.

Rubiés, Joan-Pau, "Instructions for travellers: Teaching the eye to see", *History and Anthropology* 9. (1996): 139–90.

Salkeld, Duncan, *Madness and Drama in the Age of Shakespeare*. Manchester: Manchester University Press, 1993.

Scott, William, *The Model of Poesy*. Edited by Gavin Alexander. Cambridge: Cambridge University Press, 2013.

Shakespeare, William, *Hamlet*. Edited by Harold Jenkins. London: The Arden Shakespeare, 1982.

Sherry, Richard, *A Treatise of the Figures of Grammer and Rhetorike*, 2nd ed. London, 1555.

Sidney, Philip, *Miscellaneous Prose of Sir Philip Sidney*. Edited by Katherine Duncan-Jones and Jan van Dorsten. Oxford: Oxford University Press, 1973.

Sidney, Philip *The Countess of Pembroke's Arcadia (the new Arcadia)*. Edited by Victor Skretkowicz. Oxford: Oxford University Press, 1987.

Sidney, Philip *The Poems of Sir Philip Sidney*. Edited by William A. Ringler. Oxford: Oxford University Press, 2012.

Smith, John, *The Mysterie of Rhetorique*. London, 1665.

Sowerby, Tracey A. and Hennings, Jan eds. *Practices of Diplomacy in the Early Modern World c. 1410–180*. London: Routlege, 2017.

Virgil, *Aeneid*. Translated by H. Rushton Fairclough, revised by G.P. Goold. 2 vols. Cambridge, MA, 1999–2000.

Wilson, Thomas, *The Art of Rhetorique*. 2nd ed. London, 1553.

Part II

Travel, religion, and the violence of the road

4 Arming the Alps through art

Saints, knights, and bandits on the early modern roads

Joanne W. Anderson

> Oh, you ladies, don't be startled
> When you see the swords on display
> For we don't play for real, even though
> It looks as if everybody plays in earnest!
>
> *recken spil* (1511), verses 30–33

On the late medieval stage, violence was a common trope. In the *recken spil* or *Play of Warriors*, a Shrovetide comedy from the Tyrolean Alps, the knights brandish swords in a performance of masculinity, but the illusion is highlighted from the beginning.[1] No blood will be drawn and no lives will be lost. Yet, the actual presence of these weapons in the Alps (and the men who wield them) provide an interesting pretext to the staging of such recognizable objects and performances of bodily combat, and how they might correlate to a broader visual culture of normative violence. In the late medieval and early modern periods, there was a profusion of images of military saints and knights who watched the pathways of this strategic landscape. Their formal and iconographical aspects operated in combination with the presence of numerous castles and fortresses, and also armies, armed escorts and even bandits, to create a perception of potential conflict whilst travelling through.

This chapter examines the agency of art in a landscape of travel and conflict during the fourteenth and fifteenth centuries. It begins with an overview of battles, skirmishes and bandits in Switzerland, before turning to the more focused staging ground of the *viamala* ("bad way" in Rhaeto-Romanic) in the Rhaetian Alps, a fast route to and from the mountain passes that connected the Swiss Grisons and Lombardy in northern Italy. Documentary sources tell us about the strategic importance of the route and its watchtowers for those who travelled its way, but it is also important to analyse the kind of messages one received from these buildings and their decorated surfaces whilst journeying along the *viamala*.[2] To this end, the chapter will examine the political and economic management of this landscape and how a traveller might have experienced such a construct.

This will be followed by a focused study of the imagery found on the exteriors of churches and castles that were visible along the route. These repetitive and large-scale images along the *viamala* not only signalled potential conflict but also promised protection; an understanding of how this was done enables us to comprehend how art helped in arming the Alps.

Travel and conflict in the Alps

Crossing the medieval and early modern Alps required planning, resources and the courage to cope with often-treacherous terrain, pathways and climate.[3] Banditry was also perceived as rife: without a competent guide, travellers – from solo voyagers to larger caravans – were vulnerable to a spontaneous or coordinated attack. Armies travelled through the Alps on their way to zones of conflict *en masse* or in militia groups; they camped outside market towns or large castles and brought with them an arsenal of weaponry, warhorses and soldiers.

In the early fourteenth century, the Swiss Confederacy demonstrated martial prowess against the encroachments of powerful neighbours in battles and in smaller, more effective skirmishes. Savvy tactics in direct combat and the strategic use of the landscape ensured victory for the Swiss. On November 15, 1315, in the canton of Schwyz, Austrian troops were defeated by the Confederacy on the shores of Lake Aegeri near to the Morgarten Pass. 1500 soldiers were involved in the ambush at a key point on the mountain road, making full use of the halberd to defeat the better-trained and equipped Austrian knights.[4] The battle is illustrated in the *Tschachtlanchronik* of 1470 (Bern), produced by Bendicht Tschachtlan and Heinrich Dittlinger, while Johannes Stumpf briefly describes it in his *Schwytzer Chronika* of 1554, along with a woodcut illustration that varies in composition across the editions.[5] In the former, a hand-coloured woodcut depicts the ambush on a narrow path next to a forest, with victims' body parts floating in the waters of a swamp (Fig. 4.1); the Confederates or "mountain folk" were reported to have massacred the fleeing enemy like "sheep in the shambles", thus breaking the codes of conduct enacted in warfare and opposing both cantonal and chivalric orders.[6]

The Alps continued to be the setting for warfare throughout the fourteenth century. On June 21, 1339, the battle of Laupen between Bern and Fribourg took place. It again found the Swiss Confederacy victorious thanks to the efficient brutality of their infantry and knowledge of the terrain. Fifty years later on July 9, 1386, this success was repeated at Sempach, where the Confederacy defeated Duke Leopold III of Austria in a battle that was seen as a major victory. Over 167 noblemen declared war on Switzerland, including those with alliances from Italy. The mobilization of this army of roughly 5000 men is not documented and there is no suggestion that it progressed *en masse* over the mountain terrain. However, there is evidence that smaller militia bands converged at strategic locations to take on the 6–7000 men in the Swiss Confederate army. For valleys and

Figure 4.1 *The Battle of Morgarten*, Illustration from the *Tschachtlanchronik*, 1470, Zentralbibliothek Zürich, Ms. A 120, S. 70.

villages more used to the tens rather than the hundreds or thousands, these numbers represent a substantial and significant increase in the presence of armed persons. One thing is notable in all cases: no one tried to fight in a systemic way in the Alps until modern times. The landscape was tactically dangerous for a large army: brigands and guerrilla groups were more suited to its forests, passes and narrow pathways.

Bandits are more difficult to quantify in terms of art in the late medieval Alps despite having an historical and literary tradition that dates back to at least the fourth century. For example, in the life of St Martin written by Sulpicius Serverus, the saint converts a robber in the Alps whilst taking "devious paths" on his journey from Tours to Milan.[7] According to local legend, St Martin is held by a group of brigands and threatened with beheading, but the killer blow is thwarted by one of the bandits who stops the axe with his hand before converting to the Christian faith. Later accounts from across the Alps reinforce concerns about the traveller's vulnerability to attack, with particular attention to the dangers of key routes and passes; these accounts oscillate between perception and reality, blurring the distinctions as to what exactly constituted banditry in a rural area where weapons were freely available to all.[8]

Like most of his successors in the late medieval and early modern periods, the bandit who saved St Martin remained anonymous. In the ninth century, a nameless imposter decapitated St Victor of Tomils in the Domleschg valley of the Swiss Grisons. A chapel in Tomils (consecrated in 1449) was dedicated to the headless saint with the legend representing a variant of the Italian original.[9] St Victor's relics were eventually transferred to nearby Cazis; a painted panel on the reliquary shrine showed the upright body carrying the detached head while two male brigands with pitchforks scurry away.[10] This rare depiction of bandits is telling when it comes to understanding how such criminality is typified. The brigands have used farm tools as weapons, which suggests they are outlaws whose preferred mode of combat is to attack by surprise.

More typically, the presence and impact of bandits is recorded in travel accounts and edicts from the ruling authorities.[11] What is clear from these accounts is the vital need for a trustworthy guide while travelling through the mountains. Foreigners in these climbs were vulnerable as, primarily through their appearance, they were marked as not being local to the region; indeed, crossing borders in the Alps made their appearance all the more noteworthy. Safe travel demanded money and a sensitivity to the various cultures that inhabited the Alps. Some transport routes made travellers vulnerable, especially those that passed through tight mountainous terrain; like the army skirmishes noted above, robbers benefited from the element of surprise in an environment that offered few alternatives to the principle route. This could affect routes that serviced commercial trade between major and regional centres. Shifting our locality briefly, in the mountainous terrain of southern Italy during the early fourteenth century, bandits along the key mercantile trading networks were an endemic problem for King Robert of Naples, necessitating direct action by royal edict.[12] It was an authoritarian approach to resolving conflict that has a clear resonance with the Alps where a hierarchy of rulers (including bishops and the nobility at the top) controlled the territory. However, in the case of the *viamala*, a different kind of official action was undertaken to guarantee ease of passage and safety of travellers. To a certain extent, this action operationalized the sacred and profane art on nearby churches and castles.

Arming the "Bad Way"

The *viamala* has operated as a transit route in the Rhaetian Alps since Roman times, when it was used for military purposes.[13] It leads down from the Domleschg valley, which boasts a high concentration of castles and churches leading into the Schams valley with its politically situated church of St Martin in Zillis. This route takes travellers towards the Splügen (2115 m), St Bernard (2066 m), Septimer (2310 m) and Julier (2284 m) passes, linking the Swiss Grisons to Lombardy. Any journey towards these crossings (moving north or south) negotiated the notorious gorge of the same name, a geological feature that follows the course of the Hinterrhein River. The *viamala* gorge is deep and narrow, dark and dank. It gives rise to feelings of claustrophobia and vertigo that endures along the 3.1 miles of its narrowest range. Its cultural significance is great: in 1655 the Dutch artist Jan Hackaert sketched the gorge from multiple angles as part of his larger campaign of documenting the landscape and buildings of the region, while Johann Wolfgang von Goethe paused to sketch its sheer drop in 1788 (Fig. 4.2). The local priest and chronicler, Nicolin Sererhard, who in 1742 described a local murder there, stoked the gorge's notoriety; it also featured in Mary Shelley's

Figure 4.2 Jan Hackaert, *View from the north on the lower level of the Via Mala*, 1655, pen and ink drawing, in Atlas Blaeu – Van der Hem, vol. 13, fol. 100, (pf. 48), Österreichische Nationalbibliothek, inv. no. +Z143620308 [Photo: Public Domain].

travel diaries of 1814.[14] Even today, the space challenges one's sense of safety, despite numerous improvements to the road, bridges and tunnels.

In name, the *viamala* signals its early status as a communication route rather than road (*strata*).[15] Over time, it fell into disrepair due to a lack of maintenance and parts of the passage became very dangerous or impossible to negotiate. However, it became passable once more by foot and wheeled transport from 1473; the improvements to the road included a new wooden bridge (the *Punt da Tgiern*), implemented by permission of Count Georg von Werdenberg who was reacting to complaints from the local communities about the path's desultory condition for trade and travel.[16] These measures were put down in a document, known as the *Viamala-Brief*.[17]

The *Viamala-Brief* is a lengthy document comprising nine articles that ensure clarity in the description of the work to be undertaken (including rerouting) and conditions of the agreement. The upgrade would cost each of its 60 "investors" in the cooperative 50 Rheinish Guilders (*fünfzig rinischen guldin*) with additional payments of 10 guilders to ensure its maintenance by an administration master (*rodmeister*) in the long term, as set out in article 1 and detailed in 9. Neither the investors' rank nor their trade is specified but the formulation of their names suggests their community standing and place in the social hierarchy: they are ordinary men and women of property and wealth but ultimately vassals of their lord.[18] Overall, the document encompasses both transport and commercial laws for the collective good, though its implementation did not progress without conflict.[19] What is emphasized is that the new route of the *viamala* was both economically and politically divisive. By shifting it from one side of the valley to the other, villages that lay closest to it could levy tolls, thus profiting from the veritable monopoly of the passageway; this inevitably disadvantaged travellers and villages no longer on the route.

This situation is indicative of a European approach to infrastructure maintenance and local responsibilities therein.[20] But we can also place this fifteenth-century agreement alongside earlier documents that speak to the politicization of the transport networks in the local area and how attempts to control routes could lead to potential conflict. On January 25, 1359 Emperor Carl IV issued a warning from Breslau in Poland to Rudolf IV von Werdenberg-Sargans about building new roads, imposing new customs and offering escorts or armed protection (*dhein newe strazze zolle oder gelait*). To reinforce the order, Werdenberg-Sargans is reminded of his duties as a vassal in the bishopric of Chur, of his allegiances to the Holy Roman Emperor, and of the obligations that come with the rights of such revenue streams.[21] There is sufficient evidence to suggest that Werdenberg-Sargans heeded the warning: work began and Werdenberg-Sargans's arms were on the road or it was a real prospect that they would be. Men bearing arms would carry the blazon of their lord and their traversing of the landscape was a performance of political identity and control. This kind of mobile imagery was designed to send strong signals to neighbouring lords and the bishopric, suggesting that travel along this Alpine route was anything but neutral.[22]

What distinguishes the *viamala*, alongside its reputation for danger, is the high density of fortresses, towers and castles in the connected Domleschg valley (north of the gorge towards Chur), measuring some 10 km.[23] Fourteen structures, medieval in origin, are to be found along the sides of the valley creating a veritable chain of observance points: Almens, Hochjuvalt, Innerjuvalt, Ortenstein, Hasensprung (Pratval) and Rietberg (Pratval)[24], Alt-Süns (*castrum Svnnes*) and Neue-Süns (*castrum Novi Süns*), St Lorenz Kirchenkastell (Paspels), Heinzenberg, Baldenstein, Scharans, Campbell (Sils), Ehrenfels, Hohenrätien, Obertagstein (Fig. 4.3).[25] Although many now stand in ruin, in the past they were visible and physical markers of territorial control for the local nobility, aimed squarely at each other but also signalling presence to local villagers and foreigners moving along the valley floor.[26] They perch on rocky outcrops giving the impression of an elevated realm. Their stark grey walls contrast with the green of the surrounding pine trees, yet against stretches of sheer rock face, they tend to blend in, eliding natural and manmade structures. This emphasizes that travelling through the landscape, along the route that followed a branch of the river Rhine, was not simply a matter of passive passage. It entailed the crossing of multiple, invisible and often-conflicting boundaries monitored under the watch of the castles and their lords.[27] The moving subject on the ground, walking along these routes in turn participated bodily and perceptually in the construction of that surveillance landscape.[28]

Defence is the primary function of any castle or fortress, and this is reflected in the design and materials of most of the Domleschg examples (some in their first iterations). They are simple cuboid shapes fashioned from local stone that prioritizes the vertical over the horizontal. Access is restricted to one side with sheer drops on the others. Even with such functional constraints, however, these buildings offered opportunities for display, particularly as they grew in size. Their wall surfaces were often utilized to project family identity by means of heraldic shields and scenes of fighting prowess, even if couched in the visual language of courtly leisure, such as jousting or hunting. I will return to the values of popular chivalric culture, and, by extension, how imagery that correlated to these values mediated the mountainous landscape shortly. In this next section, I want to connect the architectural chain of surveillance that enclosed the *viamala*, which promised protection but also potential conflict, to two sacred watchmen of the church: St Christopher and St George. By turning to the large-scale images of these patron saints of soldiers and travellers, it is possible to determine the contribution art made to the arming of the "bad way".

Watchmen of the Alps

Christopher was patron saint of travellers and soldiers; images of him are thus common in the Alps. These images reflect his status as one of the fourteen

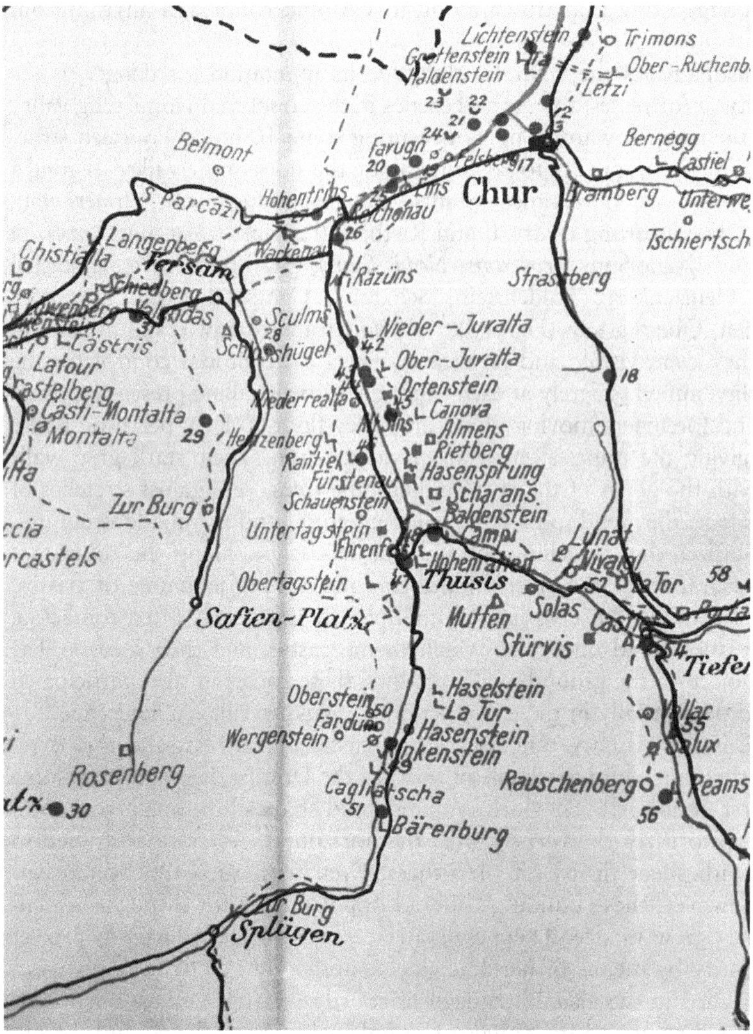

Figure 4.3 Castles on case study route, from Erwin Poeschel, *Das Burgenbuch von Graubünden* (Zürich: Orell Füssli, 1929) [Photo: Joanne Anderson].

holy helpers (*Vierzehnheiligen*) whose cult was very popular in the late medieval and early modern periods in this geographic area. The paintings of Christopher are typically found on the exterior walls of churches; they are large and simply executed, and their visibility is a key priority. Whether in town or country they typically face routes that traverse valley floors as well as the pathways that bring devotees or those seeking refuge directly to their sacred doors.[29] As such, they can be designated as popular public imagery with a particular function. Images of St Christopher are a consistent feature on the churches and chapels that line the Rhäzünser Boden, the Domleschg valley, and after exiting the *viamala* gorge into the Schams valley, creating a chain of watchmen embedded in the environment, similar to the castles discussed earlier. Three images dating from the fourteenth century illustrate how this iconography defines the route: St Paul in Rhäzüns, St Maria Magdalena in Dusch and St Martin in Zillis.

All three paintings conform to the standard mode of depiction for the saint, namely an icon figure of large proportions who is visible from a distance. Christopher carries the infant Christ on his shoulder while crossing the river, using a pole not only to steady his way but also in reference to its miraculous flowering when planted in the ground, according to his hagiography.[30] In all three cases, the painted Christopher stares out into the vast landscape, his eyes both watching and being watched by those who are passing through whether they are aware of it or not. This stands in stark contrast to the self-contained logic of other painted icons in this locality (most especially depictions of the Virgin and Child), where the gazes of protagonists do not extend far beyond the painted frames.[31] This was because Christopher was believed to offer protection against sudden death, with his salvific gaze lasting for a full day even if received from afar;[32] popular woodcuts from the time often carry an inscription that says as much.[33] Images of Christopher could be placed next to doorways in domestic residences, with the connected gaze of saint and devotee engendering a metaphorical cross on the shoulder for safe travel on leaving the home. A representative Italian example is found in the entry hall of the Palazzo Datini in Prato (1391–94). In the Alps, by contrast, Christopher was a monument in the landscape. He was a journeyman's friend who promised protection in times of adversity (the imagery had an apotropaic value), such as darkness, inclement weather or severe illness, and acted as guarantor of last rites in the absence of a priest.[34]

At the parish church of St Paul in Rhäzüns, we find a striking image of Christopher on the northern side of its east wall (Fig. 4.4). It stands at 360 cm (190 cm in width), this being an approximate height due to the loss of the lower section that would have depicted Christopher's feet in the river.[35] The church itself sits above the village and directly opposite the castle, giving the image with its perceived supernatural powers high visibility (to see and to be seen) from the route below. Significantly, there are no other paintings on this external wall to diminish this effect. Pre-dating the *Viamala-Brief*, the painting is attributed to the Waltensburg Master, an artist who, along with his workshop, dominated the region with his courtly aesthetics during the first half of the fourteenth century.[36]

Hallmark traits of the master painter can be observed in Rhäzüns, from the thick-corded hair and almond-shaped eyes to the strong linear style. However, Christopher images are of a particular type and so the artist's wont for graceful forms, as seen in nearby sites, are muted in favour of standard formulation. The saint stands square to the viewer, rather than

Figure 4.4 Church of St Paul in Rhäzüns and exterior fresco of St Christopher by the Waltensburg Master and Workshop [Photo: Joanne Anderson].

adopting the preferred three-quarters curved stance, with the Christ-child held in the nook of his left arm. Long schematic shapes and simple colours define his body – for example, the red mantle is striking in its block saturation with thick v-folds. One might be tempted to interpret red as symbolic, but in this case it simply stands out as a complementary colour to the verdant environment of this mountain valley. The vertical image is abstracted but also pushed forward to the viewer by use of bold patterns: geometrically laid out stars in red on white background plaster and red quatrefoils to enliven his robe. Only Christ's green robe, the yellow of the halo and some leaves on the walking staff provide chromatic variation. Such high visibility aesthetics are reinforced by Christopher's face: his eyebrows are arched over his staring eyes, his mouth is small and tightly closed so as not to compete with his penetrative gaze. All in all, this watchman's presence is strong and legible to all, his iconic form easily reconstructed in the mind's eye.

Further down the route, on the southern wall of Sankt Maria Magdalena in Dusch (near Paspels), a second Christopher takes over the watch (Fig. 4.5). He measures a little smaller at 295 cm in height (117 cm length), as allowed by the south wall of this mountain-side chapel which overlooks the lower valley floor; the chapel is quite defensive in its location as reinforced by its

Figure 4.5 Chapel of Sankt Maria Magdalena in Dusch and exterior fresco of St Christopher by the Waltensburg Master and Workshop [Photo: Joanne Anderson].

partner, St Lorenz Kirchenkastell on the other side of the upper valley. But the Dusch Christopher makes no less of an impact because of the recourse to general type.[37] As in Rhäzüns, Christopher is frontally positioned, looking out towards the route below. He carries the Christ-child in his left arm and a flowering staff held in his right, although this time the staff has roots. Once again, he wears a red mantle, this time over a green patterned robe, and he stands against a blue sky rather than a star-studded backdrop. Although this painting was badly damaged by weather conditions over the centuries, his strongly delineated halo, head and staring eyes reveal continuity in form and workshop practice. The small differentials simply demonstrate the room for variation within the standard formula. This points us to how one artist and his assistants could set the visual culture of a geographic area through iconographic type and style. More pointedly, it also shows how this repetition mediated any journey through the landscape.

The Christopher who watches from the west wall of Sankt Martin in Zillis is the largest of the sample, standing at a colossal 608 cm in height and 252 cm in length (Fig. 4.6).[38] It gives him increased visual presence and also speaks to the importance of this particular church in the ecclesiastical topography.[39] The composition once again is simple, likewise the execution, which comprises a blue sky background and a river below, this time populated with fish. He stands face on to the viewer allowing for a highly schematic representation of the body, like the previous two examples, with

Figure 4.6 Church of Sankt Martin in Zillis and exterior fresco of St Christopher by the Waltensburg Master and Workshop [Photo: Joanne Anderson].

bold surface detail: he wears a diamond-patterned tunic and mantle rendered in red and green. In his left hand sits Christ, who is holding a text banner (now illegible). In his right he carries the uprooted tree that will not only keep his balance but also signals his status as the patron saint of travellers.

The large yellow halo increases the size of Christopher's head, which is encased by thick locks of hair. His eyes are open and wide in this setting. They are not orientated towards the mouth of the *viamala* gorge; instead they watch towards those who approach from the direction of the Splügen pass, ready to carry them forward or watch their backs. A proverb in the Romansch language speaks to his importance to local communities and to passing travellers, including commercial traffic, "*Sontg Christoffel grand a gross porta la baselgia sen cadoss*" ("St Christopher the strong carries the church on his back").[40] He not only protects this particular church in Zillis but also, in combination with the images further up the route, he is looking out for the institution and its flock in his role as Christian soldier (*miles*). Along with the castles then, this kind of image constituted and participated in a system of watchfulness in this territory.

Before Christopher was converted to Christianity he was an outcast (known as Reprobus), a man destined to wander the wilderness far from home in search of his master. His giant size made him a weapon of kings, a hired sword, until he found the path of grace.[41] Much the same can be said of other "military" saints, although their martial might was translated into chivalric code as a means of redeeming the reputation of knights. As Peter

Burke observes, knights were so popular that a number of saints were represented as knights whose military prowess was put to work on the sacred battlefield.[42] This tacit acceptance of the feudal social order emphasizes the success of romances and other chivalric tales or songs in a time when the threat of violence – or perception of a threat – was ever present.[43] What it does not explain is how images of these warriors functioned in relation to their specific environment nor who commissioned them in the first place. Were they commissioned by the nobility, the community, or both, and to what end? How are they markers of popular culture and cultures of violence? To address these questions, we will turn to the second watchman, St George.

St George and the Dragon is one of the most popular subjects in the late medieval and early modern Alps. This martial saint is typically (though not exclusively) found on the external walls of churches alongside depictions of other protector saints. As with St Christopher, the iconography is quite standardized and conforms to chivalric code. George is on horseback lancing or bearing down upon the dragon in a mountainous or hilly landscape. The princess who requires liberation is set back from the main action, often on a rocky outcrop, praying for the success of her knight. Artists were creative in their composition of the dragon but depictions of St George emphasize his shiny armour and killer weapon. Those who looked upon this image were sent a message of saintly strength against the forces of evil but there is also a veiled warning against disruption to the established order.

Before turning to a specific image, the importance of St George can be situated in relation to the management of the route along which it lies. The *Viamala-Brief* was signed and witnessed on the feast day of St George in 1473, a significant concurrence for Count Georg von Werdenberg in name but also in terms of its wider symbolic value. George is invoked in relation to the perils of traversing the gorge (environmental or human), be it on foot, horseback or wheeled vehicle. To this perceptual picture we might add a folkloric dimension. Dragons were commonly thought to reside in mountain caves, making the *viamala*, with its dark, dank spaces, an even more daunting prospect. Johann Jakob Scheuchzer noted as late as the eighteenth century that the Swiss Grisons "is so mountainous and well provided with caves (*Spelunca*; *Drackenhöle*), that it would be odd not to find dragons there".[44] Christianity's chivalric hero, who battled one of these fearsome beasts, consequently found widespread purchase in the Rhaetian Alps; he is celebrated in local legend as the migrant soldier who converted the region to Christianity in the fourth century.

Inside the church of St Georg in Rhäzüns (Domleschg valley), we find an image of George and the Dragon (Fig. 4.7). The mural painting is spread along the upper register of the north wall of the nave; it measures 690 cm in length and 210 cm in height. Against a decorative background of red stars, laid out in (more or less) precise lines and two trees of differing species, George is depicted in the act of killing the dragon with a long lance. He is

(a)

(b)

Figure 4.7 Interior of Sankt Georg in Rhäzüns and fresco of St George and the Dragon by the Waltensburg Master and Workshop [Photo: Joanne Anderson].

dressed as a western medieval knight in complete suit of armour with a long purple tabard, rather than as a Cappadocian soldier, as described in the *Golden Legend*.[45] His spurs are neatly detailed, likewise the downward tilt of the foot in the stirrups. His halo is tucked tightly behind a raised shield that bears the cross of victory. The saddle holds the mobilized knight in place, while the noble steed wears a caparison that declares the mission of his master – green cloth replete with shields of Christ's victory of good against evil.

The lance is couched between George's right shoulder and the left side of the horse's head. Its downward horizontal thrust is halted by impact with the dragon. The blade has entered its snarling mouth and perforated the neck; blood spurts from either end. Its eye is wide open, stunned by the blow and this response is amplified by the straining tongue. The blood is depicted on the body by zig-zag strokes of the brush. In terms of composition, its body is coiled, jammed up against the corner of the nave yet balance is struck at the other end by the standing princess, who clasps her hands in the hope of her hero's victory. Although the legend presents George as killing the beast with a sword, it is more common for artists to depict his weapon as a lance. It is an elegant and precise weapon that could reinforce the martial yet civilized ideals of medieval combat and chivalry that are also linked to jousting.

George is the patron saint of this church and the painting is placed near the high altar on the proper right. It is easily the largest image inside the church and so commands the viewer's attention, particularly on account of its long horizontal format. Typically, this violent image of divine protection was displayed on the outside of buildings in the Alps in order to reach those who lived locally as well as those in transit. This would imply that such imagery was familiar in subject and representation. However, in the case of Rhäzüns these basic facts were apparently not a given or not enough for those who looked upon the painting. On inspection, the word *Gior[g]ius* can be seen painted below the girth strap of the horse. It is a surprising addition in a church wholly dedicated to the saint and his hagiography at this time, yet may find an explanation in the circumstances of patronage.[46]

The church of St Georg lies outside of Rhäzüns proper and is set on a hillock giving it a surveillance position. It was patronized by the local lords whose heraldry is prominently displayed directly below the image of George and the Dragon: a knight's helmet with upstanding festoons resting on a bi-parte shield of black and blue and white horizontal stripes (Fig. 4.8). This heraldry is also set next to a painting of the Madonna of Mercy with its donor figures representing male and female members of the family. In addition, the lord and lady of Rhäzüns are depicted in the apse, kneeling on either side of the crucifixion. The proximity of the ruling family to the image of St George by means of their chivalric symbols and avatars collapses the distance between sacred and earthly realms. In this spatial relationship, an equivalence of sorts is created between saint and lord. They

Figure 4.8 Rhäzüns Family Heraldry, fresco, Sankt Georg in Rhäzüns [Photo: Joanne Anderson].

are identifiable with one another, with the lance representing both protection and threat at a very local level.[47] The fact that the church is only a short walk from the lords' castle in Rhäzüns reinforces this family connection and their investment in imagery that shaped their identity and promoted their chivalric values.

George's conflict is aestheticized and controlled within the confines of the painting's frame and the sacred precinct of the church. But, as discussed above, he had his real-world equivalents, and this is reinforced by a battle skirmish painted directly below between George and his opponents; this is a rare inclusion in such image cycles. The lords' heraldry as public symbol was connected to the Christian *miles* but was also affected by the images of tortures visited on the saint's body by soldiers following the orders of King Dacian.[48] In the life cycle of St George, which is spread over three registers on the triumphal arch of the same church, we find numerous examples: his naked body boiled in lead; stretched round a wheel (minus the knives specified in the legend); strung up with rope that loops in illusionistic fashion over the frame of the register above; attacked with knives by a baying mob (one of whom sprinkles salt into his wounds); drinking poisoned wine and ultimately being beheaded (Fig. 4.9).

Focusing on the city and the battlefield in the fifteenth century, Valentin Groebner posits a theory of *Ungestalt* ("shapeless" or "misshapen") in relation to cultures of violence against the human body and its dismemberment.[49] Images are investigated as active agents rather than symbols in a semiotic system of reference: their representation of pain and suffering means that they are "fear inducing images on the one side and a more positively connoted 'affective piety' on the other".[50] By arming the Alps with watchmen, soldier-saints and knights, as is evident in the examples discussed so far, the threat of the *Ungestalt* is equally present yet mitigated by the beatific and almost jolly face of the saints, giving us a new way of thinking about Alpine visual culture. The repeated presence of these martyred saints in the landscape and near transit routes could influence perceptions and behaviours of those who looked upon them. This could be a passing glance or a sustained meditation on the

(a) (b) (c)

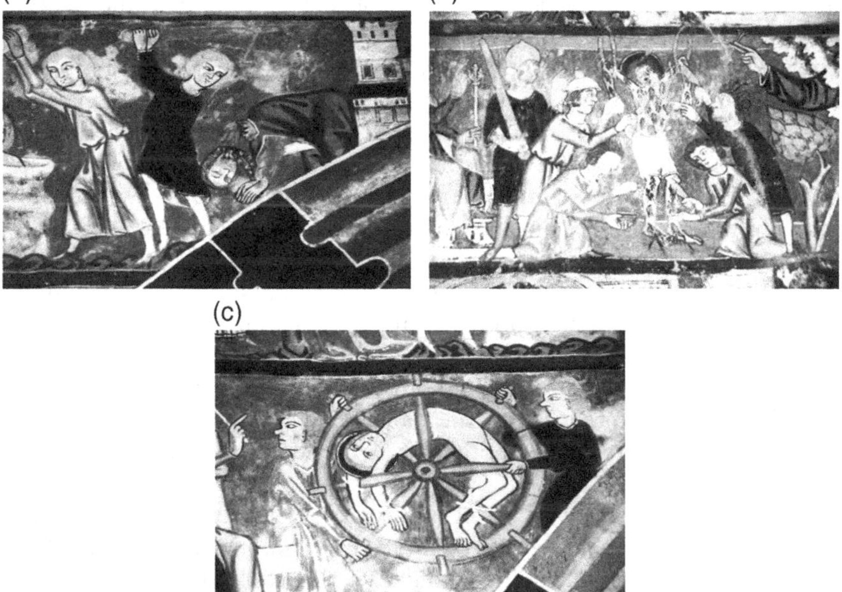

Figure 4.9 Waltensburg Master and Workshop, Scenes from the Life of St George, fresco [Photo: Joanne Anderson].

bodily forms, weapons and actions, whether in relation to the sacred narrative or the actual circumstances of the beholder's journey.

What these heroes and their oppressors in the images wear and what they carried in terms of weaponry might be thought of as material measures of perception. When a peasant looked at the shining armour and sharp lance of St George, we might ask whether they saw protection or potential attack. Similarly, when a local lord and lady gazed upon painted panoramas of battling knights, we can wonder if they only thought of chivalric tales of derring-do, or if the reality of conflict crept into their domesticated space.

Knights and hunting

The Swiss Grisons are rich in castles that, during the medieval and early modern periods, were occupied by competing noble families as expressions of their strength and power. Changes to the external and internal fabric and elemental effects have inevitably resulted in a loss of decorative surfaces or indeed complete programmes. However, what survives in the Domleschg valley is indicative of how this area participated in the pan-European phenomenon of painting palaces with romances and other imagery of courtly value. In Schloss Rhäzüns (*castello benduces et razunnes*; *castro Ruzunne*),[51] the castle of the eponymous village that sits on a rocky crop above the Hinterrhein and over the route through the valley, we find chivalric imagery that contributes to our picture of what might be called cultured or everyday violence.[52] A room on the third floor of the eastern wing is decorated with paintings that date to the end of the fifteenth century.[53] They are profane images, including the legend of Tristan and Isolde. Tristan leans on a long sword opposite Isolde and her handmaiden, Brangane in a garden replete with fruit-bearing trees and bees. The hero is labelled (*TRI – STA – ND*) to ensure the correct identification because he looks more like a battle-weary warrior than a doomed lover. What dominates, however, are the enormous heraldic arms of the Rhäzüns family displayed on either side: these are festooned helmets, each accompanied by a long lance bearing a standard.[54] The imagery parses myth and reality, giving the family an active presence in this story of frustrated love, simultaneously allowing the heraldry to make a connection between the nearby church of St Georg and thus the family's control over the land below. This control, premised on the clan system and the fluid borders that separate one territory from another, is depicted on the outside of their building.

On the exterior of the west tower of the castle is a bear hunt set against a backdrop of trees with large leaves (painted ca 1350; Fig. 4.10). To the left, we see a group of men, including a knight who drives his lance towards the rump of the bear. He is accompanied by two fellow hunters, one on foot carrying his lance upright, the other on horseback holding a hooded falcon. The noble men are attended by a pack of hounds, and they encircle the bear.

Figure 4.10 Schloss Rhäzüns with exterior fresco of Hunting Scenes [Photo: Joanne Anderson].

All the animals reveal their snarling teeth, in a scene of high-pitched aggression and courtly recreation. This is confirmed by the heraldry on the far right of the scene, that of the Rhäzüns lords. It is a visual declaration of chivalric culture and aestheticized violence that is clearly visible to those passing along the route, even today.

Further down the road, towards the *viamala* gorge, in Fürstenau, similar subjects were added to the fourteenth-century residence of the Bishop of Chur, known as the Haus Stoffel.[55] A stag hunt is depicted in the north part, along the easterly wall. A man rides from the left behind the herald who blows a horn to drive the beast – already agitated by the hounds – towards the man holding a lance. This imagery is rich in chivalric topoi and appropriate to such noble residences. It is self-reinforcing in terms of rank and also imposes order on the more volatile world beyond the protective walls. Animals are tamed or killed; valour, skill and strategy are celebrated. In this way they are active agents in shaping how people managed their environment and subjects.

These images demonstrate how violence can be romanticized, but they also show how violence can be aestheticized. A skirmish between knights is depicted in the few lines from a late medieval play that prefaces this chapter: viewers saw how violence could be enacted, controlled and aestheticized for entertainment. It is not that far from the popular treatises on swordplay that construct the ideal fighting body. An iconography of experience in relation to the fighting body is present in the earliest known combat manual in Europe.[56] Dating to *c.* 1300, this fencing book (*Fechtbuch*) from Franconia is characterized by typologies and repetition of bodily forms that help us to understand the art of combat by means of a series of agreed and orchestrated moves. This pictorial realism invested in a work of pragmatic literature gives weight to the historical data of images.[57] From iconography/iconology to phenomenology, the experience of fighting is contained in the pictorial layers giving the images a psychological thrust.[58] In the contained environment of such books, images bear a relationship to the text, equivocal or not. They are owned, perused and applied. They might be shelved, stored or gifted. The imagery is enlivened by means of leafing through pages, slowly or rapidly, with new combinations made possible by non-sequential selections. While this might stand in contrast to the public nature of mural paintings, especially those on external walls, where imagery is open, flat and static, it has bearing on the relationship that I have suggested between the images and the mobile subject who viewed it. Villagers and townspeople had access to images, to staged performances, and to the displays of arms at the castles of the local nobility or on the livery worn by servants. While any correlation must be treated cautiously, it is possible that such performances mapped onto the sacred and profane imagery depicted in this landscape of transit in times of peace and conflict, giving rise to recognized forms of aestheticized combat.[59]

Arming the Alps: a final paradox

Art, in presenting sacred and secular archetypes of martial life, mediated a culture of itinerant violence in the Alps. On the walls of churches, castles, fortresses and town gates, the saints stood watch guarding against the threat of outlaws and foreigners yet also reinforced the possibility of that threat.

The surviving visual and material record, namely the images and buildings discussed throughout this chapter, provide the basis for a reconstruction of how travel through the landscape was a mediated experience. Some travellers only saw exterior walls from a distance or when they passed close by en route. Others stopped, and even took a detour inside buildings to look at the imagery. They could perceive these buildings as places of physical and not just spiritual refuge: the imagery they gazed at could be the life cycle of St George or the episodes from chivalric tales. However, such static works of art or architecture were not the only presence in the landscape or near the roads. For travellers, there were armies and bandits whose violence was feared, even if not always directly encountered.

For the *viamala* there is no record of endemic robbery, only the odd sensational crime that became part of its mythology, as perpetuated in the works of Nicolin Sererhard and Johann Jakob Scheuchzer, who both described it as a terrible place of murder.[60] The route's organized repair in 1473 was motivated by a need to increase merchant footfall, creating a monopoly as a *via mercatorum*. The very reason for its improvement was exactly to ensure that travellers might go about their business without fear of natural disaster or violent confrontation and to increase profit for the surrounding communities and the overlord. Into this pragmatic, mercantile approach to the route, we must place the imagery that stood guard on the external walls of churches and castles at either end of the corridor for over one hundred years. The composition of the images' eyes and actions sent signals about watchfulness by heavenly and temporal knights who themselves were not afraid to wield weapons or to take the painful blow on the road to their final destination of salvation. However, what results from this investment is a paradox. Against a backdrop of military skirmishes and jostling for position by lords, knights and the freemen of cantonal Switzerland, the "bad way" actually was the safest. Art thus played its part in arming this territory of the Alps in a most distinctive and enduring way.

Notes

I thank the editors for the generous invitation to contribute to this volume and for all their guidance and encouragement.

1 See Klaus Amann and Max Siller, "Urban Literary Entertainment in the Middle Ages and the Early Modern Age: The Example of Tyrol," in *Urban Space in the Middle Ages and the Early Modern Age*, eds. Albrecht Classen and Marilyn Sandidge (Berlin and New York: Walter de Gruyter, 2009), 505–35, esp. 528–35.

2 For a recent study of how routes define landscapes and shape perceptions, see *Routes, Roads and Landscapes*, ed. Marie Hvattum, Brita Brenna and Janike Kampevolde Larson (Farnham: Ashgate, 2011).

3 On attitudes towards mountains and specifically the Alps, see Martin Korenjak, "Why Mountains Matter: Early Modern Roots of a Modern Notion," *Renaissance Quarterly* 70 (2017): 179–219, esp. 179–80. For the premodern Alps, see Arno Borst, "Alpine Mentalität und europäischer Horizont im Mittelalter," *Schriften des Vereins für Geschichte des Bodensees und seiner Umgebung* 92 (1974): 1–46 and Josef Riedmann, "Vie di comunicazione, mezzi di trasporto," in *Communicazione e mobilità nel Medioevo. Incontri fra il Sud e il Centro dell'Europa (secoli XI-XIV)*, ed. Siegfried de Rachewiltz and Josef Riedmann (Bologna: il Mulino, 1997), 109–34.

4 The tactics employed by the Swiss are discussed in Brian Todd Carey, *Warfare in the Medieval World* (Barnsley: Pen and Sword Military, 2006), 4–5, and 177–201.

5 Johannes Stumpf, *Schwytzer Chronika: aus der grossen in ein Handbüchel zusammengegozen, in welcher begriffen ist … der Eydgenossen zeyt, harkummen … bis 1546* (Zurich: Froschauer, 1554).

6 Image in Karl von Elgger, *Die Kämpfe am Morgarten in den Jahren 1358 and 1798* (Einsiedeln: Verlag Carl & Nicolaus Benziger, 1868), 11. John of Winthertur described the battle in around 1340, quoted in *Warfare in the Medieval World*, 184.

7 Jacobus de Voragine, *The Golden Legend. Readings on the Saints*, trans. William Granger Ryan, 2 vols. (Princeton: Princeton University Press, 1993 and 1995), 2:292–300, esp. 293. For the earliest manifestation of the cult of the bandit, and the role of folklore in the transformation of pagan bandit into Christian martyr, see Andrea Giardina, "Il Culto del bandito nella tarda

antichità romana," in *Psicopatologia cultura e pensiero magico*, ed. Goffredo Bartocci (Naples: Liguori Editore, 1990), 207–28. He notes that encounters between saints and bandits (*latrones*) are a common *topos*.

8 On the problems of travel in the Alps but also the challenges for historians in quantifying and qualifying banditry, see Timothy Reuter, *Medieval Polities and Mentalities* (Cambridge: Cambridge University Press, 2006), 38–71. See also, Scott G. Bruce, *Cluny and the Muslims of La Garde-Freinet: Hagiography and the Problem of Islam in Medieval Europe* (Ithaca: Cornell University Press, 2015), 19–20. For the classic albeit much disputed study of "social banditry", see Eric Hobsbawm, *Bandits* (London: Weidenfeld & Nicholson, 1969).

9 The legend of Victor of Tomils is included in the *Proprium Sanctorum Antiquissimi Episcopatus Curiensis* (Chur: Embsii: Ex Officina Typographica Bartholomaei Schnell, 1646), b. 28.

10 All the local noble families have their heraldry displayed on the chasse: Schauenstein, Planta, Flugi von Aspermont, Mohr. Illustrated in Poeschel, *Die Kunstdenkmäler des Kantons Graubünden*, 186.

11 Banditry as organized crime is discussed in Julius R. Ruff, *Violence in Early Modern Europe 1500–1800* (Cambridge: Cambridge University Press, 2001), 216–47.

12 Eleni Sakellariou, *Southern Italy in the Late Middle Ages: Demographic, Institutional and Economic Change in in the Kingdom of Naples, c. 1440–c. 1530* (Leiden and Boston: Brill, 2012), 161–62.

13 For a concise overview of the Roman occupation and its lasting impact, see Andrew Beattie, *The Alps: A Cultural History* (Oxford: Oxford University Press, 2006), 25–32. Alpine transport routes from antiquity to modernity are discussed in Uta Lindgren, *Alpenübergänge von Bayern nach Italien, 1500–1850: Landkarten – Strassen – Verkehr* (Munich: Deutsches Museum & Hirmer Verlag, 1986).

14 For Hackaert's contribution to topographical knowledge of Switzerland, see Sven Stelling-Michaud, "Die Via Mala im Jahre 1655, wie sie Hackaert sah und zeichnete," *Anzeiger für schweizerische Altertumskunde* 38 (1936): 261–73. The treachery of the *viamala* is noted by Nicolin Sererhard in his 1742 history and chronicle of the area, *Einfalte Delineation aller Gemeinden gemeiner dreyen Bünden nach der Ordnung der Hochgerichten eines jeden Bunds, ihren Nammen, Nachbarschafften, Höfen, Situationem, Landsart, Religion u. Land-Sprach nach ... beschrieben ... im Jahr 1742 ...* vol. 2 (Chur: Antiquariats-Buchhandlung am Riegerungs-Platz, 1872), 2:33 and 35–36.

15 Riedmann, 116. Riedmann suggests that before *strata* was used to designate a public communication route, the preferred term was *via*.

16 Armon Planta, *Verkehrswege im alten Rätian* (Chur: Terra Grischuna Buchverlag, 1985), 75.

17 The original document is held in the Gemeindearchiv Thusis (doc. 3). It was first published in full by R. Wagner, "Der Viamala-Brief vom Jahre 1473," *Zeitschrift für das Gesammte Handelsrecht* 30 (1884-85): 60–68 but quoted in part by Carl Ulisses von Salis Marschlins, "Bemerkungen auf einer Reise durch einen Theil des Canton Graubünden," *Der Neue Sammler: ein gemeinnütziges Archiv für Graubünden* 4 (1808): 153–92 (163).

18 The exception is the named housewife (*hußfrow*), Elsy Sin; there is only one other woman mentioned, Anna Tschörn. The preposition "*von*" is absent from nearly all names suggesting that most were not of noble rank. The nobiliary particle principally applies to Georg, Grafe von Werdenberg and towards the end of the document, the *Junkherrn* Joachim von Castelmur. See Wagner, 64.

19 For the communal composition of Graubünden and the village conflicts that arose over control of alps and passes, see Yoshihisa Hattori, "Community, Communication, and Political Integration in the Late Medieval Alpine Regions," in *Communities and Conflicts in the Alps from the Late Middle Ages to Early Modernity*, ed. Marco Bellabarba, Hannes Obermair and Hitomi Sato (Bologna: Il Mulino, 2015), 13–38 (esp. 19–21).
20 See Marcella Lorenzini, "Infrastructure Financing in the Early Modern Age. The Beginning of a 'Little Divergence,'" in *Infrastructure Finance in Europe: Insights Into the History of Water, Transport, and Telecommunications*, eds. Youssef Cassis, Giuseppe De Luca and Massimo Florio (Oxford: Oxford University Press, 2016), 61–80 (esp. 63 and 80).
21 "*so wolt wir den byschof von Kur vnd sein gotzhaus beschirmen vnd bei seinen rechten vnd gewonheiten genedichlich behalten,*" in *Bündner Urkundenbuch*, edited through the Historisch-Antiquarische Gesellschaft von Graubünden, introduction by Elizabeth Meyer-Marthaler and Franz Perret, vol. 6 (Chur: Bischofberger & Staatsarchiv Graubünden, 2010), 6:255, 268.
22 The practice of maintaining armed men in livery in order to retain local position continued beyond pre-modern times, see Ruff, 45–46.
23 This concentration is representative of the region as a whole. See Otto P. Clavadetscher and Werner Meyer, *Das Burgenbuch von Graubünden* (Zürich: Orell Fussli, 1984). For the Domleschg circle and districts of Hinterrhein and Imboden, see esp. 115–88.
24 The lords of Rietberg held the castle until 1349. It then fell into the possession of the Bishop of Chur who used it for strategic gain in the valley (political and economic) until the Planta family acquired it in the seventeenth century. More conflict arose, this time between Catholics and Protestants, leading to the murder of Pompeius Planta in 1621. See Clavadetscher and Meyer, 126.
25 St Lorenz is defined as a church-castle; a hybrid structure that blurs the boundaries between sacred and profane architecture. It sits above the rocky outcrop that supports Schloss Ortenstein in Tomils, a doubling up of watching structures.
26 Most were vassals of the Bishop of Chur, who held ultimate yet frustrated authority in the region, with rights of possession interchanging between the nobility with relative frequency. The Werdenberg-Sargans family was dominant during the fourteenth and fifteenth centuries, along with the free lords of Rhäzüns, Vaz and Belmont. See Benjamin R. Barber, *The Death of Communal Liberty: A History of Freedom in a Swiss Mountain Canton* (Princeton: Princeton University Press, 1974), 31–33.
27 A point also observed by Barber in relating landscape to political freedom (80).
28 Hvattum, 3.
29 There are sometimes pictures of Christopher on the interior walls, including the remnants of one giant inside Chur cathedral, but this is more common to late medieval Italy and England.
30 Christopher proves his worth by carrying people across a dangerous river, with his most important "client" being Christ, whose small but leaden body symbolizes the weight of the world and its creation. See de Voragine, 2:10–14.
31 For discussion of sacred embodiment and style, see Megan Holmes, *The Miraculous Image in Renaissance Florence* (New Haven and London: Yale University Press, 2013), 179–207.
32 Suzannah Biernoff, *Sight and Embodiment in the Middle Ages* (Basingstoke: Palgrave Macmillan, 2002), 133–34. She argues for the participation of the body in the act of seeing that constitutes medieval vision, which is "optical, carnal and redemptive."
33 A hand-coloured woodcut from 1423 states in Latin, "In whatsoever day you see the likeness of St Christopher, In that same day you will from death no evil

blow incur". It is preserved in the John Rylands Library, University of Manchester in the *Laus Virginis* (Bohemian) manuscript of 1417.

34 For a brief but thoughtful study on Christopher imagery and its perceptual influence on devotional practices in the fifteenth century, through the example of Umbria, see Goffredo Bartocci, "Medieval enchantment techniques: St Christopher and the Siren. Images in Cultural Psychiatry," *World Cultural Psychiatry Research Review* 6 (2011): 84–92.

35 Alfons Raimann, *Gotische Malereien in Graubünden* (Disentis: Destertina Verlag, 1983), 353–54.

36 For the Waltensburg Master see, *Der Waltenburger Meister in seiner Zeit*, ed. Horst F. Rupp and Simona Boscani Leoni (Lindenburg: Verlag Joseph Fink, 2015). Costs are unknown for the painting in Rhäzüns. However, it would have cost substantially less than the one painted for the Palazzo Datini in Prato. Niccolò di Pietro Gerini asked for four gold florins for his Christopher, a work that was eventually completed by assistants in 1394. See Anne Dunlop, *Painted Palaces. The Rise of Secular Art in Early Renaissance Italy* (University Park, PA: The Pennsylvania State University, 2009), 21.

37 Raimann, 310.

38 *Ibid.*, 427. The upper parts of the mural painting are original (albeit restored) while the lower sections were "refreshed" *c.* 1619. For the problems of conservation of external wall paintings, see Simona Boscani Leoni, *Essor et fonctions des images religieuses dans les Alpes. L'example de l'ancien diocèse de Coire (1150–1530 env.)* (Bern: Peter Lang, 2008), 85–94.

39 Zillis was an episcopal See in the ninth century. It is worth noting the dedication of the church to another soldier-saint, St Martin of Tours, whose own legend has him encountering a bandit in the Alps. See n. 7.

40 From the foreword by Christoph Simonett, in Ernst Murbach, *Zillis. Die Romanische Bilderdecke der Kirche St. Martin* (Zürich and Freiburg: Atlantis Verlag, 1967), 10.

41 de Voragine, 2:10–11.

42 Peter Burke, *Popular culture in early modern Europe* (Aldershot: Scholar Press, 1994), 157–59. Burke identifies four hero types (saint, warrior, ruler and outlaw) that are transformed according to societal need (150).

43 *Ibid.*, 174.

44 Johann Jakob Scheuchzer, *Itinera per Helvetiae Alpinas Regiones facta annis 1702–1711* (Leiden: Van der Aa, 1723). Scheuchzer's infamous dragon expedition took place in 1706. He drew upon local witness and folk tales in the construction of his typologies, allowing these accounts to override proto-enlightenment sensibilities.

45 de Voragine, 1:238–40.

46 The fine identity line between St George and an average knight is also found in the *Casa delle Guardie* in Sabbionara d'Avio (Trentino), where the battling knight bears no trace of a halo, even in the presence of the dragon. See, *Le vie del Gotico. Il Trentino fra Trecento e Quattrocento*, ed. Laura del Prà (Trento: Provincia Autonoma di Trento, Servizio Beni Culturali, 2003), 552–54.

47 George may have been a titular saint given that the last lord of the family was Georges Brun von Rhäzüns.

48 de Voragine, 1:238–42, esp. 240–41.

49 Valentin Groebner, *Defaced. The Visual Culture of Violence in the Late Middle Ages*, trans. Pamela Selwyn (New York: Zone Books, 2004). Chapter 4 discusses performance practice in images and their relation to religious theatre.

50 *Ibid.*, 101.

51 *Codex Diplomaticus*, ed. Theodor von Mohr, 2 vols (Chur: Leonhard Hiss, 1848 and 1852–1854), 1:80 and 94), 2:53.
52 See René Girard, *Violence and the Sacred*, trans. Patrick Gregory (London: Continuum, 2005) who claimed (in 1977) that cultural order could be maintained through sacrifice, offsetting the potential for male-driven violence.
53 Erwin Poeschel, *Die Kunstdenkmäler des Kantons Graubünden*, vol. 3 (Basel: Birkhäuser, 1940), 3:74–77.
54 Later restructuring of the rooms has led to some losses of the pictorial surfaces.
55 Fürstenau is in fact a tiny city, the market rights being conferred in 1354 by Charles IV, Holy Roman Emperor. The repair and rerouting of the *viamala* in 1473 impacted on its trade.
56 *Le Livre de L'Art du Combat (Liber de arte dimicatoria). Commentaires et exemples*, critical edition of Royal Armouries MS. 1.33, ed. Franck Cinato and André Surprenant (Paris: CNRS Éditions, 2009), XLIII–LXI.
57 Jens Peter Kleinau, "Visualised Motion: Iconography of Medieval and Renaissance Fencing Books," in *Late Medieval and Early Modern Fight Books. Transmission and Tradition of Martial Arts in Europe (14th-17th Centuries)*, ed. Daniel Jaquet, Karin Verelst and Timothy Dawson (Leiden and Boston: Brill, 2016), 88–116.
58 David Freedberg, *The Power of Images. Studies in the History and Theory of Response* (Chicago: University of Chicago Press, 1989).
59 For similar observations although with different conclusions, see John Astington, *Stage and Picture in the English Renaissance: The Mirror up to Nature* (Cambridge: Cambridge University Press, 2017). External wall paintings, as form of public art, during this period probably had greater symbiosis with staged performances than the luxury textiles discussed by Astington.
60 Scheuchzer, 440.

Bibliography

Amann, Klaus and Max Siller. "Urban Literary Entertainment in the Middle Ages and the Early Modern Age: The Example of Tyrol." In *Urban Space in the Middle Ages and the Early Modern Age*. Edited by Albrecht Classen and Marilyn Sandidge, 505–35. Berlin and New York: Walter de Gruyter, 2009.

Anon. *Proprium Sanctorum Antiquissimi Episcopatus Curiensis*. Chur: Embsii: Ex Officina Typographica Bartholomaei Schnell, 1646.

Astington, John. *Stage and Picture in the English Renaissance: The Mirror up to Nature*. Cambridge: Cambridge University Press, 2017.

Barber, Benjamin R. *The Death of Communal Liberty: A History of Freedom in a Swiss Mountain Canton*. Princeton: Princeton University Press, 1974.

Bartocci, Goffredo. "Medieval enchantment techniques: St Christopher and the Siren: Images in Cultural Psychiatry." *World Cultural Psychiatry Research Review* 6 (2011): 84–92.

Beattie, Andrew. *The Alps: A Cultural History*. Oxford: Oxford University Press, 2006.

Biernoff, Suzannah. *Sight and Embodiment in the Middle Ages*. Basingstoke: Palgrave Macmillan, 2002.

Borst, Arno. "Alpine Mentalität und europäischer Horizont im Mittelalter." *Schriften des Vereins für Geschichte des Bodensees und seiner Umgebung* 92 (1974): 1–46.

Boscani Leoni, Simona. *Essor et fonctions des images religieuses dans les Alpes. L'example de l'ancien diocèse de Coire (1150–1530 env.)*. Bern: Peter Lang, 2008.

Bruce, Scott G. *Cluny and the Muslims of La Garde-Freinet: Hagiography and the Problem of Islam in Medieval Europe*. Ithaca: Cornell University Press, 2015.

Burke, Peter. *Popular Culture in Early Modern Europe*. Aldershot: Scholar Press, 1994.

Carey, Brian Todd. *Warfare in the Medieval World*. Barnsley: Pen & Sword Military, 2006.

Cinato, Franck and André Surprenant. *Le Livre de L'Art du Combat (Liber de arte dimicatoria). Commentaires et exemples*, critical edition of Royal Armouries MS. 1.33. Paris: CNRS Éditions, 2009.

Clavadetscher, Otto P. and Werner Meyer. *Das Burgenbuch von Graubünden*. Zürich: Orell Fussli, 1984.

Dunlop, Ann. *Painted Palaces: The Rise of Secular Art in Early Renaissance Italy*. University Park, PA: The Pennsylvania State University, 2009.

Elgger, Karl von. *Die Kämpfe am Morgarten in den Jahren 1358 and 1798*. Einsiedeln: Verlag Carl & Nicolaus Benziger, 1868.

Freedberg, David. *The Power of Images: Studies in the History and Theory of Response*. Chicago: University of Chicago Press, 1989.

Giardina, Andrea. "Il Culto del bandito nella tarda antichità romana." In *Psicopatologia cultura e pensiero magico*, ed. by Goffredo Bartocci, 207–28. Naples: Liguori Editore, 1990.

Girard, René. *Violence and the Sacred*. Translated by Patrick Gregory. London: Continuum, 2005.

Groebner, Valentin. *Defaced: The Visual Culture of Violence in the Late Middle Ages*. Translated by Pamela Selwyn. New York: Zone Books, 2004.

Hattori, Yoshihisa. "Community, Communication, and Political Integration in the Late Medieval Alpine Regions." In *Communities and Conflicts in the Alps from the Late Middle Ages to Early Modernity*. Edited by Marco Bellabarba, Hannes Obermair and Hitomi Sato, 13–38. Bologna: Il Mulino, 2015.

Holmes, Megan. *The Miraculous Image in Renaissance Florence*. New Haven and London: Yale University Press, 2013.

Hvattum, Marie, Brita Brenna and Janike Kampevolde Larson, eds. *Routes, Roads and Landscapes*. Farnham: Ashgate, 2011.

Kleinau, Jens Peter. "Visualised Motion: Iconography of Medieval and Renaissance Fencing Books." In *Late Medieval and Early Modern Fight Books. Transmission and Tradition of Martial Arts in Europe (14th-17th Centuries)*. Edited by Daniel Jaquet, Karin Verelst and Timothy Dawson, 88–116. Leiden and Boston: Brill, 2016.

Korenjak, Martin. "Why Mountains Matter: Early Modern Roots of a Modern Notion." *Renaissance Quarterly* 70 (2017): 179–219.

Lindgren, Uta. *Alpenübergänge von Bayern nach Italien, 1500–1850: Landkarten, Strassen, Verkehr*. Munich: Deutsches Museum & Hirmer Verlag, 1986.

Lorenzini, Marcella. "Infrastructure Financing in the Early Modern Age. The Beginning of a 'Little Divergence.'" In *Infrastructure Finance in Europe: Insights Into the History of Water, Transport, and Telecommunications*, edited by Youssef Cassis, Giuseppe De Luca and Massimo Florio, 61–80. Oxford: Oxford University Press, 2016.

Meyer-Marthaler, Elizabeth and Franz Perret, eds. *Bündner Urkundenbuch*. vol. 6. Chur: Bischofberger & Staatsarchiv Graubünden, 2010.

Mohr, Theodor von, ed. *Codex Diplomaticus*. 2 vols. Chur: Leonhard Hiss, 1848 and 1852–1854.

Murbach, Ernst. *Zillis. Die Romanische Bilderdecke der Kirche St. Martin*. Zürich and Freiburg: Atlantis Verlag, 1967.

Planta, Armon. *Verkehrswege im alten Rätian*. Chur: Terra Grischuna Buchverlag, 1985.

Poeschel, Erwin. *Die Kunstdenkmäler des Kantons Graubünden*. Basel: Birkhäuser, 1940.

Prà, Laura del, ed. *Le vie del Gotico. Il Trentino fra Trecento e Quattrocento*. Trento: Provincia Autonoma di Trento, Servizio Beni Culturali, 2003.

Raimann, Alfons. *Gotische Malereien in Graubünden*. Disentis: Destertina Verlag, 1983.

Reuter, Timothy. *Medieval Polities and Mentalities*. Cambridge: Cambridge University Press, 2006.

Riedmann, Josef. "Vie di comunicazione, mezzi di trasporto." In *Communicazione e mobilità nel Medioevo. Incontri fra il Sud e il Centro dell'Europa (secoli XI-XIV)*, ed. by Siegfried de Rachewiltz and Josef Riedmann, 109–34. Bologna: il Mulino, 1997.

Ruff, Julius R. *Violence in Early Modern Europe 1500-1800*. Cambridge: Cambridge University Press, 2001.

Rupp, Horst F. and Simona Boscani Leoni, eds. *Der Waltenburger Meister in seiner Zeit*. Lindenburg: Verlag Joseph Fink, 2015.

Sakellariou, Eleni. *Southern Italy in the Late Middle Ages: Demographic, Institutional and Economic Change in in the Kingdom of Naples, c. 1440– c. 1530*. Leiden and Boston: Brill, 2012.

Scheuchzer, Johann Jakob. *Itinera per Helvetiae Alpinas Regiones facta annis 1702–1711*. Leiden: Van der Aa, 1723.

Sererhard, Nicolin. *Einfalte Delineation aller Gemeinden gemeiner dreyen Bünden nach der Ordnung der Hochgerichten eines jeden Bunds, ihren Nammen, Nachbarschafften, Höfen, Situationem, Landsart, Religion u. Land-Sprach nach ... beschrieben ... im Jahr 1742 ....* Chur: Antiquariats-Buchhandlung am Riegerungs-Platz, 1872.

Stelling-Michaud, Sven. "Die Via Mala im Jahre 1655, wie sie Hackaert sah und zeichnete," *Anzeiger für schweizerische Altertumskunde 38* (1936): 261–73.

Stumpf, Johannes. Schwytzer Chronika: aus der grossen in ein Handbüchel zu-sammengegozen, in welcher begriffen ist ... *der Eydgenossen zeyt, harkummen ... bis 1546*. Zurich: Froschauer, 1554.

Voragine, Jacobus de. *The Golden Legend. Readings on the Saints*. Translated by William Granger Ryan. 2 vols. Princeton: Princeton University Press, 1993 and 1995.

5 Between hermits and heretics

Maronite religious renewal and the Turk in Catholic travel accounts of Lebanon after the Council of Trent

Robert John Clines

Over the course of the 1590s, two Jesuits, Diego de Salazar and Girolamo Dandini, travelled to the Levant and spent the majority of their time there with the Syriac Maronite Church of Antioch, a Catholic community nestled in the mountains of Lebanon. The Maronites had embraced Catholicism following the establishment of the Crusader County of Tripoli in 1109.[1] Their relationship with Rome was solidified when Patriarch Jeremiah II participated in the Fourth Lateran Council of 1215, called by Pope Innocent III (r. 1198–1216).[2] Despite the Mamluk conquest of Tripoli in 1289, the union between Rome and the Maronites remained intact, as evidenced by the Maronite delegation at the Fifth Lateran Council (1512–1517).[3] Union was demonstrated again in 1577, when Pope Gregory XIII (r. 1572–1585) sent Jesuit missionaries after Maronite Patriarch Mihail ar-Ruzzy had convinced the pontiff that a series of Ottoman incursions had threatened the Maronites' autonomy. The Jesuit Giovanni Battista Eliano subsequently conducted two missions to the Maronites between 1578 and 1582, which brought the Maronites closer into theological and liturgical union with Rome.[4] Likewise, in 1584, Gregory established the Jesuit-run Maronite College, which allowed Maronite seminarians to receive their training in Rome.[5]

However, when Diego de Salazar arrived there in 1590 on pilgrimage and Girolamo Dandini arrived in 1596 to continue the work that Eliano had suspended in 1582, both men found a community that could hardly be considered Catholic beyond their continued expression of loyalty to Rome. In their eyes, the Maronites' failings in practice were the result of their separation from the rest of the Catholic world due to the political and religious climate of the Ottoman Empire, which caused both of them to fear that the community could be in danger of apostasy. However, the Maronites' centuries-long loyalty to Rome in the face of such obstacles was proof enough of their faith and Dandini saw them as "some relic of the ancient hermits that – if you read the most excellent authors – lived apart from the community of man, and lived in the greatest numbers in the deserts of Syria and Palestine"; they could therefore be part of the Catholic Church's effort to renew itself.[6]

Salazar and Dandini highlighted the complexity of the Maronites straddling the Catholic and Ottoman worlds; they saw the Maronites as anciently pious Christians who were in danger of apostasy, but who could be brought in step with the reforms of the early modern church. Likewise, Salazar's and Dandini's reflections on their religiously-motivated travels draw attention to how contemporaries believed securing the Maronites' Catholic future – like all Catholic futures – centred on reformers travelling to and immersing themselves in the Maronites' cultural milieux. These reflections recognize the community's desire to be Catholic and how European travellers grappled with local cultural practices that could not be eradicated entirely but needed to be amended before the Maronites could truly be considered Catholic. In turn, these two Jesuits confronted perceived cultural excess (in this case the perception that Islam was grounded in sexual depravity, vanity, and social license); its conflicted relationship with the Maronites' piety; and the acculturative challenges that early modern Catholicism faced as it reformed from within, came to blows with its religious rivals, and expanded globally. Jesuit travel is thus imbued with tensions caused by mobility and anxiety to (re)establish Catholic conformity.

The Turkish mirror: the cultural entanglements of Eastern Rite Catholicism and Islam

Diego de Salazar and Girolamo Dandini travelled to the Levant for very different reasons. Salazar, who had entered the Society of Jesus in 1560, arrived in the Levant in 1590 as a pilgrim on behalf of King Philip II of Spain.[7] The product of that voyage is Salazar's *Libro de las peregrinaciones del Catholico Rey Don Philippe Segundo*, which exists only in manuscript fragments.[8] For his part, Girolamo Dandini, who entered the Society of Jesus in 1579, was sent to Lebanon in 1596 by Pope Clement VIII to guide the Maronites in a synod.[9] In addition to his letters sent home while on mission and his final report written in 1597, his *Missione apostolica al patriarca e maroniti del Monte Libano* was published in 1656 but circulated widely beforehand in manuscript form.

In spite of these two authors having different motives for travelling to the Levant, their accounts each elucidate the Maronites' struggles to maintain their Catholic identity in the Ottoman Empire. Undergirding their explorations of the Maronites are the challenges that the two Jesuits believed the Catholic Church faced in its global effort to renew itself.[10] Part of this was exploring the cultural and religious mores of Muslims, whom they invariably called Turks, thereby conflating Muslims' religious, political, ethnic and cultural identities, and allowing the Jesuits to tie Islam more demonstrably to the widespread fear of the Ottoman Empire as a direct threat to Catholic Europe.[11] Through their explorations of the Maronites and their cultural milieux, both Jesuits outlined what they saw as the most significant differences between Christendom and the Muslim world. In turn, they attempted to pinpoint the ways in which Christians could succumb to the threat of Islam and how this might be prevented. As part of this, the pair evaluated the ways in which Muslims' religious and cultural practices compared to Catholic orthodoxy. This elucidates their desire to explore the deep-seated disparities in cultural practice, as well as the Europeans' fascination with the seemingly foreign nature of Islam.[12]

There are some differences in tone between the two Jesuits' texts, but the overall topic is much the same. Salazar painted Islam as a schismatic form of Christianity rather than a different faith, such as his description of mosques as "the church of the Turks, where they perform their ceremonies and some form of sermons". Likewise, Muslim worshippers "were seated on certain benches and in a certain order". There was also preaching, a familiar sight for any churchgoer.[13] Salazar furthered the "broken Christian" paradigm by explaining that mosques had "certain men, or boys" who sang in an "acute and sonorous voice" to call people to prayer, not an uncommon occurrence in the post-Tridentine church, which supported sacred music and saw communal ritual as integral to the social fabric of a true Catholic community.[14] Salazar continued these types of evaluations when he described Ramadan:

> Recently, that is, at the new moon of July, the Turks began to make their Lent, when that Moor or Turk who oversees the mosque cries out from the tower at certain times of day and night. And he notifies the people from the said tower with a flickering light almost all night long. Their fasting practice is to not eat all day, until they see the stars of the night, when they eat endlessly and abundantly.[15]

By referring to mosques as Muslim churches and Ramadan as the Islamic Lent, Salazar exposed both the importance of upholding Christian religious practice as well as the Ottomans' ability to threaten it by offering a corrupted (but not wholly unfamiliar) alternative. Rather than painting Muslims as foreigners with no points of comparison in Christendom, Salazar depicted them as deviant Christians to demonstrate that apostasy to Islam was as threateningly normal as the temptation to worship in the all-too-familiar heretical churches that peppered the landscape of Protestant Europe.

While Dandini also relied on quasi-Christian descriptions of Islam in *Missione apostolica*, such as likening Muslim prayer beads to the rosary, in comparison to Salazar he more forcefully condemned Islam for its apparent "stunning liberty, and license to sin, from which comes every grave wickedness".[16] Dandini furthered this in the chapter entitled "On their Greed and Garbage", in which he explained that Muslims had a wild appetite for money and more readily sought it than their own salvation. And oftentimes they strove to stomp underfoot Christian communities to fuel their greed, rendering them all the more base:

> Therefore, the reason the Turks pursue us Christians so much, move us mortals, and continue wars [against us], is not for what others have

> estimated: for adventure, faith, or religion. Rather, they do it for their greed of money and unbridled appetite for possessions, and, even more so, to stretch the boundaries of their vast empire. Therefore, this vice together with the sensual, the carnal, reigns far too licentiously over every one of those people without any restraint.[17]

Two of Dandini's beliefs are exposed in this invective: first, the foundation of Islam is vice; second, the expression "us Christians" stresses Dandini's view that Catholics and Eastern Rite Christians should be partners in the defence of an ecumenical Christendom against Islam, a purported foreign scourge driven by Ottoman license and tyranny.[18] Whereas Salazar was not a missionary to the Maronites and therefore did not necessarily see the apparent threats that Islam presented to Christian minority groups to the same extent as Dandini, Dandini saw Islam not as an aberrant form of Christianity but as a false religion that threatened the Maronites and aimed to strike at the heart of Christendom itself.[19]

Dandini also discussed Ottoman oppression in his exploration of the ways in which clothing legislation legally marked one's place in Ottoman society: "those of noble blood or for other dignity, wear the great turban", whereas others wore smaller ones, as the size of one's turban served as a marker of social class. Likewise, Christian subjects of the Sultan wore white shirts or other colours on rare occasions. But they never wore black or green: European Catholics tended to wear black; green was the colour of the descendants of Mohammed and was a colour far too lofty for a commoner to don. The Maronites were therefore limited in their clothing options as dressing in black would have been a declaration of political autonomy from the Ottomans and dressing in green would have been a severe insult to the Muslim establishment. In either case, the way they dressed reflected their minority status in the Ottoman world.[20]

Dandini's account of Ottoman clothing restrictions and Muslim religious customs reflects a preoccupation with outer appearance as a mode of cultural evaluation, especially concerning perceived differences. As Jan-Petter Blom has suggested, "the organization of ethnic identities does not depend on cultural diversity *per se* ... but rather on the assignment of particular social meanings to a limited set of acts".[21] Blom's point is reflected in Ottoman clothing restrictions and their ramifications for the material lives of the Maronites and for how Christians would have perceived such restrictions as attacks on their material lives. Dandini saw them as legislative tyranny through which the Ottomans stratified society along religious lines, as they used clothing to establish impermeable, hierarchical boundaries between themselves and others as well as between their subjects and foreigners. While such practices were common throughout Catholic and Protestant Europe, Dandini positioned them here as a threat to Europe itself; he presented the idea that, if the Ottomans should conquer Europe, it too would be subject to such cultural, religious, and sartorial oppression.

While Dandini agreed with his predecessor in Lebanon, Giovanni Battista Eliano, that the Maronites themselves often embraced these restrictions on dress as evidence of their poverty and living martyrdom,[22] Dandini also viewed them as an Ottoman edict that needed to be resisted. Not only were Muslims and Orthodox Christians pressuring the Maronites to convert, he thought, but these Catholics were also culturally fettered in their own development by Ottoman political oppression, right down to how they could dress. Given why Dandini was in the Levant (to lead the Maronites in Catholic reform), these observations elucidate a preoccupation with saving the Maronites from apostatizing; if one wanted to raise one's social standing, at least in theory, all one had to do was change clothes.

Dandini's effort in this regard provides insights into how he and Salazar viewed the Ottomans' mores and laws – religious and secular – as direct threats to the Maronites. In these texts, the Ottomans remained the ultimate enemy: they lacked morality and religious rectitude, and this often manifested itself not only in religious practice but also in social and material customs. Salazar presented Islam as a schismatic form of Christianity whereas Dandini vilified it as a godless cult of depravity: while their tactics were slightly different, both strove to fashion the Ottomans as immoral, foreign, heretical barbarians who were responsible for the Maronites' plight. Nevertheless, both remained hopeful that the Maronites could be pulled back through proper instruction and, as the next section discusses, their own piety. When Salazar's and Dandini's descriptions of Muslims are tied to their exploration of the Maronites' mores, they lay bare their belief that the Ottomans were a direct threat to the Maronites and the Catholic faith as a whole. This points to the authors' desire to reconcile the universal vision of the church with the local peculiarities that needed to be tackled before Catholic reform could take place. The problem of Ottoman incursions and the threat of cultural transformation were challenges to the Jesuits' efforts, and these travel accounts served to expose this. But essential to this as well was emphasizing that, despite these threats, the Maronites could reform themselves just as if they were in Catholic Europe.

Maronite piety and the defence of Christendom against oriental vice

Throughout their accounts, Dandini and Salazar praised the Maronites for their austerity and piety that they believed had originated in late antiquity. At the same time, they reminded their readers that the Maronites' Catholic identity was on the verge of collapse due to centuries of external pressures. On the one hand, these descriptions served as proof that the Maronites were good Catholics. On the other hand, Dandini's and Salazar's texts are laden with evaluations of, and references to, what they saw as the dire state of Catholicism; they did this by laying bare that (in the guise of the Maronites) true Christians ought to return the faith to its antique origins and resist the various types of threats that all Catholics faced.

Much of the descriptions hinge on the Maronites clinging to Catholicism despite being surrounded by Muslims and non-Catholic Christians such as the Greek and the Syriac Orthodox. For example, rather than purely seeing Ottoman clothing restrictions as oppressive, the Jesuits complicated the Maronites' dress by also positioning it as evidence of their devotion, such as Salazar's description of four Maronite archbishops, who wore "large turbans on their heads", which were "not white like those of the Turks, but of a blue colour". Furthermore, they were "very humble and poor" and they "go about very poorly dressed". While the Ottomans forced the Maronites to dress in a manner that would at first appear to be unchristian, Salazar was sure to couple this description with assurances to the reader that the priests and monks who served the people were all "subject to the High Pontiff and the Roman Church".[23] Likewise, Dandini explained that he felt great piety reverberating from Patriarch Sarkis ar-Ruzzy, who was also dressed in a turban.[24] When read against his descriptions of Ottoman-dictated dress codes, Dandini's exploration of Sarkis's appearance and piety despite his "oriental" dress demonstrates Dandini's cautious optimism: while the Maronites might have been under pressure from the Ottomans, they nevertheless desired to be devout Catholics. In some sense, their perseverance in the face of heterodoxy and their embrace of such sartorial oppression were evidence of just how Catholic they were.

This praise for the Maronites continued in descriptions of their liturgical life. Salazar recounted that their liturgical language was Syriac and they performed "particular ceremonies that they have preserved according to their own mode and use", such as saying mass in the evenings shortly before sunset. Customs that often had no equivalent in the Latin rite were, for Salazar, evidence that the Maronites had retained them from the early days of Christianity.[25] Much like Salazar, Dandini believed that the Maronite liturgy fit within the framework of Catholicism because of its antiquity even though it was "in many ways different from our Latin [rite]". Likewise, Dandini's evaluation of the patriarchal monastery, which he described as "very beautiful", captures the essence of the depth of the Maronites' piety in the face of external pressures. While Dandini lamented that it was "dark and poorly maintained" with "little splendour" because the Ottomans' "insatiable greed" prevented the Maronites from maintaining it, he believed that the Maronites embraced an architectural martyrdom not seen since early-Christian house-churches, anchorite caves, and pillars.[26]

Both men also admired the austerity of the monks. Salazar explained that during Lent "not only do [the monks] not eat eggs, dairy etc., but they also abstain from fish and from drinking wine; and their wine is the best and tastiest in all of Mount Lebanon".[27] Despite their ostensibly peculiar practices and customs, the monks' ability to abstain from seemingly delectable wine for forty days rendered them, as far as Salazar was concerned, good Catholics. This is evident when coupled with Salazar's description of Muslim fasting, mentioned above: when "the Turks began to make their Lent ... their fasting practice is to not eat all day, until they see the stars of the night, when they eat endlessly and abundantly".[28] Whereas apparently not a single Muslim could make it a full day without stuffing themselves full of culinary vice, these pillars of the Maronite community fasted uninterruptedly for forty days. Likewise, despite the fact that they did not wear vestments with "a maniple, stole, and chasuble similar to ours" when saying mass, they nevertheless radiated a piety that reflected the antiquity of their liturgy.[29] And, as I noted in the introduction to this chapter, because these monks lived in the remotest parts of the mountains, dressed humbly and simply, and practiced strict dietary austerity, Dandini was convinced that these monks "could be some relic of the ancient hermits" who served as moral exemplars for the rest of the community.[30]

While Salazar and Dandini explained that Islam surely perpetuated it, the monks' destitution was partially "caused by their ancient constitution, by which these holy hermits and servants of God strove to flee from idleness, and procure food with their own industry, and spend other times performing manual labour".[31] Their generosity also struck Dandini, as they welcomed not only Christians but even Muslim guests into their monastery.[32] They did not perform the Latin rite, and they struggled to confess to the Jesuits because, as Salazar explained, they spoke "no Latin, so I had to guess or draw out (as they say) with discretion what was said".[33] Yet Dandini and Salazar recognized that the Maronites' ancient piety enabled them to resist the perfidy that surrounded them and to embrace poverty, something that European Catholics would do well to emulate. Such an idea – that Europeans could learn something about being good Catholics from poor Levantine Christians on the precipice of apostasy – renders the process of salvific universalism a dialogic exchange between culturally diverse members of one faith rather than a dictate from Rome.[34]

However, despite the Maronites' ancient piety and the lessons that European Catholics might glean from it, Salazar and Dandini reminded their readers that the Ottomanisation of the Mediterranean was a direct threat to the Maronites' Catholic identity and thus a potential challenge to Christendom itself. When these two men were in Lebanon, the reality of the Battle of Lepanto's failure to curb Ottoman advances had already set in, and the Ottomans were pushing further into Habsburg territories in Eastern Europe while also expanding their naval presence in the Indian Ocean World.[35] It was not just the Ottomans themselves but "the Turk" as a metaphor for moral baseness and the erosion of civil society that was increasingly seen as a challenge to Catholic Europe and its burgeoning overseas empires.[36]

In this regard, the Jesuits muddled differences between Europeans and Maronites as a means of exposing "the Turk" as a potentially unstoppable tide that would envelop Christendom. They did this to prove that proper reform – as well as the thorough removal of any stain of "oriental" vice – would be necessary to reverse the process of Ottoman acculturation that had long plagued the Maronites and could even soon strike Europe. This explains why Dandini wrote that the Maronites were "like the Italians" in terms of physical appearance,[37] and Maronite women were "most modest, and the mirror of honesty", and their manner of dress "is quite like what is used in our country, being long, going to the ground, with their shoulders and chest most diligently covered".[38] By giving the Maronites European characteristics and a reticence that reflected the ideal cultural mores of Catholic Europe while also reminding the reader that they were imperilled in their surroundings, Dandini used the Maronites as a cultural paragon to compel his audience to admire and perhaps imitate them. Europe too, it would seem, could succumb to Islam if the proper measures were not in place.[39] Given that fears of turning Turk and sliding into vice weighed heavily on the early modern consciousness, the notion of Europeans being victimized by such a cultural threat was very real – both for Dandini and his audience.[40]

Despite lessons that Europeans might learn from the Maronites, there were troubling elements of Maronite customs, which suggested that Turkisms had seeped into the Maronite community; inevitably, these would arrive on the doorsteps of Europe. For example, Maronite women wore bracelets and headbands that resembled what Dandini called Turkish accessories – clear signs of vanity for Dandini – and, on occasion, men wore Turkish-style trousers. The Maronites also tended to dine while seated on a carpet rather than use fixed furniture; they also used no cutlery, which, to Dandini, resembled Ottoman barbarity rather than European civility.[41] Exposure to Islam or something as seemingly innocuous as drinking coffee struck at the heart of European preoccupations with ostentation, commercial cosmopolitanism, and the proliferation of foreign goods.[42] Anxieties about Catholic Europe succumbing to Ottoman license through conspicuous consumption were thus being projected onto the Maronites' material lives.[43]

Dandini understood that Maronite men and women, regardless of how Catholic they might have desired to be, remained exposed to Muslim mores in ways that the secular clergy and the monks were not and were therefore on the frontlines of combatting Ottomanisation. Dandini never overtly condemned the Maronites for their seemingly uncivilized manners; he tended to describe rather than evaluate or pass judgement. However, when read against his excoriation of the Ottomans' moral baseness and their proclivity for oppressing religious minorities in other parts of the work – as well as the pervasive European fear of Islam as an infectious disease – his explorations of the Maronites take on new meaning. Dandini feared that the Maronites had long been, and continued to be, under attack due to their presence within the Ottoman world; Catholic Europe could also succumb to the Ottomans' growing empire of heresy and vice should customs like wearing ornate bracelets and eating without cutlery while seated on carpet become popular in Europe.

This fear of potential flirtation with Ottoman mores illuminates Dandini's perception that, in addition to Islam, the Ottomans' material culture oppressed and corrupted the Maronites. Daniel Vitkus has argued that Europeans saw Mediterranean travel, cross-cultural exchange, conspicuous consumption, and moral licence as inextricably linked; the very acts of exchange and consumption, whether it be of clothing, luxury goods, or ideas, could "bring disease to the body politic" and eventually lead to the slow erosion of a pure Christian society.[44] Vitkus's discussion of anxieties concerning the spread of vice through consumerism points to a larger fear of cultural contamination that pervaded the early modern European consciousness and that manifested itself in sumptuary laws.[45] As religious purification and the eradication of worldliness became principal aspects of reform,[46] political leaders and high-ranking churchmen became obsessed with safeguarding the frontiers of the faith against religious, political, and cultural dangers; the consensus was that the Ottomans were the gravest threat.[47] For Dandini, the Maronites were evidence that all Catholics were under attack from the two-headed monster of vanity and heresy. Thus, the Maronites were a reminder that the deterioration of Catholic purity was never too far from home. Yet if the Maronites could be reformed, they might serve as a satellite in the church's efforts to evangelize other Eastern Rite Christians as well as prove that Catholicism was no longer on the defensive and in fact could persist in the face of its enemies, Protestant, Muslim, or Jewish.

Global missionary Catholicism as pious unity against the Turk

Despite the notion that the Maronite community could be integral to combating Islam, Dandini and Salazar understood that the reality of things was more complicated. Dandini discovered numerous inefficiencies and imperfections in practice, particularly among the clergy. He found priests unable to differentiate between venial and mortal sins; as a result, the penance issued never seemed requisite to the sin, rendering confession ineffective. Even more troublesome was the fact that "at times they give Holy Communion to some who have not confessed their sins".[48] Likewise, Maronite priests, despite their devotion, "never preach the Word of God to the people, nor to children, and stupid men teach Christian doctrine".[49] All of this created a devoutly ignorant community that was entrenched in false belief and action – and thus at risk of damnation – because of poor clerical practice.[50] Diocesan and parochial reform of ill-trained clergy remained a constant struggle in post-Tridentine Europe as well, so Dandini's exposition of the need to institute reform on this level, especially regarding the education of ignorant clergy, would have been a quite familiar call for Europeans struggling to reform their own dioceses.[51]

Likewise, Dandini listed numerous challenges that he believed prevented the Maronites from being fully orthodox.[52] Many of these errors were theological, but, for Dandini, several exposed that the Ottomans' purported vice had already contaminated the community. One was particularly grave: "because of adultery, or for other reasons and infirmities, one can repudiate one's wife and take another", which he attributed to "Turkish violence and tyranny".[53] Because the patriarch refused to approve divorces, Maronite men often paid a fee to the Ottoman emir, who ordered the patriarch to carry out the divorce. One such divorce occurred while Dandini was in Lebanon. He explained that Patriarch Sarkis was saddened by having to participate in such a scandal. Sarkis granted the divorce, albeit reluctantly, lest his refusal anger the emir and bring more hardship to the community.[54] Such acts were not uncommon in religio-imperial borderlands: men and women left home for the Ottoman Empire and converted to Islam to obtain a divorce or to avoid being married off or enclosed in religious houses.[55] But for Dandini, this abandonment of one's Catholic familial duties was clearly a Turkism, a practice tied to effeminate sexual license, abject depravity, and allure of Ottoman vice.[56] Dandini played on the widely employed trope of Ottoman sexual predation to argue that Maronite chastity was under attack.[57] In turn, Dandini believed that divorce would not be possible if Maronites were freed from the Ottomans' yoke.

But there were larger implications as well. For Dandini, Maronite husbands' ability to circumvent the authority of the patriarch and obtain a divorce from the Ottomans demonstrated that the patriarch was unable to prevent the spread of apostasy on his own. Thus, the only means of saving the Maronites in this world and the next was institutional and clerical reform, seminary education, catechetical instruction, and uninterrupted communication with Rome. Salazar reminded the reader of the great work of previous Jesuit missionaries like Eliano and explained that more missions were necessary. And much as Jesuits conceptualized educational activities as key to reform throughout the Catholic world, Salazar believed that the Maronites were showing signs of reform because of the training Maronite priests had received at the Maronite College, which had opened in Rome in 1584.[58]

Dandini, like Salazar, explored education as one of the many elements necessary for reform. In one sense, the infrastructure that the Jesuits had set up to instruct Maronites in the 1570s and 1580s, such as the establishment of the Maronite College in 1584, was integral to reform, as it would allow Maronite youths to be instructed in Rome and subsequently dispatched to Lebanon to serve the spiritual needs of their community And this, considering the cooperation of the community, should have been easy to maintain. Dandini remained confident that graduates of the Maronite College, who were "extremely knowledgeable of Roman matters", would be welcomed into the community and that the Maronites would maintain their place within the church as a result.[59] Such a revolution in clerical training would render Maronite reform both a Roman initiative and a grassroots effort.

Likewise, frequent local synods – like the one Dandini conducted in Lebanon in 1596 – would ensure that the Maronites did not abandon reform efforts. As Dandini explained in his 1597 final report, the synod's purpose was to fix the Maronites' errors concerning doctrine and the performance of the sacraments.[60] While much of this centred on challenges unique to the Maronites, Dandini realized that reforming the Maronites was not any different from reform elsewhere. Catholic Europe faced many obstacles after the Council of Trent concluded in 1563 that could only be addressed via instructional reform and local synods. The notion that the synod conducted under Dandini's leadership functioned much like every other local Catholic synod the world over is apparent in the canons themselves, such as the requirement that bishops visit their respective dioceses at least once a year. This suggests that Dandini's desire to reform the Maronites was in step with the larger post-Tridentine efforts to confront religious laxity and ensure that proper Catholic reform regardless of locale was contingent upon cooperation between the church's apostolic ministers, the local clergy, and the community itself.[61]

This was an important move toward uniting all Catholics with Rome beyond solely recognizing papal authority. The synodal decrees elucidate the reality that for all Catholics – the Maronites included – true bonds could only be achieved once the faithful believed that they were participating in the purification and renewal of the church. This confirms again the importance of long-distance networks of communication between Rome and local communities, local desires to participate in the life of the church, and the institutional church's ability to promote reform as a collaborative effort. Through the profession of faith and the Maronites' eagerness to adhere to the synodal decrees, the synod textually constructed a roadmap for what global early modern Catholic religious culture might look like. The synod was a means of presenting the Maronites as centuries-old Christians while also exposing that Catholic renewal hinged on broad-spectrum participatory engagement and confrontation of local conflicts regardless of where a given community lived.[62] It was also a message to the Catholics of Europe and beyond that local cooperation was as integral to achieving Catholic universalism as were Trent and any subsequent councils or papal decrees.

In this regard, one needs only look to the Maronite patriarchs' desire to receive papal approval upon their election, their resubmission to Rome after each papal conclave, and the exchange of letters between Maronite prelates and Catholic leaders that stress mutual admiration and desires to work together for the betterment of the faith. While the challenges that they faced in the Ottoman world might have rendered the Maronites different from other Catholics, such cooperation between Rome, the patriarchs, and the Maronite community suggests that reform was a dialogue between the institutional church and local Christians regardless of locale and cultural differences. In other words, it reflected the larger reality of Catholic reform the world over.

Conclusion

Dandini and Salazar elucidated that attempting to reform the Maronites – and by extension the church as a whole – by simply commanding them to adhere to Trent would be a fool's errand, as the Maronites would almost assuredly return to error unless a full apparatus of continuous renewal were put into place. Salazar and Dandini thus saw efforts to properly reform the Maronites as threefold: travel to and immerse oneself in the community, understand the unique conflicts and obstacles to reform that the community faced, and rely on local communities' desires to maintain lines of communication with Rome. This was no different from how reform was to take place elsewhere. After all, despite their "oriental" tendencies, the Maronites repeatedly acknowledged papal authority, they had sent delegates and seminarians to Rome for centuries, and they had welcomed the Jesuits into their community. Likewise, given Europeans' anxieties concerning their own proclivity for "turning Turk" and the ease by which conversion could take place, the Maronites' Catholic experience was hardly a world apart from those of the streets of Rome, German villages and hamlets, English priest-holes, or bi-confessional spaces in France.[63]

These travel accounts betrayed the peculiarity of the Maronites by showing that despite their unique situation under the Ottomans, they nevertheless were devout Catholics, purportedly since late antiquity, and desired to be a part of the larger Catholic ecumene. By exploring where the Maronites fit within both the Catholic and Ottoman worlds, Salazar's and Dandini's travel accounts illustrate that Catholic reform was more complicated than Tridentifying local communities, regardless of where in the world they found themselves. In other words, Dandini's and Salazar's experiences with the Maronites elucidated that the process of Catholic renewal on a global scale could never be the hegemonic, homogeneous endeavour of the papacy and its agents. Rather, it had to be a dialogue between Rome and local communities that hinged on a mutual confrontation of the unique challenges that each community faced.

On a certain level, orthodoxy was orthodoxy; as the decrees of Dandini's synod illustrate, the Council of Trent certainly mattered. But as European Catholics saw more and more of the world, encountered new forms of religious difference, and established missions worldwide, it became obvious that constructing a global faith required amending expectations in the face of geographic isolation, cultural difference, and religious conflict. In sum, as Salazar's and Dandini's accounts attest, Catholic renewal required recognizing that true universality was not replacing local peculiarities and traditions with a totalized Tridentine Catholicism as defined by a hegemonic pontiff and his curia. Rather, travelling across the globe to spread Catholicism had to include coming into conflict with the myriad challenges to those efforts, whether they were the religio-imperial aims of the Ottoman Empire, the perseverance of indigenous religious traditions in Asia and the Americas, or the reform efforts of the various Protestant confessions of Europe. This reality rendered early modern global Catholicism an ongoing dialogue between church leaders in Rome, evangelizers, and local communities, all of whom had to reconcile their idiosyncrasies within one another.[64]

Notes

1 Kamal S. Salibi, "The Maronites of Lebanon Under Frankish and Mamluk Rule (1099-1516)," *Arabica* 4 (1957): 288–96.
2 Robert W. Crawford, "William of Tyre and the Maronites," *Speculum* 30 (1955): 222–28.
3 Robert J. Wilkinson, *Orientalism, Aramaic, and Kabbalah in the Catholic Reformation: The First Printing of the Syriac New Testament* (Leiden: Brill, 2007), 11–28.
4 ARSI, Gall. 95 II. fol. 23r. On the Maronites' call for assistance in 1577 and the subsequent mission, see Charles A. Frazee, *Catholics and Sultans: The Church and the Ottoman Empire, 1453–1923* (Cambridge: Cambridge University Press, 1983), 137–39; Matti Moosa, *The Maronites in History* (Syracuse, NY: Syracuse University Press, 1986), 245–55; Robert John Clines, "Wayward Leadership and the Breakdown of Reform on the Failed Jesuit Mission to the Maronites, 1577–1579," *Journal of Early Modern History* 22 (2018): 215–37.
5 Pierre Raphael, *Le rôle du Collège Maronite Romain dans l'orientalisme aux XVIIe et XVIIIe siècles* (Beirut: Université Saint Joseph, 1950); Nasser Gemayel, *Les échanges culturels entre les Maronites et l'Europe: Du Collège Maronite de Rome (1584) au Collège de `Ayn-Warqa (1789)* (Beirut: Impr. Y. et Ph. Gemayel, 1984).
6 Girolamo Dandini, *Missione apostolica al patriarca, e maroniti del Monte Libano e sua pellegrinazione à Gierusalemme alla santità di N.S. Alessandro VII* (Cesena: Neri, 1656), 75–77.
7 ARSI, Tolet. 37. fols. 271v-275r. On Salazar, see "DE SALAZAR, Diego Marañon de Porres," in *Monumenta Promixi Orientis*, vol. III, ed. Sami Kuri (Rome: Institutum historicum Societatis Iesu, 1994), 355-56 (hereafter MPOIII). See as a parallel Houari Touati, *Islam and travel in the Middle Ages*, trans. Lydia G. Cochrane (Chicago and London: The University of Chicago Press, 2010), 45–78.
8 See Sami Kuri's introduction to the document in MPOIII.146–47 for more on the fragmentation of Salazar's account.
9 Anna Foa, "DANDINI, Girolamo," *Dizionario biografico degli italiani,* (Rome: Istituto della Enciclopedia italiana, 1960-present), 32:423–24; Carmelo Capizzi, "Un gesuita italiano di fine Cinquecento per i maroniti," *Studi e ricerche sul-l'oriente christiano* 1 (1978): 19–36.
10 Robert Bireley, *The Refashioning of Catholicism, 1450-1700: A Reassessment of the Counter Reformation* (Washington, D.C.: Catholic University Press, 1999); R. Po-Chia Hsia, *The World of Catholic Renewal, 1540-1770*. 2nd ed. (Cambridge: Cambridge University Press, 2005).
11 Mustafa Soykut, "The Development of the Image 'Turk' in Italy Through 'Della

Letteratura De Turchi' of Giambattista Dona," *Journal of Mediterranean Studies* 9 (1999): 175–203.

12 Nancy Bisaha, "'New Barbarian' or Worthy Adversary? Humanist Constructs of the Ottoman Turks in Fifteenth-Century Italy" (185–205), and Daniel J. Vitkus, "Early Modern Orientalism: Representations of Islam in Sixteenth- and Seventeenth-Century Europe" (207–30), in *Western Views of Islam in Medieval and Early Modern Europe: Perception of Other*, ed. David R. Blanks and Michael Frassetto (New York: Palgrave Macmillan, 1999).

13 MPOIII.151.

14 *Ibid.*

15 *Ibid.*, 163–64.

16 Girolamo Dandini, *Missione apostolica*, 42.

17 *Ibid.*, 43.

18 The belief in Christian ecumenism being the key to the ultimate defeat of the Ottomans is rooted in the period surrounding the lead up to the fall of Constantinople in 1453 when the Catholic and Orthodox churches attempted to reconcile their differences at the Council of Florence. See Deno J. Geanakoplos, "The Council of Florence (1438-1439) and the Problem of Union between the Greek and Latin Churches," *Church History* 24 (1955): 324–46; Joseph Gill, *The Council of Florence* (Cambridge: Cambridge University Press, 1961); Martin Anton Schmidt, "The Problem of Papal Primacy at the Council of Florence," *Church History*, 30 (1961): 35–49; Henry Chadwick, *East and West: The Making of a Rift in the Church: From Apostolic Times until the Council of Florence* (Oxford and New York: Oxford University Press, 2003).

19 Eric R. Dursteler, "Fearing the 'Turk' and Feeling the Spirit: Emotion and Conversion in the Early Modern Mediterranean," *Journal of Religious History* 39 (2015): 484–505.

20 Dandini, *Missione apostolica*, 35–39.

21 Jan-Petter Blom, "Ethnic and Cultural Differentiation," in *Ethnic Groups and Boundaries: The Social Organization of Culture Difference*, ed. Fredrik Barth (Boston: Little Brown, 1969), 74–85 (76).

22 ARSI, Gall. 106. fols. 157v-58r.

23 MPOIII.153.

24 Dandini, *Missione apostolica*, 55.

25 MPOIII.153–54.

26 Dandini, *Missione apostolica*, 55.

27 MPOIII.153–54.

28 *Ibid.*, 163–64.

29 Dandini, *Missione apostolica*, 79.

30 Dandini, *Missione apostolica*, 75–77.

31 *Ibid.*, 77.

32 *Ibid.*, 78.

33 MPOIII.154.

34 *Ibid.*, 153–54. For more, see Peter Iver Kaufman, *Augustinian Piety and Catholic Reform: Augustine, Colet, and Erasmus* (Macon, GA: Mercer University Press, 1982), 111–42.

35 Andrew C. Hess, "The Battle of Lepanto and Its Place in Mediterranean History," *Past & Present*, 57 (1972), 53–73.

36 Palmira Johnson Brummett, *Ottoman Seapower and Levantine Diplomacy in the Age of Discovery* (Albany, NY: State University of New York Press, 1994); Giancarlo L. Casale, *The Ottoman Age of Exploration: Spices, Maps and Conquest in the Sixteenth-Century Indian Ocean* (Oxford: Oxford University Press, 2010).

37 Dandini, *Missione apostolica*, 65.
38 *Ibid.*, 67.
39 Joyce de Vries, "Fashioning the Self in Early Modern Europe: Gender, Consumption, and Material Culture," *Journal of Women's History* 23 (2011): 187–97; Maria Giuseppina Muzzarelli and Antonella Campanini, eds., *Disciplinare il lusso: la legislazione suntuaria in Italia e in Europa tra Medioevo ed età moderna* (Rome: Carocci, 2003).
40 Lucia Rostagno, *Mi faccio turco: esperienze ed immagini dell'Islam nell'Italia moderna* (Rome: Istituto per l'Oriente C.A. Nallino, 1983); Giovanni Ricci, *Ossessione turca: in una retrovia cristiana dell'Europa moderna* (Bologna: Il mulino, 2002); Daniel J. Vitkus, *Turning Turk: English Theater and the Multicultural Mediterranean, 1570–1630* (New York: Palgrave Macmillan, 2003); Marina Formica, *Lo specchio turco: immagini dell'altro e riflessi del sé nella cultura italiana d'età moderna* (Roma: Donzelli, 2012); Valentina Oldrati, "«Remota Causa Removetur Effectus». Cattività, Gioventù e Apostasia All'Islam Nelle Fonti Dell'Arciconfraternita per La Redenzione Dei Cattivi Di Palermo," *RiMe. Rivista Dell'Istituto Di Storia Dell'Europa Mediterranea* 16, no. 2 (2016): 123–63.
41 Dandini, *Missione apostolica*, 63–68.
42 Kristen G. Brookes, "'Inhaling the Alien: Race and Tobacco in Early Modern England," in *Global Traffic: Discourses and Practices of Trade in English Literature and Culture from 1550 to 1700*, ed. Barbara Sebek and Stephen Deng (New York: Palgrave Macmillan, 2008), 157–78.
43 On seemingly innocuous goods causing moral turpitude, see Ina Baghdiantz McCabe, *Orientalism in Early Modern France: Eurasian Trade, Exoticism, and the Ancien Régime* (Oxford: Berg, 2008), 163–230.
44 Daniel J. Vitkus, "Poisoned Figs, or 'The Traveler's Religion': Travel, Trade, and Conversion in Early Modern English Culture," in *Remapping the Mediterranean in Early Modern English Writings*, ed. Goran V. Stanivukovic (New York: Palgrave Macmillan, 2007), 41–57 (43).
45 While sumptuary laws were self-imposed throughout Europe, they nevertheless point to the same anxieties that the Jesuits highlighted in the Maronite experience in Lebanon and the fear that unfettered cultural vice through consumerism could cause the deterioration of civil society. Alan Hunt, *Governance of the Consuming Passions: A History of Sumptuary Law* (Basingstoke: Macmillan Press, 1996); Catherine Kovesi Killerby, *Sumptuary Law in Italy 1200–1500* (Oxford: Oxford University Press, 2002).
46 Wolfgang Reinhard, "Reformation, Counter-Reformation, and the Early Modern State: A Reassessment," *The Catholic Historical Review* 75 (1989): 383–404; Ute Lotz-Heumann and Matthias Pohlig, "Confessionalization and Literature in the Empire, 1555–1700," *Central European History* 40 (2007): 35–61.
47 Daniel Goffman, *The Ottoman Empire and Early Modern Europe* (Cambridge: Cambridge University Press, 2002); Paula S. Fichtner, *Terror and Toleration: The Habsburg Empire Confronts Islam, 1526–1850* (London: Reaktion Books, 2008).
48 Dandini, *Missione apostolica*, 87.
49 *Ibid.*
50 On confession and preaching, respectively, see Wietse de Boer, *The Conquest of the Soul: Confession, Discipline, and Public Order in Counter-Reformation Milan* (Leiden: Brill, 2001); Emily Michelson, *The Pulpit and the Press in Reformation Italy* (Cambridge, MA: Harvard University Press, 2013).

51 Francesco C. Cesareo, *A Shepherd in Their Midst: The Episcopacy of Girolamo Seripando (1554–1563)* (Villanova, PA: Augustinian Press, 1999).
52 Dandini, *Missione apostolica*, 98
53 *Ibid.*, 91.
54 *Ibid.*, 92.
55 Eric R. Dursteler, *Renegade Women: Gender, Identity, and Boundaries in the Early Modern Mediterranean* (Baltimore: Johns Hopkins University Press, 2011); Stephen Ortega, *Negotiating Transcultural Relations in the Early Modern Mediterranean: Ottoman-Venetian Encounters* (Aldershot: Ashgate, 2014), 77–104.
56 On the gendered aspects of conversion to Islam and its ramifications for traditional European gender roles, see Eric R. Dursteler, "To Piety or Conversion Prone? Gender and Conversion in the Early Modern Mediterranean" (21–40), and Chloë Houston, "'I Wish to Be No Other but as He': Persia, Masculinity, and Conversion in Early Seventeenth-Century Travel Writing and Drama," in *Conversions: Gender and Religious Change in Early Modern Europe*, ed. Simon Ditchfield and Helen Smith (Manchester: Manchester University Press, 2017), 216–35.
57 Ronald C. Jennings, "Divorce in the Ottoman Sharia Court of Cyprus, 1580–1640," *Studia Islamica* 78 (1993): 155–67; Bernadette Andrea, *Women and Islam in Early Modern English Literature* (Cambridge and New York: Cambridge University Press, 2007).
58 MPOIII.153.
59 Dandini, *Missione apostolica*, 108.
60 For Dandini's final report, see ARSI, Gall. 95 II. fols. 226r–29v.
61 On the larger struggle with reforming Catholic episcopacies, see Francesco C. Cesareo, "The Episcopacy in Sixteenth-Century Italy," in *Early Modern Catholicism: Essays in Honour of John W. O'Malley, S.J.*, ed. Kathleen M. Comerford and Hilmar M. Pabel (Toronto: University of Toronto Press, 2001), 67–83.
62 Dandini, *Missione apostolica*, 97–98.
63 On fear of Islam, see Vitkus, *Turning Turk*. On religious change in Post-Reformation Europe, see Marc R. Forster, *The Counter-Reformation in the Villages: Religion and Reform in the Bishopric of Speyer, 1560–1720* (Ithaca: Cornell University Press, 1992); Eamon Duffy, *The Stripping of the Altars: Traditional Religion in England, c.1400-c.1580* (New Haven, CT: Yale University Press, 1992); Keith P. Luria, "Separated by Death? Burials, Cemeteries, and Confessional Boundaries in Seventeenth-Century France," *French Historical Studies* 24 (2001), 185–222.
64 On the challenges of early modern missionary Catholicism in the Levant, see Megan C. Armstrong, "Spiritual Legitimisation?: Franciscan Competition over the Holy Land (1517–1700)", in *The Frontiers of Mission: Perspectives on Early Modern Missionary Catholicism*, ed. Alison Forrestal and Seán Alexander Smith (Leiden: Brill, 2016), 159–79.

Bibliography

Andrea, Bernadette. *Women and Islam in Early Modern English Literature*. Cambridge and New York: Cambridge University Press, 2007.

Armstrong, Megan C. "Spiritual Legitimisation?: Franciscan Competition over the Holy Land (1517–1700)." In *The Frontiers of Mission: Perspectives on Early*

Modern Missionary Catholicism. Edited by Alison Forrestal and Seán Alexander Smith, 159–79. Leiden: Brill, 2016.

Baghdiantz McCabe, Ina. *Orientalism in Early Modern France: Eurasian Trade, Exoticism, and the Ancien Régime*. Oxford: Berg, 2008.

Bireley, Robert. *The Refashioning of Catholicism, 1450–1700: A Reassessment of the Counter Reformation*. Washington, D.C.: Catholic University Press, 1999.

Bisaha, Nancy. "'New Barbarian' or Worthy Adversary? Humanist Constructs of the Ottoman Turks in Fifteenth-Century Italy." In *Western Views of Islam in Medieval and Early Modern Europe: Perception of Other*. Edited by David R. Blanks and Michael Frassetto, 185–205. New York: Palgrave Macmillan, 1999.

Blom, Jan-Petter. "Ethnic and Cultural Differentiation." In *Ethnic Groups and Boundaries: The Social Organization of Culture Difference*. Edited by Fredrik Barth, 74–85. Boston: Little Brown, 1969.

Boer, Wietse de. *The Conquest of the Soul: Confession, Discipline, and Public Order in Counter-Reformation Milan*. Leiden: Brill, 2001.

Brookes, Kristen G. "'Inhaling the Alien: Race and Tobacco in Early Modern England." In *Global Traffic: Discourses and Practices of Trade in English Literature and Culture from 1550 to 1700*. Edited by Barbara Sebek and Stephen Deng, 157–78. New York: Palgrave Macmillan, 2008.

Brummett, Palmira Johnson. *Ottoman Seapower and Levantine Diplomacy in the Age of Discovery*. Albany, NY: State University of New York Press, 1994.

Capizzi, Carmelo. "Un gesuita italiano di fine Cinquecento per i maroniti." *Studi e ricerche sull'oriente christiano 1* (1978): 19–36.

Casale, Giancarlo L. *The Ottoman Age of Exploration: Spices, Maps and Conquest in the Sixteenth-Century Indian Ocean*. Oxford: Oxford University Press, 2010.

Cesareo, Francesco C. *A Shepherd in Their Midst: The Episcopacy of Girolamo Seripando (1554–1563)*. Villanova, PA: Augustinian Press, 1999.

Cesareo, Francesco C. "The Episcopacy in Sixteenth-Century Italy." In *Early Modern Catholicism: Essays in Honour of John W. O'Malley, S.J.* Edited by Kathleen M. Comerford and Hilmar M. Pabel, 67–83. Toronto: University of Toronto Press, 2001.

Chadwick, Henry. *East and West: The Making of a Rift in the Church: From Apostolic Times Until the Council of Florence*. Oxford; New York: Oxford University Press, 2003.

Clines, Robert John. "Wayward Leadership and the Breakdown of Reform on the Failed Jesuit Mission to the Maronites, 1577–1579." *Journal of Early Modern History 22* (2018): 215–37.

Crawford, Robert W. "William of Tyre and the Maronites".*Speculum 30* (1955): 222–28.

Dandini, Girolamo. *Missione apostolica al patriarca, e maroniti del Monte Libano e sua pellegrinazione à Gierusalemme alla santità di N.S. Alessandro VII*. Cesena, 1656.

Ditchfield, Simon and Helen Smith, eds. *Conversions: Gender and Religious Change in Early Modern Europe*. Manchester: Manchester University Press, 2017.

Duffy, Eamon. *The Stripping of the Altars: Traditional Religion in England, c. 1400–c. 1580*. New Haven, CT: Yale University Press, 1992.

Dursteler, Eric R. *Renegade Women: Gender, Identity, and Boundaries in the Early Modern Mediterranean*. Baltimore: Johns Hopkins University Press, 2011.

Dursteler, Eric R. "Fearing the 'Turk' and Feeling the Spirit: Emotion and Conversion in the Early Modern Mediterranean." *Journal of Religious History 39* (2015): 484–505.

Fichtner, Paula S. *Terror and Toleration: The Habsburg Empire Confronts Islam, 1526–1850*. London: Reaktion Books, 2008.

Foa, Anna. "Dandini, Girolamo." *Dizionario biografico degli italiani*. Rome: Istituto della Enciclopedia italiana 32 (1986): 423–24.

Formica, Marina. *Lo specchio turco: immagini dell'altro e riflessi del sé nella cultura italiana d'età moderna*. Roma: Donzelli, 2012.

Forster, Marc R. *The Counter-Reformation in the Villages: Religion and Reform in the Bishopric of Speyer, 1560–1720*. Ithaca: Cornell University Press, 1992.

Frazee, Charles A. *Catholics and Sultans: The Church and the Ottoman Empire, 1453–1923*. Cambridge: Cambridge University Press, 1983.

Geanakoplos, Deno J. "The Council of Florence (1438–1439) and the Problem of Union Between the Greek and Latin Churches." *Church History 24* (1955): 324–46.

Gemayel, Nasser. *Les échanges culturels entre les Maronites et l'Europe: Du Collège Maronite de Rome (1584) au Collège de `Ayn-Warqa (1789)*. Beirut: Impr. Y. et Ph. Gemayel, 1984.

Gill, Joseph. *The Council of Florence*. Cambridge: Cambridge University Press, 1961.

Goffman, Daniel. *The Ottoman Empire and Early Modern Europe*. Cambridge: Cambridge University Press, 2002.

Hess, Andrew C. "The Battle of Lepanto and Its Place in Mediterranean History." *Past & Present 57* (1972): 53–73.

Hsia, R. Po-Chia. *The World of Catholic Renewal, 1540–1770*. 2nd ed. Cambridge: Cambridge University Press, 2005.

Hunt, Alan. *Governance of the Consuming Passions: A History of Sumptuary Law*. Basingstoke: Macmillan Press, 1996.

Jennings, Ronald C. "Divorce in the Ottoman Sharia Court of Cyprus, 1580-1640." *Studia Islamica 78* (1993): 155–67.

Kaufman, Peter Iver. *Augustinian Piety and Catholic Reform: Augustine, Colet, and Erasmus*. Macon, GA: Mercer University Press, 1982.

Kovesi Killerby, Catherine. *Sumptuary Law in Italy 1200–1500*. Oxford: Oxford University Press, 2002.

Kuri, Sami, ed. *Monumenta Promixi Orientis*, vol. III. Rome: Institutum historicum Societatis Iesu, 1994.

Lotz-Heumann Ute and Matthias Pohlig. "Confessionalization and Literature in the Empire, 1555–1700." *Central European History 40* (2007): 35–61.

Luria, Keith P. "Separated by Death? Burials, Cemeteries, and Confessional Boundaries in Seventeenth-Century France." *French Historical Studies 24* (2001): 185–222.

Michelson, Emily. *The Pulpit and the Press in Reformation Italy*. Cambridge, MA: Harvard University Press, 2013.

Moosa, Matti. *The Maronites in History*. Syracuse, NY: Syracuse University Press, 1986.

Muzzarelli, Maria Giuseppina and Antonella Campanini, eds. *Disciplinare il lusso: la legislazione suntuaria in Italia e in Europa tra Medioevo ed età moderna*. Rome: Carocci, 2003.

Oldrati, Valentina. "«Remota causa removetur effectus»: cattività, gioventù e apostasia all'Islam nelle fonti dell'arciconfraternita per la redenzione dei cattivi di Palermo." *RiMe. Rivista dell'istituto di storia dell'Europa mediterranea 16*, no. 2 (2016): 123–63.

Ortega, Stephen. *Negotiating Transcultural Relations in the Early Modern Mediterranean: Ottoman-Venetian Encounters*. Aldershot: Ashgate, 2014.

Raphael, Pierre. *Le rôle du Collège Maronite Romain dans l'orientalisme aux XVIIe et XVIIIe siècles*. Beirut: Université Saint Joseph, 1950.

Reinhard, Wolfgang. "Reformation, Counter-Reformation, and the Early Modern State: A Reassessment." *The Catholic Historical Review 75* (1989): 383–404.

Ricci, Giovanni. *Ossessione turca: in una retrovia cristiana dell'Europa moderna*. Bologna: Il mulino, 2002.

Rostagno, Lucia. *Mi faccio turco: esperienze ed immagini dell'Islam nell'Italia moderna*. Rome: Istituto per l'Oriente C.A. Nallino, 1983.

Salibi, Kamal S. "The Maronites of Lebanon Under Frankish and Mamluk Rule (1099–1516)." *Arabica 4* (1957): 288–96.

Schmidt, Martin Anton. "The Problem of Papal Primacy at the Council of Florence." *Church History 30* (1961): 35–49.

Soykut, Mustafa. "The Development of the Image 'Turk' in Italy Through 'Della Letteratura De Turchi' of Giambattista Dona." *Journal of Mediterranean Studies 9* (1999): 175–203.

Touati, Houari. *Islam and travel in the middle ages*. Translated by Lydia G. Cochrane. Chicago and London: The University of Chicago Press, 2010.

Vitkus, Daniel J. "Early Modern Orientalism: Representations of Islam in Sixteenth- and Seventeenth-Century Europe." In *Western Views of Islam in Medieval and Early Modern Europe: Perception of Other*. Edited by David R Blanks and Michael Frassetto, 207–30. New York: Palgrave Macmillan, 1999.

Vitkus, Daniel J. "Poisoned Figs, or 'The Traveler's Religion': Travel, Trade, and Conversion in Early Modern English Culture." In *Remapping the Mediterranean in Early Modern English Writings*. Edited by Goran V. Stanivukovic, 41–57. New York: Palgrave Macmillan, 2007.

Wilkinson, Robert J. *Orientalism, Aramaic, and Kabbalah in the Catholic Reformation: The First Printing of the Syriac New Testament*. Leiden: Brill, 2007.

6 Avoiding conflict in the early modern Levant

Henry Blount's adaptations in Ottoman lands

Eva Johanna Holmberg

One of the most celebrated British travel accounts of the seventeenth century, *A Voyage Into the Levant* (1636), was written by Henry Blount (1602–1682). It went into seven editions before 1670. Blount's *A Voyage* consists of his itinerary, detailing the eleven months it took him to travel through the Ottoman Empire, from Venice to Constantinople and Cairo and then back to Venice, and a second part discussing the Ottoman Empire from various perspectives. The latter part contains a more general discussion of its armies, religion, moralities, and minority populations.[1] A gentleman traveller, Blount had toured the Levant in the spring of 1634. Before his departure, he received his BA from Trinity College, Oxford, along with a legal education and work experience with Gray's Inn, one of the four Inns of Court of London; his formative years shaped the extent to which he kept his potential London readers and future patrons firmly in mind while abroad.[2] He had written some poems and already travelled in Spain, France and Italy,[3] but an expedition to the Ottoman lands was still a brave, arduous, and dangerous enough endeavour to warrant publishing a book about it. It was also a feat that authorized Blount to make wide-ranging observations about the Ottoman state, claiming to rely only on knowledge gained through his own travel experiences and not the biased views of previous authors. Formal and firm diplomatic relations had only been established a few decades previously by William Harborne and the Levant Company merchants who had followed in his footsteps. In *A Voyage Into the Levant* (1636), Blount portrayed the Ottomans and their civilization as quite different from his own, but still worthy of observation, particularly because of this difference.

Two recent contributions to scholarship about Blount have both looked at his text as a performance of identity. According to Gerald MacLean, the early seventeenth century saw a new kind of Levant traveller emerging, the kind who travelled in order to write an account of it. MacLean writes that "for Blount, travel into the Ottoman Empire and writing about it were aimed at exposing incrusted prejudices and directed at future possibilities".[4] He highlights the ways in which British travellers like Blount "performed" the East and how encountering Ottomans made Britons more conscious of their own national identity, expressing what he calls a sense of "imperial envy",[5] a stance towards the East far more conflicted and insecure than the imperial positions that they adopted in later periods.

Another useful approach is expounded by Sabine Schülting, who has explored Blount's account as "strategic improvisation".[6] Even while admitting that Blount's tone is rather "self-congratulatory" when presenting his encounters and performances among the Ottomans, Schülting proceeds by assessing the success rate of Blount's performances, arguing that, "throughout his text, Blount shows a surprising ability to adopt the perspective of cultural 'others' and to empathize with their point of view".[7] She further claims that "Blount's self-fashioning results from his situation of powerlessness, the permanent threats English travellers were facing in the Ottoman Empire and England's fervent fascination with the power and wealth of the 'Turks'".[8] It needs to be pointed out that it was customary and almost trope-like to recount threats and travel dangers in travel accounts, whether in the West or in the East. Blount's powerlessness is also relative: the only account we have is both retrospective in nature and one-sided, carefully catered to his English readers and conscious of the perspective it provided them.

Building on these performative perspectives, but taking a slightly modified approach, I will explore how Blount utilized the Ottomans, their customs, and dispositions in his own presentations of himself. Thus, I, too, will be examining the writing, or even the "performance" of the self, but I will focus more on the way in which this writing was first and foremost aimed at Blount's readership in the English literary and cultural context. In doing this, I will follow MacLean's suggestion that, for Blount, writing about the East was about building a career.[9] In the following pages, I will also argue that, despite his programmatic stance against book knowledge and academic study in the preface of his book, Blount in fact carefully operated and participated in the discursive world woven by contemporary travel advice literature. We may have taken Blount's criticisms of previous books and reliance on his own eyewitnessing too much at face value. Despite his protestations to the contrary, Blount's travel account owed much to preceding and near contemporary travel advice literature and previous travel accounts. By analyzing closely the episodes where Blount describes his conflicts with Ottomans and explains how he successfully managed such situations, I will show how instead of showing signs of cultural sympathy and understanding towards Turks, Blount used his self-proclaimed expertise in Ottoman dispositions to better represent himself as knowledgeable, quick-witted, and better able to "read the room" in his encounters with the Ottomans. These episodes, full of the potential for conflict, showed Blount's skills, adaptability and "sprezzatura" – an effortless confidence favoured by followers of Baldassare Castiglione's *Cortegiano*[10] – all in a favourable light. All these qualities were desirable for a gentleman seeking to better his position back home.

The shape of the *Voyage* and its author

Blount's critical attitude to received knowledge has been interpreted positively by contemporaries and scholars alike. Robert Harley commended the "facetiousness of wit so peculiar to him", which gained respect during his years at Oxford. Anthony Wood (1632–1695) praised Blount in his *Athenae Oxoniensis* as "a Gentleman of a very cler judgment, great experience, much contemplation (tho not of much reading)".[11] In 1942, Boies Penrose included Blount on his list of "urbane travellers", describing such attributes as wit, genius, and "quickness of parts" that dated back to Blount's university years.[12] Blount's attitude has been described both as "hubris" and as a new kind of Baconian attitude towards empirical evidence.[13] Indeed, Blount stressed in the preface that his book relies more on his own experiences than on reading previous books. Convinced of his unbiased perspective, he claims to have received his information "from [his] own eye, not dazled with any affection, prejudice, or mist of education, which preoccupate the mind, and delude it with partial ideas, as with a false glass, representing the object in colors and proportions untrue".[14] Blount's views on sensory perception, imagination, and eyewitness accounts followed the intellectual currents of his day, amalgamating ideas from the empirical sciences, the humanist study of statecraft by the likes of Jean Bodin, and commonly held views of bodily and spiritual characteristics, which shall be discussed below.[15] The attack on books and university education might also be a response to new ideals regarding an aristocratic education, which had begun to emphasize practical and physical accomplishments over scholarly achievements.[16]

In his preface, Blount conceded that his travels were but of limited duration but tried to turn this into an asset. He defended his own perspective as a "passenger"[17] and a short-term observer by arguing that inhabiting a place for too long actually creates more bias. In the second part of his book, Blount claims that "usually quick glances, take in the most eminent peeces" and that, like the colour scarlet, "better discerned by a passing eye, then a fixt; of this nature I esteeme the morall points of behauiour".[18] In his opinion, "a new comer apprehends them with a judgment fresh, and sincere, which further familiaritie corrupts with affection or hatred, according as it meets a disposition conforme or contrary".[19] Later, he and other early seventeenth-century travellers would be criticized for this by Sir Paul Rycaut (1629–1700),[20] who dismissed the authors of his generation as mere "ingenious travellers" and short-term passengers and, hence, of being incapable of providing authoritative knowledge about the Ottomans, as his own *The Present State of the Ottoman Empire* (1670) aimed to do.[21]

MacLean suggests Blount's critiques of university education and book knowledge were directed against Richard Knolles, famous for his 1603 *Generall Historie of the Turkes,* who Blount saw as the biggest competitor in the field and as the epitome of one who had just accumulated book knowledge on the "Turks" and their history.[22] Another competing text was *Relation of a Journey begun An. Dom. 1610* by the poet, translator of Ovid, and later colonialist Sir George Sandys (1577–1644). Julia Schleck has read Sandys's *Relation* as a plea for the better governance of the Ottoman lands and as a book more focused on ancient history and heritage and thus suited for aristocratic readers. In comparison, Blount's book is much more focused on politics and statecraft.[23] Both perspectives, each in its own way, served the dual goals of helping construct their authors' credibility and shape the gentlemanly and aristocratic identities of their readers, who were more inclined to favour short tours instead of arduous and lengthy stays in foreign lands and textual treatment of foreign lands that catered to their (assumed) more refined tastes.

Knowing how to travel

Dangers and conflicts were common topics in travel accounts. Quick-witted or not, in the lands of the Ottomans, it was believed that Englishmen like Blount were in need of protection, patronage, and goodwill from both the diplomatic representatives of their own country and from Ottoman officials. Travellers were also encouraged to adapt to the divergent "customs of the country" without charging into easily avoidable conflicts. A failure to do so, authors warned their readers, would put them at risk. Despite such precautions, conflicts and dangers also provided readers of travel accounts with exotic excitement. Travellers repeatedly complained of attacks by "wild Arabians", of the wicked inhabitants of Jerusalem; of the annoyance caused by Ottoman cavalry soldiers, Spahies, and their footmen, Janissaries; or of being charged levies for their safe passage. An oft-repeated accusation was that the Ottomans called European travellers names or beat them up. Travellers' fears also included being sold off into slavery or forcefully converted to Islam.[24] Most of Blount's travel incidents seem to have occurred during his passage with Ottoman soldiery through the war-ridden Balkans,[25] through rocky landscapes in Greece, and on ships to Ottoman Egypt and back. The dangers took the form of the Ottoman people themselves, of unruly, drunken, and haughty soldiers; potential slave snatchers; or local administrators and guards who suspected him of spying. Blount's clashes with locals provided him with both chances to explore Ottoman dispositions and to show his readers his own management and reconciliation of these tricky situations.

As has been convincingly argued by MacLean, Blount seems to have followed the "travel precepts" of Fynes Moryson presented in the third book of the latter's *Itinerary* (1617). Blount mentions carefully regulating the amount of time spent in the company of fellow Englishmen, preferring Ottoman khans, Jews, and Eastern Christians as his travel companions and hosts while on ships or travelling with caravans. This, Blount asserts, helped him investigate the "arms, religion, and social customs" of the "Turks" without the undue risks incurred by other people's mistakes.[26]

Fynes Moryson had borrowed most of his travel precepts from Justus Lipsius, whom he admired greatly. John Stradling's adapted translation[27] of Lipsius' well-known travel advice book, *A direction for trauailers* (1592), had advised travellers to be "friendlie to al, familiar to a few, and speake but sildome".[28] Men of "noble and virtuous natures" desired to travel, unlike those of "base and badder minds", who remained "glued to their home they carrie (with the sluggishe and slowfooted snaile) their howses on theyr backs".[29] In comparison, the great, "haught and heavenlie spirited men" were only happy and well at ease when they were on the move. These same valiant men travelled to "profite, and inrich themselues with experience, and true wisedome" *and* to profit and benefit their country. Their great desire was "to seek, to search, to learne, and to attaine to true pollicie and wisedome", returning home better versions of themselves.[30] Lipsius had stressed the urgent need for experienced, practised, and well-educated men to travel, those assumed to resemble the intended readers of Blount's book. Blount could here find justifications for his preferred knowledge making, which owed to the tradition of humanists, who had tried to navigate the balance between book learning and firsthand experience longer than him.

Similarly to Lipsius, Blount directed his words at gentlemen and aristocrats who, similarly to him, thought highly of their own wit. Like Odysseus, Lipsius/Stradling writes, the ideal travellers came to know "many mens maners and saw many cities", especially by observing with one's own eyes the "sundry rites, manners, pollycies and gouernments", comparing them and then putting this knowledge to good use. While books represented a good starting point for acquiring knowledge, they could not compensate for empirical sensory experience, "lerning…is obtained either by the eare, or by the eie". Delving into books was not enough, even for a scholar, for "the things we heare, lesse cause the mind, and sences to arise: Then do the thinges that presently are subiect to the eyes".[31] Blount found this idea useful in defending his own position, dismissing his predecessors' portrayals of the Ottomans as mere prejudiced opinions, and also when justifying his own travels in order to better study Ottoman manners and policies firsthand.

For Lipsius, in addition to observing foreign manners and government, a man was to also refine his own manners in the process and behave kindly and courteously "towardes all kinde and conditions of men", restraining himself from imitating their gestures like a player on a stage, and likewise avoiding the "naturall" faults and vices of foreign nations.[32] For Lipsius, too, dissembling was necessary, but only to keep safe. A man should be

> in countenance…as courteous as you can, and as your state wil beare; in talk as affable as you shall see cause; but keepe your minde secret vnto your selfe, till you come to those, whose heartes are as yours.[33]

Blount wrote in his preface words to the same effect: travelling alone allowed him to avoid difficult questions and to become "all things to all men which let me into the breasts of many".[34] This idea was later adopted by James Howell for the second edition of his largely derivative *Instructions*, where he advises the traveller to "bee reserv'd in concealing his own Religion, so he must be a πολύτροπος [*polytropos*][35] he must become all to all in point of morall conversation".[36] This need was especially pertinent when travelling with people of different religions.

"Turkish" dispositions and "moralls"

How did Blount then become "all things to all men" and use his management of dangerous situations in the Ottoman Empire to portray his resourcefulness? Like his contemporaries, Blount thought that human beings were shaped by time, culture, and geographical location. He compares this process to glass blowing, in which men and their "morall parts" are glasses, the finished product and the education and laws of a specific country "are the moulds wherein they are blowne to this, or that shape", perhaps intending to imply that men and their morals were somewhat malleable and fragile. "The spirit of a man" was, however, "the metal", the most important component.[37] His assumption that Ottomans were made of different stuff than Englishmen made them a useful example in his discussions of governance and customs, but it also meant that they served as backdrops for Blount's presentation of himself.

Blount began his discussion of Ottoman dispositions by considering whether some men were "naturally borne, some for slavery, others to command". Although sceptical that nature would have such intentions, he was still willing to claim that "differences of spirit" can sometimes be detected in "whole Nations". In the Ottomans, Blount sees a natural tendency for world domination.[38] According to Blount, a great empire often "exercises" or demonstrates its attributes via "mighty employments", "admirable examples", and "great victories", in such a way that accustoms them to a greatness of spirit, too.[39]

There were, however, thorns in this Ottoman rose garden as well. Just as every human body is prone to specific illnesses according to its frame and constitution, so, too, following the humoral logic of his time, Blount associated the magnanimous "Turks" and their mindset with their own particular weaknesses or "errors":

> The magnanimous are apt to bee corrupt with an haughty Insolency, though in some sort generous: this is the Turkish way, remorcelesse to those who beare up, and therefore mistaken for beastly; but such it is not; for it constantly receives humiliation with much sweetnesse: This to their honour, and my satisfaction, I ever found: I had almost hourely experience hereof, *which my unsoiled successe makes mee not blush to remember*.[40]

Here, Blount carves out an expert position for himself, listing his successes in correctly reading and interpreting Ottoman behaviours and mental states. In mere months, he claims to have grown so accustomed to the "Turkish spirit" that he could easily anticipate and respond to Ottoman actions. He had become an expert interpreter of their dispositions, the tone of their voices, and the meaning of their gestures; at the same time, nowhere in his narrative does Blount claim much knowledge of the languages he encountered in the Ottoman Empire.[41]

It was relatively commonplace in travel writing of the time to reflect on the differences between bodily practices, divergent expectations about deference and politeness, and struggles with foreign status rituals.[42] Famous conflicts over gestures and ceremony have included Polish aristocrats refusing to bow to the Sultan in Constantinople and Ambassador Macartney's refusal to kowtow to the Chinese emperor, both of which have been discussed in histories of colonialism and imperialism. How much space was there for improvisation for travellers of an earlier generation? And how was this forced humility turned to an advantage without seeming weak?

Blount wrote that "Turks" had no rivals in "courteous" salutations, illustrating this by mentioning how the "Turks" held their hands on their breasts or exchanged "Salaum Aleeks" and "ever sweetning their conuersation with such accent of pronunciation, and so much respectiue gesture, as savours of a gentle Genius, from that rudeness, whereof they are accused".[43] In Ottoman Egypt, Blount tells how he learned to take his shoes off at the right time, "as the rest had done", followed by "bowing often, with my hand upon my breast" and of letting his host perfume him, sprinkle him with rose water, and cover him with a "sheete of Taffaty".[44]

Deciphering foreign cultural codes and gestures was just as vital as an ability to dissimulate when needed. Blount describes asking too many prying questions in Rhodes, where he was intent upon inquiring about tombs. It turned out that the tombs belonged to a "Captain Basha", slain the year before in an attack by two English ships. Blount used the situation to advertise his skills in dissimulation. He faked an identity, first claiming to be a Scot and then explaining why he did not think it was a serious lie.[45] Blount's faked Scottishness saved him only temporarily, however. The next day a "Turk" came up to him, spoke to him in Italian, and communicated with a gesture that he was "on to him":

> yet the third day, in the morning, I prying up, and downe alone, met a Turk who in Italian told me ah! are you an Englishman, and with a kinde of malicious posture, laying his fore-finger under his eye, me thought he had the lookes of a designe, he presently departed, I got to my Galleon, and durst goe to land no more. The next morne we departed for Alexandria of Egypt.[46]

At the time Blount was writing his narrative, treatises on sign language and the meaning of gestures had begun to appear, including John Bulwer's *Chirologia or the Naturall Language of the Hand* (1620). Also Castiglione thought that using correct gestures was essential for a courtier.[47] Similarly, contemporary manuals for quick language acquisition provided advice on customs and mores, though perhaps not so much on gestures.[48] Blount, without the requisite language skills, nonetheless easily claimed expertise in understanding people's demeanours, gestures, and both threatening and benevolent acts. The Ottoman gestures that he encountered also included friendly ones, presented as rewards for his quickly adopted and correctly displayed acts of humility. He reported being lovingly and respectfully treated by Ottoman sailors, who showed themselves to excel

> not only in ordinary civility, but with so ready service, such a patience, so sweet, and gentle a way, generally through them all, as made me doubt, whether it was a dreame, or reall.[49]

For Blount, the key to understanding Ottoman customs, courtesy, and gestures lay in understanding the nature of the Ottoman people. He explained many incidents by attributing to the "Turks" a proud and haughty disposition.[50] In doing so, he was by no means unique. "Turks" personified ambiguous, contradictory, often conflicted qualities, which had been portrayed both in early modern dramatic performances and prose literature; in the words of Matthew Dimmock, "Turks" managed "to represent for European men everything they imagined themselves not to be".[51]

Blount returned to the issue of Ottoman "Turkish" character more fully in the second part of his book. Again, Blount drew from his own experiences, which now serve his depiction of Ottoman peoples and represent his own resourcefulness – making this section of the book peculiarly auto-ethnographic in nature.[52] As has already been mentioned, the "Turks" were described as being "haughty and insolent" because of their "magnanimousness", but they often combined such characteristics with generosity. Blount mentions one example where he was riding with two "Turks" in the Balkans an hour ahead of a larger caravan when they met with four "Spahy-Timariots":

> They were at dinner, and seeing by my head, I was a Christian, they called to me; I not understanding what they would, stood still, till they menacing their weapons, rose, and came to me, *with lookes very ugly*; I smiling met them, and taking him who seemed of most port, by the hand, layed it to my forehead, which with them is the greatest signe of love, and honour, then often calling him Saltanum, spoke English, *which though none of the kindest*, yet gave I it such a sound, as to them who understood no further, *might seeme affectionate, humble, and hearty*; *which so appeased them, as they made me sit and eate together, and parted loving*. [my emphasis][53]

Here, Blount is sharing an inside joke with his readers, while at the same time making the Ottomans seem unaware of his secret and subtle mockery. In an effort to further contrast his own cautious comportment with that of another traveller's, Blount then tells of how the Spahies severely beat up a merchant from Ragusa, who had forgotten "how the place had changed his condition".[54] Blount continues by reporting that only two hours after this incident, he was confronted by six (or seven) more Spahies, who threatened him with his own dagger. Again, Blount claims that his adaptability and skills in dissimulation saved him. He describes the "Turks" trying to interrogate him about his travel motives and routes and how he misled them with his vague answers.[55] It was customary to warn Christian travellers about carrying weapons or displaying their wealth through ostentatious apparel. The Ottoman lands were supposed to be "a geography of fear" for Christians, just as Catholic lands had been only a short while ago for Protestants.

The book includes several instances where Blount manages to follow local customs and to escape threatening situations by showing humility. He takes his shoes off in correct places, thanks people for their kindness and hospitality, and avoids all kinds of dangers. All of these experiences, we are led to believe, have given Blount the confidence to "read" "Turkish" dispositions without making mistakes. He tells readers that after having "daily successe in the like kinde" of happenings, he grew so "confident of the Turkish nature" that he did not doubt himself, unless he had to deal with drunkards or mercenaries. Back home in mid-seventeenth-century England, a nation increasingly plagued by sectarian divisions and political conflict, such skills could later prove quite useful for Blount. After his travels, Blount gained fame as both a teetotaller and a coffee drinker, probably prompted by his journey, or possibly also by his experiences in the Inns of Court. In dealing with drunkards, a fierce confidence and effortless navigation of strangers' "structures of feeling" he wanted to impart was more useful than either book knowledge or impeccable language skills.

The key to Blount's successful management of his Ottoman encounters was in understanding that the "fashions of other countreys" were despised by "Turks".[56] Blount claims to have learned this in the Turkish inlands, where, if appearing dressed at all like a Christian, he was "tufted like an Owle among other birds"; just as unhappy owls were attacked by other birds, so, too, was he at the time.[57] When complaining about this treatment, he was told that the "Turks" did not despise him as such, but instead they hated all novelties. This was not a question of civility but of *institution*. The "Turks" had to "estrange themselves" from all Christian fashions, customs and people because they desired "perpetuall hostilitie" with them. They likewise had to "disgrace" all novelties, including Christian dress.

Blount's conclusion was that the "Turkish way" was not entirely or essentially barbarous, but – as he had already mentioned in his preface – just another kind of "civilitie". The Ottoman Empire provided Blount with a good setting to effectively display his intellectual tool box: his quick wit and resourcefulness in the face of adversity and conflict, a "sprezzatura" suitable for a courtier, and a situational awareness useful for an intelligencer/diplomat/gentleman about town. Answering haughty with haughty or a more forthright display of military bravery would have gotten him into trouble and was not something his university education had prepared him for.[58]

Stephen Greenblatt quotes Keith Thomas when he proposes that Protestants rarely valued "sprezzatura" because it would hide all their hard work from view.[59] Blount's careful writing of himself seems to hover somewhere in between, engaging in fake humility in Ottoman lands but making sure his readers understood what he was doing. He wants to make his experiences and successes in dissimulation and adaptation visible to readers and to advertise his knowledge of "Turkish" humours and dispositions. For Blount, such expertise and skills, as I am suggesting, revealed his great suitability for a more prestigious office; he was someone who could deal with the haughtiest and most difficult of men, including the most outlandish rules of conduct and ceremony. In addition, Blount carefully tried to hide his own literary knowledge of Ottoman news, books, and apodemic texts from view and take credit for knowledge that he had clearly attained from reading other books about travels.

"Treating with men with several humours"

Just before Blount set out on his journey in 1633, a small volume of travel advice had been published in London. It contained a list of travel instructions on what to observe and report back home and two short letters of travel advice by famous men: Robert Devereaux, the late Earl of Essex, and Sir Philip Sidney. Similar justifications for travel/travel advice had been published before in response to prominent attacks against the usefulness of travel, which included most notably Roger Ascham's *The Scholemaster* (1570) and Bishop Joseph Hall's *Quo Vadis? A Iust Censure of Travell* (1617). The publisher, Benjamin Fisher, stated in the preface to the reader that travel experience was essential for all men intending to enter public service and that limiting oneself to the study of maps and books, even if useful to prevent moral and spiritual corruption by the outside world, also guaranteed continuing ignorance:

> Our sedentary Traueller may passe for a wise man, as long as hee converseth either with dead men by reading; or by writing, with men absent. But let him once enter on the stage of publike imployment, and hee will soone find, if he can bee but sensible of contempt, that he is vnfit for Action. *For ability to treat with men of seueral humours, factions, and Countries; duly to comply with them, or stand off, as occasion shall require, is not gotten only by reading of books, but rather by studying of men.* [my emphasis][60]

Similar themes abound in John Stell's dedicatory preface in the English translation of probably the most influential European account of the Ottoman Empire, Nicolas de Nicolay's *Voyage into Turkie* (1585). Here, Stell praised experience and activity over sedentary knowledge of maps and books because "perfect prayse of wisedome and learning, is not to be sought for in bookes, but to be gotten by verie vse and practise". Stell further pointed out that wise men should be

> conuersant in this worlde as in a Theater of mans life, and by due triall are taught the diuersities of countries, the differences of peoples manners, the examples of life, and manifold thinges besides, which Trauellers doe comprehend with their eie, and compasse by their wit.[61]

A studious and noble-born man should not remain home like a snail in his shell but aspire to

> attaine to the sight of great matters, wherwith they shal neuer be acquainted, otherwise than by the booke or the map, which differ as much in mouing the imagination and other faculties of the mind, as the representation of a thing in a glasse or a paynted table, doth vary from the thing represented?[62]

We can recognize here praise for seeing things with one's own eyes and for personal experience, the importance of sensory experience, and even tacit criticism of the sluggards lurking in both the work of Lipsius and later that of Fynes Moryson.[63] Blount had stated his interest in Ottoman civilization and its characteristics, the diverse non-Muslim peoples living in the empire, the success and manner of the Ottoman armies, and his desire to see Gran Cairo with his own eyes. He travelled there after spending only five days in Constantinople. Blount wrote that "intellectual complexions have no desire so strong, as that of knowledge...as that of humane affaires", which sounds very similar to Fisher's "studying of men" and Sidney's advice. It is possible that Blount had read Lipsius, Sidney, or Moryson, but more than anything it was quite probable that in planning his journey to the Ottoman Empire, he had read both apodemic texts and Nicolay as well as other previous travel accounts, all of which praised active engagement with the world over merely perusing books and maps. A similar emphasis on the importance of eye-witness accounts and the usefulness of brave engagement with men of diverse customs and humours to the state can be found in all of them. It makes sense that someone intending to write a book about the Ottomans would have read pre-existing texts and then criticize, utilize, and even either accidentally or wilfully imitate them, followed by an attack on all previous attempts at depicting the Ottomans as false, distorted, or blinded by "zealous ignorance".

In order to learn how to treat men of different humours, Blount had to observe people who were least like himself, people whose institutions and dispositions differed the most from the ones he was used to. Blount acquainted himself with all of these aforementioned aspects of the Ottoman Empire, using his own travel experiences as exempla, which provided a sounding board and served as proof for his general observations. For Blount, the Ottomans became like a language, the tone of which he had learned to read and decipher with the help of his senses, especially the ocular senses, but not to speak. In addition, he had learned which set of gestures would have a desired effect on Ottomans from all walks of life.

The ability to read tones and gestures, to listen, and to remain silent and submissive are all useful skills when in a precarious situation or when a traveller's sole purpose is to observe and not to interfere. Blount's line of communication was similar to that of a spy and an intelligencer, and some have suggested that he was one:[64] regardless, this perspective certainly confirms the fact that he was not seeking to communicate with the Ottomans so much as with his readers in England.

After the publication of his book in 1636, Blount gained the patronage of Charles I, becoming a gentleman pensioner of his court, a position that gave him the right to make an occasional appearance at court in addition to a small pension.[65] He was then knighted in 1639, but probably mostly for reasons other than his services in providing useful information about Ottoman harbours and armies. The adaptability that he so confidently describes as serving him well during his travels may also have been useful later in his life, when Blount switched sides during the Interregnum and then back again after the Restoration; he aligned himself with different rulers and successfully avoided conflicts and dangers until the end of his long life. In retrospect, we can see how he used his discussions of the nature of the "Turks" to promote his own abilities and that similar dissimulations were put to practice in his own life. *A Voyage Into the Levant* would then have been written not only in order to produce knowledge about the Ottoman Empire but *especially* to advertise the intellectual and practical qualities and skills of its author.

Notes

1 Henry Blount, *A Voyage Into the Levant* (London, 1636), 61.

2 Nabil Matar, "Blount, Sir Henry (1602–1682)," *Oxford Dictionary of National Biography* (Oxford: Oxford University Press, 2004), http://www.oxforddnb.com/view/article/2687. For Blount's life, see also Gerald MacLean, "Ottomanism before Orientalism," in *Travel Knowledge. European "Discoveries" in the Early Modern Period*, ed. Ivo Kamps and Jyotsna Singh. (Basingstoke and New York: Palgrave Macmillan, 2001), 85–96 and MacLean, *The Rise of Oriental Travel: English Visitors to the Ottoman Empire, 1580–1720* (Basingstoke and New York: Palgrave Macmillan, 2004). For the early modern Inns of Court and social formation of the literati, see Michelle O'Callaghan, *The English Wits: Literature and Sociability in Early Modern England* (Cambridge: Cambridge University Press, 2007); Jayne Elisabeth Archer, Elizabeth Goldring and Sarah Knight, eds, *The Intellectual and Cultural World of the Early Modern Inns of Court* (Manchester: Manchester University Press, 2011).

3 On September 23, 1629, a passport was granted for Francis and John Mannock from Suffolk and "another for Henry Blunt and William Blount, gentlemen, of the Countie of Oxford to travill into the partes beyond the Seas for the space of three yeares next after the date hereof, and to take with them their Truncks of Apparell etc. not prohibited". Underneath is a memorandum, which explains that each of them were allowed one servant and that the passes were issued in November. The acts of the Privy Council for 1629–30, 138. The decision to travel then was probably due to the new-found/looming peace with France and Spain. See Kevin Sharpe, *The Personal Rule of Charles I* (New Haven and London: Yale University Press, 1992), 67–69.

4 MacLean, *Looking East: English Writing and the Ottoman Empire Before 1800* (Basingstoke and New York: Palgrave Macmillan, 2007), 175. Note MacLean's discussion of Bishop Henry King's appraisal of Blount's book and its profitability with respect to future travellers and how to "deal successfully in Ottoman lands," 178–79.

5 For imperial envy, see MacLean, *Looking East*, 20–23.

6 See Schülting, "Strategic Improvisation: Henry Blount in the Ottoman Empire," in *Early Modern Encounters with the Islamic East*, ed. Sabine Schülting, Sabine Lucia Müller and Ralf Hertel (London and New York: Routledge, 2012), 67–83 (70–71).

7 *Ibid.*, 74.

8 *Ibid.*, 78.

9 See MacLean, "Ottomanism before Orientalism," 94.

10 On the European reception of Castiglione's work, see Peter Burke, *The Fortunes of the Courtier* (Cambridge: Polity Press, 1995). On the many gendered meanings and practices of sprezzatura in Castiglione and Della Casa's *Galateo*, see Harry Berger, *The Absence of Grace. Sprezzatura and Suspicion in Two Renaissance Courtesy Books* (Stanford: Stanford University Press, 2000).

11 See MacLean, *The Rise of Oriental Travel*, 118, 119, wherein MacLean lists several near contemporary characterisations of Blount, ranging from Samuel Pepys' "the famous travellor & water drinker" to Charles Gildon's calling Blount "the Socrates of the age".

12 Boies Penrose, *Urbane Travelers 1591–1635: Biographical sketches of Fynes Moryson, John Cartwright, Thomas Coryate, William Lithgow, George Sandys, Sir Thomas Herbert, Sir Henry Blount* (Philadelphia: University of Pennsylvania Press, 1942), 216.

13 See MacLean, *The Rise of Oriental Travel*. MacLean has suggested that during his time in London and at Gray's Inn, Blount might even have met Bacon. It has been demonstrated that Blount had contacts with both political and literary elites of his time.

14 Blount, 3, 4.

15 For similar views, see Fynes Moryson in the third book of his *Itinerary* (1617), "Lesse doth it moue the minde that beates the eares, Then what before the faithfull eye appeares" (3:1–2).

16 Clare Howard, in *English Travellers of the Renaissance* (London: John Lane, 1914), 104, cites the letter of the Earl of Newcastle to his pupil the prince, the future King Charles II, advising him not to be "too studious, for too much contemplation spoils action, and Virtue consists in that". The ideal was to be learned and wise, but to hide it discreetly. At the same time, noblemen were increasingly losing interest in university education, and the ideals of courtly comportment were being transmitted and received more from France than Italy, where fencing and dancing was more in vogue than lectures and scholarly pursuits; *ibid.*, 106–07. See also Michèle Cohen, *Fashioning Masculinity:*

National Identity and Language in the Eighteenth Century (London: Routledge, 1996).

17 *Ibid.*, 2. On Blount's "self-styled role as a passenger" and perhaps being an "ideal spy" for his monarch, see McLean, "Ottomanism before Orientalism," 92–94.

18 Blount, 61.

19 *Ibid.*

20 Sir Paul Rycaut made a distinction between himself, a long-term resident and scholar, and occasional travellers, who did not know what they were writing about. On Rycaut, see Sonia Anderson, *An English Consul in Turkey: Paul Rycaut at Smyrna, 1667–1678* (Oxford: Clarendon Press, 1989).

21 Paul Rycaut, *The Present State of the Ottoman Empire* (London, 1670), "The Epistle to the Reader," 1.

22 MacLean, *The Rise of Oriental Travel*, 127.

23 Julia Schleck, *Telling True Tales of Islamic Lands: Forms of Mediation in English Travel Writing, 1575–1630* (Selinsgrove: Susquehanna University Press, 2011), 51. Jonathan Haynes has suggested that "Sandys thinks historically, Blount philosophically; Sandys prizes identity and continuity, Blount difference and change; Sandys' stance is public and rhetorical, Blount's private, scientific, alienated". In addition, Haynes has claimed (based on their published outputs) that Sandys was celebrated, while Blount was a failure as an author because the *Voyage* is practically the only thing he ever published. However, Gerald MacLean has shown that the positive reception of Blount's work by his contemporaries testifies to the contrary. See Jonathan Haynes, "Two Seventeenth Century Perspectives on the Middle East: George Sandys and Sir Henry Blount," *Alif: Journal of Comparative Poetics* 3 (1983): 4–22. See also MacLean, *Looking East*, 185–86.

24 Fynes Moryson, in *Unpublished Chapters of Fynes Moryson's Itinerary* (1903) 1:63–5, lists all the dangers, exploitations, and abuses he encountered during his journey. Thomas Gainsford attributed the worsening of the status and safety of Britons to Ottoman revenge on Edward Barton, the former ambassador, because his overly intimate relationship with Murat III's grandmother. See Gainsford, *The Glory of England* (1618), 191–92.

25 Blount, 8.

26 These precepts were an amalgamation of advice gathered from predecessors in the *ars apodemica* genre of the sixteenth century. MacLean convincingly demonstrates which of them seem to have been taken up by Blount. See MacLean, *The Rise of Oriental Travel*, 128. Moryson had clearly gleaned some of these precepts from Lipsius and other writers in the apodemic tradition. For a discussion of women's engagement with *ars apodemica*, see Gábor Gelléri's chapter in this volume. See also Natalya Din-Kariuki's chapter in this volume on how a travel journal characterizes its subject in harmony with travel advice literature.

27 The translation was made and adapted by John Stradling to fit the travel aspirations of the young Earl of Bedford.

28 Justus Lipsius, *A Direction for Trauailers* (London, 1592), C3v.

29 *Ibid.*, A2v.

30 *Ibid.*, Br.

31 *Ibid.*, B4v.

32 *Ibid.*, C2r.

33 *Ibid.*, C3v. He talks of dissembling and counter-dissembling on pages C3r–C3v. A traveller should thus have "frons aperta, lingua parca, mens clausa". See also Mareile Pfannebecker, "'Lying by Authority': Travel Dissimulations in Fynes Moryson's Itinerary," *Renaissance Studies* 31 (2017): 569–85.

34 Blount, 5.

35 Greek, "A man of many devices" shapes and shifts, the first epithet by which Homer refers to Odysseus in Odysseia.
36 James Howell, *Instructions and Directions For Forreine Travell* (London, 1650), 130–31. I thank Jan Stenger for deciphering the almost illegible Greek word in Howell's text and for helping me contextualize it.
37 Blount, 96.
38 "Thus, if ever any race of men were borne with *Spirits* able to beare downe the world before them, I thinke it to be the *Turke*"; Blount, *A Voyage Into the Levant* (1636), 97.
39 Lipsius similarly believed that great places gave rise to noble thoughts; Blount, 97.
40 Blount, 97–8.
41 For language learning in early modern England, see John Gallagher, *Learning Languages in Early Modern England* (Oxford: Oxford University Press, 2019).
42 On status ritual, see Richard Handler, "Erving Goffman and the Gestural Dynamics of Modern Selfhood," *Past & Present* 203 (2009): 280–300.
43 Blount, 107. Compare with William Biddulph, who also included a detailed account of Ottoman gestures and salutations; see Biddulph, *The Travels of certaine Englishmen* (1609), 66–8.
44 Blount, 42.
45 *Ibid.*, 32–3.
46 *Ibid.*, 33.
47 See Castiglione, *The Courtier from the Italian of Count Baldassare Castiglione / done into English by Sir Thomas Hoby Anno 1561; With an Introduction by Walter Raleigh. Baldassarre Castiglione, conte, 1478–1529* (London: David Nutt, 1900), 1:24.
48 According to Pamela Allen Brown, the same gesture in Sicily and Calabria means both "you are full of shit" and "I'm on to you". Charlotta Wolff, from Helsinki University, has informed me that in France it could mean "and you want me to believe that?" and can be accompanied or replaced by the phrase "mon œil". In France and Greece, however, this would then mean "I am alert. I'm looking. You can't fool me". In Italy and Spain, though, it is a friendlier warning, meaning "Be alert, that guy is clever". Henry Blount had clearly been found out. I want to thank John Gallagher for information on early modern language learning manuals.
49 Blount, 75.
50 *Ibid.*, 97.
51 See Nabil Matar, *Islam in Britain: 1558–1685* (Cambridge: Cambridge University Press, 1998); MacLean, *Looking East*, 7–8; Matthew Dimmock, *New Turkes: Dramatizing Islam and the Ottomans in Early Modern England* (Aldershot: Ashgate, 2005), 207. These characteristics included being "inconstant", "unbelieving", "cruel", "circumcised", or "covetous", all of which were perceived as negative qualities and could be used to characterize women, Jews, deviants and criminals, bankers or anyone who the writer wanted to morally undermine. For examples in Ottoman histories, see Anders Ingram, *Writing the Ottomans. Turkish History in Early Modern England* (Basingstoke and New York: Palgrave Macmillan, 2015).
52 On Mary Louise Pratt's idea of colonial subjects adopting the rhetoric of the colonizers and engaging in "auto-ethnography" in the context of seventeenth-century Asia, see Anna Winterbottom, "Self-Fashioning and Auto-Ethnography: Samuel Baron's Description of Tonqueen (1686)," *Journeys* 14 (2013): 85–105.
53 Blount, 97–98.
54 *Ibid.*, 98.
55 *Ibid.*, 98–99.

56 On clothing as a marker of status and religion in the Levant, see Robert Clines's essay in this current volume.

57 *Ibid*., 99–100. I want to thank Professor Katharine Hodgkin for pointing this out to me, when I was about to interpret the passage differently.

58 Lack of space prevents the further exploration of the gendered nature of Blount's performance, but it seems that he was trying to tread the fine line between tactical submissiveness and showing his masculinity in unfavourable light. For the embodied and gendered performance of masculinity with gestures, see Miranda Fay Thomas, *Shakespeare's Body Language: Shaming Gestures and Gender Politics on the Renaissance Stage* (London: The Arden Shakespeare, 2019), introduction.

59 Stephen Greenblatt, *Renaissance Self-Fashioning: From More to Shakespeare* (Chicago: University of Chicago Press, 1980), 190. "Sprezzatura, which sets out to efface all signs of 'hard work and application,' is a cult of the 'too easy,' a kind of aesthetic magic". It makes sense that Anglicans like Joseph Hall would have distrusted such a stance; Hall would have mistrusted any traveller who had learned too many foreign mannerisms.

60 F. B. [B.F.], *Profitable instructions* (1633), preface, A2–A3v. The passage continues by claiming that scholars still make the best travellers, "Yet this ever holds true; The best scholler is fittest for a Traueller, as being able to make the most vseful obseruation: Experience added to learning, makes a perfect Man" (A3v). In the same book, Philip Sidney advised travellers (and interprets Homer's advice) to study men's mores, meaning "the certainnesse if true discerning of mens minds both in virtue, passion and vices" (88), and to observe human affairs, meaning by it "their Religion, Policies, Lawes, bringing vp of Children, discipline both for warre and peace and such like" (89).

61 John Stell, in Nicolay, *The Navigations into Turkie* (London, 1585), 2–2v.

62 John Stell adds a reference to *The Poet*, an emphasis on eyesight that is also taken up by Blount, who seems fascinated by the sensory world of glasses, eyes, and their connection to both the mind and imagination, "That which we heare with outwarde eare, moues not the minde so much, As that which beames of rowling eyes, atteine vnto and touch". He ends by praising travellers who learn practical skills, like languages, manners, religion and policy and put such knowledge to good use, "Onely this I say, that the greatest commendation and praise of a traueller is, not onelye to talke by knowledge roundly of such famous Cities, Countries, people, and other straunge obiectes as he hath seene; but also to speake their language redily, to learne their manners desirously, to know their religion perfectly, their pollicie throughly, their gouernment absolutely, and all thinges els, that hee may imploy himselfe about profitably". Stell, 2r–2v.

63 "But it is the part of an industrious man, to act their affaires in the world, tho sluggards lie by the fire". Moryson, 3:5.

64 On Blount being "an ideal spy", see MacLean, "Ottomanism before Orientalism", 93.

65 Gerald Aylmer advises that gentlemen pensioners should not be seen "as royal officials rather than as part-time attendants at court". See Gerald Aylmer, *The King's Servants: The Civil Service of Charles I 1625–1642* (Oxford: Clarendon Press, 1989), 27.

Bibliography

Anderson, Sonia. *An English Consul in Turkey: Paul Rycaut at Smyrna, 1667–1678*. Oxford: Clarendon, 1989.

Archer, Jayne Elisabeth, Elizabeth Goldring and Sarah Knight, eds. *The Intellectual and Cultural World of the Early Modern Inns of Court*. Manchester: Manchester University Press, 2011.

Aylmer, Gerald. *The King's Servants. The Civil Service of Charles I 1625–1642*. New York: Columbia University Press, 1961.

Berger, Harry. *The Absence of Grace. Sprezzatura and Suspicion in Two Renaissance Courtesy Books*. Stanford: Stanford University Press, 2000.

Blount, Henry, *A Voyage Into the Levant*. London, 1636.

Brotton, Jerry. *This Orient Isle: Elizabethan England and the Islamic World*. London: Penguin Press, 2016.

Cohen, Michèle. *Fashioning Masculinity: National Identity and Language in the Eighteenth Century*. London: Routledge, 1996.

Dasent, John Roche, ed. *Acts of the Privy Council of England Volume 30, 1599–1600*. London, 1905. *British History Online* http://www.british-history.ac.uk/acts-privy-council/vol30.

Dimmock, Matthew. *New Turkes: Dramatizing Islam and the Ottomans in Early Modern England*. Aldershot: Ashgate, 2005.

F. B. [B.F.]. *Profitable Instructions*. London, 1633.

Gallagher, John. *Learning Languages in Early Modern England*. Oxford: Oxford University Press, 2019.

Greenblatt, Stephen. *Renaissance Self-Fashioning: From More to Shakespeare*. Chicago: University of Chicago Press, 1980.

Handler, Richard. "Erving Goffman and the Gestural Dynamics of Modern Selfhood." *Past & Present 203* (2009): 280–300.

Haynes, Jonathan. "Two Seventeenth Century Perspectives on the Middle East: George Sandys and Sir Henry Blount." *Alif: Journal of Comparative Poetics 3* (1983): 4–22.

Howard, Clare. *English Travellers of the Renaissance*. London: John Lane, 1914.

Howell, James. *Instructions and Directions For Forren Travell*. London, 1650.

Ingram, Anders. *Writing the Ottomans: Turkish History in Early Modern England*. Basingstoke and New York: Palgrave Macmillan, 2015.

Lipsius, Justus. *A Direction for Trauailers*. London, 1592.

MacLean, Gerald. "Ottomanism before Orientalism." In *Travel Knowledge: European "Discoveries" in the Early Modern Period*, 85–96. Edited by Ivo Kamps and Jyotsna Singh. Basingstoke and New York: Palgrave Macmillan, 2001.

MacLean, Gerald. *The Rise of Oriental Travel: English Visitors to the Ottoman Empire, 1580-1720*. Basingstoke and New York: Palgrave Macmillan, 2004.

MacLean, Gerald. *Looking East: English Writing and the Ottoman Empire Before 1800*. Basingstoke and New York: Palgrave Macmillan, 2007.

MacLean, Gerald and Nabil Matar. *Britain and the Islamic World, 1558–1713*. Oxford: Oxford University Press, 2011.

Matar, Nabil. *Islam in Britain, 1558–1685*. Cambridge: Cambridge University Press, 1998.

Matar, Nabil. "Blount, Sir Henry (1602–1682)." *Oxford Dictionary of National Biography*. Oxford: Oxford University Press, 2004. http://www.oxforddnb.com/view/article/2687.

Moryson, Fynes. *An Itinerary*. London, 1617.

Moryson, Fynes. *Shakespeare's Europe: Unpublished Chapters of Fynes Moryson's Itinerary: Being a survey of the Condition of Europe at the End of the 16th Century*, edit edby Charles Hughes. London: Sherratt & Hughes, 1903.

Nicolay, Nicolas de. *The Nauigations, peregrinations and voyages, made into Turkie*. London, 1585.

O'Callaghan, Michelle. *The English Wits: Literature and Sociability in Early Modern England*. Cambridge: Cambridge University Press, 2007.

Penrose, Boies. *Urbane Travelers 1591–1635: Biographical Sketches of Fynes Moryson, John Cartwright, Thomas Coryate, William Lithgow, George Sandys, Sir Thomas Herbert, Sir Henry Blount*. Philadelphia: University of Pennsylvania Press, 1942.

Pfannebecker, Mareile. "'Lying by Authority': Travel Dissimulations in Fynes Moryson's Itinerary." *Renaissance Studies 31* (2017): 569–85.

Rycaut, Paul. *The Present State of the Ottoman Empire*. London, 1670.

Schleck, Julia. *Telling True Tales of Islamic Lands: Forms of Mediation in English Travel Writing, 1575–1630*. Selinsgrove, Susquehanna University Press, 2011.

Schülting, Sabine, "Strategic Improvisation: Henry Blount in the Ottoman Empire." In *Early Modern Encounters with the Islamic East*, 67–83. Edited by Sabine Schülting, Sabine Lucia Müller and Ralf Hertel. London and New York, Routledge, 2012.

Sharpe, Kevin. *The Personal Rule of Charles I*. New Haven and London, Yale University Press, 1992.

Thomas, Miranda Fay. *Shakespeare's Body Language. Shaming Gestures and Gender Politics on the Renaissance Stage*. London: The Arden Shakespeare, 2019.

Winterbottom, Anna, "Self-Fashioning and Auto-Ethnography: Samuel Baron's Description of Tonqueen (1686)." *Journeys 14* (2013): 85–105.

Part III

War, diplomacy, and dissimulation

7 Ambassadors as travellers in Italy in the second half of the fifteenth century

Paul M. Dover

Conflict and travel combine uniquely in the role of the ambassador. Ambassadors have alternatively been mediators, averters, and abettors of conflict. Even more so than today, in the Middle Ages and Renaissance, ambassadors were regarded by their sovereign authorities primarily as agents to prepare and ensure readiness for the next conflict. At the same time, ambassadors have also always been travellers. Throughout most of history, their travel has been time-consuming, uncomfortable, and often dangerous; ambassadors, like merchants and pilgrims, were who they were largely because they travelled. Much of the theoretical and legal framework that has bounded the practice of diplomacy has been predicated on the assumption that the ambassador was an individual who had travelled from his domicile and was the living extension of his home prince or polity. Travel is so central to diplomacy, in fact, that significant changes in diplomatic praxis necessarily shape the experience of the ambassador as a traveller. Such was the case with the advent of the "new diplomacy", which first appeared in fifteenth-century Italy, where an important contributing factor was the relative ease of travel within the peninsula at the close of the Middle Ages.

The two outstanding tenets of this "new diplomacy" were the appearance of residency in diplomatic representation and of growing permanence in diplomatic institutions. These emerged within a space that shared common political and cultural languages and practices, in response to pressing needs of representation, negotiation, and information gathering. As Isabella Lazzarini has recently demonstrated in her comprehensive examination of Italian Renaissance diplomacy, these developments had multiple origins and to see their development as a straightforward path to "modern" diplomacy is unduly teleological.[1] Nonetheless, these twin transformations were deeply consequential and become clearly evident in the years after 1450. An ordinary network of resident embassies between the principal Italian states was in place by about 1480. Ambassadors increasingly resided for long periods at courts in Italy and occasionally beyond, with the expectation that their office would be filled by another upon their departure. These resident ambassadors maintained a regular stream of correspondence, as they not only attended to urgent political business but also acted as conduits for information gathering.

Alongside resident ambassadors, short-term extraordinary envoys remained important diplomatic operatives. These individuals tended to be of higher social station than resident ambassadors, who often emerged from the secretariat of Italian states. The service of such high-ranking individuals was typical of the "old diplomacy", and Italian states continued to dispatch them to negotiate treaties, send greetings to new sovereigns, make formal requests, and attend celebrations. But their activity now served to supplement the bulk of the diplomatic work carried out by resident ambassadors, who attended to the far less glamorous day-to-day work of a diplomatic functionary. Italian states concurrently developed chanceries and offices dedicated to the drafting, retrieval, and storage of the rapidly increasing volume of diplomatic correspondence.[2] The implementation of these changes was not universal across the diverse collection of Italian regimes, nor was there anywhere a relentless march toward modern forms. By 1500, however, the basic outlines of this "new diplomacy" were firmly in place and widely associated with Italy.

This essay will examine the travel of Italian ambassadors in the second half of the fifteenth century and will suggest that travel and the well-established and geographically confined network in which this diplomacy was conducted were important considerations in the transformations described above. Despite the associated vagaries and hardships, the familiarity and predictability of travel within Italy were important factors making possible the advent of the new diplomacy.

Varieties of diplomatic travel and sources

It is harder than one might imagine to ascertain detailed information about the experience of diplomatic travel in the second half of the fifteenth century. Regulations and legislation concerning travel were largely concerned with its expense and what proportion of these costs was to be borne by the state. The statutory record governing embassies and diplomatic practice contains very little pertaining to travel; diplomatic treatises of the period also seldom cover the vagaries of travel and are more concerned with qualities of the ambassador in service; the protocol of the court; and the everyday practice of representation, negotiation, and information gathering. The best source for details on diplomatic travel is the correspondence of the ambassadors themselves. These reports are among the few largely unmediated and nearly real-time accounts of travel that we have from the late medieval and early Renaissance period.

Even among the large volumes of extant fifteenth-century correspondence, however, the information furnished about travel is surprisingly thin. Within Italy, at least, the routes between the most important courts were well established and dotted with locales geared to provide services to travellers.

Ambassadors stayed at the same inns, generally in urban centres or near postal stations, and passed familiar landmarks that served as mile markers along the route. Ambassadors did not regularly write letters en route to and from their assigned destinations, unless there was a political crisis or climatological event delaying their arrival, or if they learned something of sufficient importance along the way that merited immediate reporting. When travelling from northern Italy to Rome or Venice, they usually had commissions to stop in other cities en route, especially Bologna, Florence, and Siena, to consult with the governments there, to meet with ambassadors stationed at these locales, or to deliver correspondence. Especially for travel within Italy, ambassadors generally wrote very little about the physical circumstances of their journeys. This changed when extraordinary events intervened, as for example when Pier Candido Decembrio, sent by Duke Francesco Sforza as an envoy to Naples in January 1457, described the difficulties of reaching the city amid terrible weather and in the wake of a devastating earthquake that had hit southern Italy a month earlier:

> When I left Milan to come here I did not believe that I would find such obstacles. I do not want to repeat the difficulty of the journey, such that one has never seen a more painful one on account of the mud, the continuous rain and flooded rivers, which had overflowed their banks beyond any measure; it is enough for me to write of the terrible news of the earthquake, which has left this whole region desolate, such that one sees no one left alive.[3]

Pier Candido then credits an astrologer with recommending he delay his departure for Naples, which certainly prevented him from being caught in the "recent disaster" (*ultimum terribilium*).

Speed and conditions of travel in the fifteenth-century Mediterranean were far closer to those of the Roman Empire than to those of our own age. The wheels of statecraft could turn only as fast as couriers could carry the letters. In practice, however, speed of communication was less important than the consistency of travel times, for ambassadors and couriers alike. The exigencies of travel still mattered, especially for the extraordinary envoys sent for specific, temporally limited purposes. But the new emphasis on nearly daily reportage by resident ambassadors, largely in the pursuit of information gathering, meant that the secure, quick, and reliable travel of couriers and their paper cargo was now a top concern.

When travelling by land, ambassadors invariably travelled on horseback. If they had to, ambassadors could travel very quickly, in some cases as quickly or quicker as a courier might. In May 1471, Agostino Rossi, sent by Duke Galeazzo Maria Sforza of Milan to negotiate the renewal of the Italian League in Rome, travelled so quickly that he left three horses by the side of the route as he rode night and day.[4] This was faster even than a *cavallaro*, a courier on horseback. The Milanese ambassador Tommaso da

Rieti in March 1454 reported from Aix in France that he had ridden 440 miles in ten days, "on the road [I] beat the hide off several horses. This I did in order to be here on time and obey the orders that Your Lordship wrote to me".[5] He arrived on the same day as a letter carrier who had left the same day that Tommaso had.

Good horses were expensive to purchase and to feed, and much of the legislation pertaining to embassies in the fifteenth century has to do with the serious financial burden of providing horses. A large retinue of horses was a means of conferring prestige on both the receiving and sending party.[6] The Milanese ambassador in Venice in September 1471 noted that the Venetian embassy of four men that was being readied to congratulate the new Pope Sixtus IV was to include 120 horses.[7] Such massive deployments of horses were rare, however, and were generally associated with extraordinary missions designed primarily to make a splash.

The number of horses included in an embassy was usually much smaller. When the Milanese ambassador Gian Pietro Panigarola travelled to Burgundy in 1475, he had eight horses in his train.[8] Venetian authorities allowed between eight and twelve horses, before an act of September 1500 limited embassies inside Italy to eleven.[9] In the case of resident ambassadors, the horses assigned to an embassy were nearly always immediately repatriated upon the arrival of the envoy. He would not need more than one or two horses for his daily activity at court.

Although the Italian travel network was sophisticated for its day when compared with most of the rest of Europe, it was still susceptible to physical degradation as a result of weather and overuse. The road system remained largely that built by the Romans and most of the byways remained unpaved, subject to floods, landslides, and ubiquitous mud. Travel by carriage was almost nonexistent in the fifteenth century; it would not be until the end of the sixteenth century that Italy would see large-scale initiatives to build the first roads that were passable for long distances by wheeled vehicles.[10]

Diplomatic journeys frequently involved a combination of land and water travel. River travel was more predictable than that by sea and was an attractive option in some of Italy's river valleys, for example, along the Po. It was not uncommon for ambassadors travelling to Ferrara or Venice, or to the riparian cities of Lombardy in the Po River valley, to travel by river barge for part of the way. Passage by sea might be preferable when travelling to the relatively remote kingdom of Naples – Lorenzo de' Medici took this route during his famous voyage to Naples in 1479. But in most cases, unless a convenient port and vessel were readily available, travel by land was almost always preferable to that by sea.

Arrivals and departures

The arrival of an embassy into the receiving city or court was the most formalized and carefully choreographed portion of an ambassador's journey and had been regulated by similar brands of ceremony across Western Europe stretching back into the Middle Ages.[11] It was customary for an embassy, as it drew near to its destination, to send forward news of its imminent arrival. The advance warning served as a courtesy, but it was also designed to allow the host to prepare a suitably ceremonial entrance. In some cases, the host might ask the ambassadors to delay their entry.[12] Generally, a welcoming party of officials, courtiers, and other eminent individuals, along with trumpeters and jesters, would then ride out from the town and accompany the newly arrived embassy to the town gates or to the house or *osteria* where the members of the embassy were to reside. Catherine Fletcher has reconstructed with some precision the route that was customarily taken by ambassadors arriving at the papal court, showing just how much the papal master of ceremonics Paris de Grassi sought to control the logistics and protocol of the final few miles of foreign ambassadors' travel into Rome. Grassi strove, not always with success, to dictate the precise route of the approach as well as the pomp associated with it.[13]

In most cases, the first, formal audience with the local prince would not take place until the following day at the earliest. It was then that an ambassador would present his credentials, deliver an oration, and perhaps share his letter of instruction, thus laying out the purpose of his mission. Interestingly, the ceremony and ritual associated with the arrival of extraordinary ambassadors were almost never extended to a newly arrived resident ambassador. The lack of such formalities when a new resident arrived and took another's place in his office indicates the recognition of an inherent continuity in the office of resident ambassador; it simply was not necessary to reprise the ceremonial of a formal entry. Indeed, resident ambassadors, while naturally wishing to meet their replacements as soon as practicable, were also often part of the entourages that welcomed special envoys – an indication that they were, in addition to being representatives from abroad, an intrinsic part of the court community of the receiving state.[14]

The departure of an ambassador was always preceded by the formal granting of leave, by both the sending and receiving parties. Leaving before acquiring such permission was a significant transgression in protocol. The Venetian Senate applied harsh penalties to ambassadors who did so.[15] Should a host not wish to see an ambassador leave, he could delay matters by refusing to grant an audience. More common in the dispatches of the period, however, are requests by the ambassador to his sender for release from service in order to return home. Such complaints were far more common among the ambassadors of republican states, fearful that time away would prevent them from attending to their economic and political concerns at home.[16] Pier Filippo Pandolfini, sent as Florentine ambassador to Rome in 1487, wrote pleading letters to the First Secretary in Florence, Niccolò Michelozzi, "One day here seems like a month...I am compelled and need to return [to Florence] on account of my private affairs, which, I am ashamed to say, are in considerable disarray. And every day that I stay here they get worse". [17] Back in Rome three years later on another mission, he lamented that he had no desire to become a *romanescho* himself.[18]

Dispatches and couriers

Part of the hardship lamented by the likes of Michelozzi had to do with paperwork. Among the most important components of any ambassador's entourage, which had to be taken into consideration when undertaking any travel, were the embassy's chancery and archive of papers. This is because an embassy, and perhaps especially a resident embassy, whose primary function was the provision of information, was ultimately judged on the paper that it produced.

The process of composing dispatches was closely linked to the schedule and availability of couriers. Within the Italian geopolitical space, a great many couriers were delivering letters across the most travelled routes at any given time. In addition to their own couriers, ambassadors might share couriers with friendly colleagues and used institutionalized postal services that maintained regular routes with standard stations along the way. Both the Duke of Milan and the King of Naples established postal routes to connect the most important locales on the peninsula, used also by ambassadors of states friendly to Milan or Naples. In a letter of Duke Francesco Sforza to his ambassador in Naples, Antonio da Trezzo, the duke indicated that he kept fifty couriers employed in running the route between Milan and Naples, spending more than a thousand *lire* a month on them. For this investment he expected to receive from his ambassador at least two letters a week.[19] Under the direction of the chancellor Cicco Simonetta, the Sforza established a system of relays on the Rome–Milan route, with horse-changing stations every forty miles.[20] Cicco's declared desire was a reliable 84-hour route between the two cities.[21] Pieces of fifteenth-century diplomatic correspondence often begin with details about which courier delivered the last letter received and which will be carrying the current one. Details on the dates of the letters most recently received were usually also included. Letters written on different days might be sent or received on the same day with a single *staffetta* (courier relay).[22] Ambassadors often hurried to complete a letter so that the missive could travel with an available courier.[23]

Italian ambassadors relied on a combination of systems for the delivery of correspondence. In some cases, they could rely on pre-existing networks: ambassadors arranged to have their letters included in the mail bags of Italian bankers and merchants whose networks of communication stretched across Italy. It was also not uncommon to see ambassadors employ more than one of these vectors at once, in order to ensure that a particularly important letter would arrive at its intended destination.

A network of couriers was essential for the employment of the resident embassies to take hold, and these men (widely known as *cavallari* and *cavalcanti*) occasionally make their way into the dispatches of the period. Access to trusted couriers was a considerable asset to ambassadors – in 1487, Giovanni Lanfredini, stationed in Rome, thanked Lorenzo de' Medici's secretary, Niccolò Michelozzi, for supplying him with a steady supply of dedicated couriers. Among other things, he was grateful to spend less time coding his dispatches, given he had carriers whom he could trust.[24] We learn little about these men themselves, as they were commonly referenced solely by somewhat cryptic names.[25] Extended discussions of couriers appear only when there was a significant problem with their service, such as a pay dispute or the detention of a courier carrying an important letter. Technically, couriers carrying letters enjoyed all of the privileges and protections afforded to ambassadors. But this rule was frequently violated. Tommaso da Rieti, writing to Francesco Sforza from Tours in March 1454, informed the duke that "Pietro da Sesto, courier, has reached me here with letters from Your Lordship, which he delivered all opened. He says he was detained for seven days by the Duke of Savoy, and it is very likely that they have deciphered the letters if they have anyone who knows how to do it".[26] In Milan in 1488 the resident ambassador of the King of Naples complained that the couriers of the royal post were receiving ill treatment at the hands of Ferrarese officials while crossing Estense territory.[27]

A "tyranny of distance"?

Such complaints are a reminder that travel exposed ambassadorial and postal personnel to physical and financial pressures. These risks increased with distance. In his compelling discussion of the circulation of information in the early modern Mediterranean, Fernand Braudel borrowed from the work of Pierre Sardella (who himself relied on data from Marino Sanudo's *Diarii*), as well as from a selection of other sources, to examine travel times radiating from Venice as a hub. Braudel placed Venice at the centre of an information map, surrounded by a series of concentric rings, representing the speed with which one could expect a letter from Venice to travel.[28] It is telling that all of northern and central Italy fell within the inner ring, representing a voyage of less than one week from Venice. The political space contained by this innermost ring was almost wholly Italian and (in the fifteenth century as in Braudel's sixteenth) was characterized by the circulation of large volumes of information, generated by communication between commercial, political and cultural contacts. These were part of a broader "intensification of contacts" that characterized late medieval and early Renaissance Italy.[29] New institutions and services catered to this waxing volume of contacts – the Milanese, for example, had a state postal system in operation by the end of the fourteenth century.[30] Relying on an extensive data, Federigo Melis has demonstrated that within Italy in the fifteenth century, one might have reasonable expectations of quick and reliable communication times between the major towns. He has given us a picture of the *viabilità postale* of the Italian peninsula, in many of its particulars, down to the relative cost and duration of such services. From Milan, average times ranged from four days to Venice to sixteen to Naples.[31] For Florence, the numbers stretched from five days to Rome to twelve to Naples.[32] For Rome, it was from four days to Naples to eleven to Milan.[33] Melis' figures also show how unpredictability in travel times soared once one ventured beyond Italy into France, Spain, the Empire, and across the Mediterranean, but did not include travel times for couriers running diplomatic correspondence. Given the increasing sophistication of the networks of diplomatic couriers, there is reason to believe that they would regularly exceed the average speeds that he provides. This, along with the established and familiar rhythms of travel within the Italian peninsula, meant that Italian statesmen could rely on reasonably speedy and consistent communication, an important spur to the advent of resident ambassadors and of continuous diplomatic activity between states.

Several hundred years' worth of commercial, political, and cultural exchanges between these major Italian locales had created a human space within which the dispatch and receipt of men and news could be undertaken with considerable confidence in the time that would be needed.[34] Jacopo Trotti, the resident ambassador of Duke of Ferrara, Ercole d'Este, in Milan for many years, would call letters he received from Ferrara that took four days or more to get to Milan "old" (*vecchia*).[35] As distances grew inside of Italy, the level of uncertainty rose in such two-way communication but remained within a manageable margin of unpredictability.

We can see how these dynamics of distance inside Italy worked by looking at the fluid interplay of travel and diplomacy for one extended resident embassy of the second half of the fifteenth century. For most of the period between 1450 and 1494, the Marquis of Mantua maintained a resident ambassador at the court of the Sforza in Milan.[36] The well-travelled route between Milan and Mantua could be negotiated in two to three days on most occasions by a traveller on horseback.[37] Zaccaria Saggi, the resident ambassador of the Marquis of Mantua at the Milanese court for twenty years, travelled between Milan and Mantua frequently, and when he found himself delayed, he invariably felt compelled to tell the Marquis why. Customarily, it was weather, or more accurately, its effect on the roads, that slowed him down. In November 1478, he wrote from Cremona that, in the wake of incessant rain, he would risk drowning were he to continue on to Milan. It seemed to him it might be "a thousand years" before he reached Milan. In the event, Saggi chose not to heed the warnings and ventured forward and indeed almost drowned as a result: he fell into an inundated ditch with his mule on top of him – "thus bathed, I arrived here", he quipped.[38]

Resident ambassadors like Zaccaria, however, travelled the route infrequently in comparison to the couriers who ran letters back and forth. By nearly every measure of the day, the communication links between Milan and Mantua were secure, efficient, and timely. Travel in the winter, predictably, was marginally slower than during the spring and summer months. The Gonzaga marquises employed a team of couriers to ferry letters, money, and merchandise between the two cities.

The importance of news played a part in how fast it travelled. The distance between Milan and Mantua is ninety miles, at the limit of what a horse can travel in one day. In his chronicle of Mantua, Andrea Schivenoglia was so surprised by the arrival of a courier from Milan some fourteen hours after his departure that he made mention of the episode.[39] This look at the admittedly short route between Milan and Mantua is suggestive of communication within Italy. The Gonzaga marquises employed these ambassadors in the hope of a turnaround time of between three and six days in the exchange of correspondence; that is, that they could expect to have his envoy receive instructions formulated in response to news that was only a few days old. The Gonzaga expected to receive prompt information from their ambassadors in Milan, and they had matching expectations that their responses would be received promptly.

Travel by resident ambassadors

Resident ambassadors, whose offices were largely defined by their sedentary nature, were nonetheless forced to travel within the confines of their positions. There might be travel within the host city itself: for masses in churches some distance from court, to the residences of influential nobles, or to the gates of the city to welcome newly arriving ambassadors. When the Duke of Milan Galeazzo Maria Sforza was assassinated in the Church of St. Stephen in 1476, he was flanked by two resident ambassadors, from Mantua and Ferrara, who had travelled to the service with him and were splattered with blood from the duke's wounds.[40] Giovanni Lanfredini, Lorenzo de' Medici's man in Naples, regularly travelled with King Ferrante, who came to value the Florentine as a political sounding board and personal companion.[41] Resident ambassadors, who had among their most important duties the systematic gathering of news and gossip, occasionally travelled away from the central court in this pursuit. In Rome, for example, where cardinals oversaw social centres that amounted to satellite courts of the Curia, ambassadors would travel between the households of the most prominent members of the college.[42]

Travel simultaneously compromised and enhanced the demands of information gathering: it took ambassadors away from their usual sources but also exposed them to others. Travel necessarily upset routine, and complaints about the physical and financial burdens of travel among ambassadorial and postal personnel were frequent and numerous. Some Italian princes and their courts, while largely fixed, did still move en masse with some regularity. The Dukes of Milan, for example, while usually based at their castle in Milan, frequently relocated to their many residences and hunting lodges across the *Milanese*, both to pursue leisure and to make their presence felt in the provinces of their fractious Duchy, much like other European monarchs in their respective territories. Duke Galeazzo Maria Sforza spent as much as a third of his days in the *contado*, at ducal residences in Pavia, Vigevano, Abbiate, and Galliate.[43] Such travel was a source of considerable inconvenience to ambassadors in Milan – it meant not only the relocation of the diplomat but also of his secretary and personal archive. By then it was customary for ambassadors to carry personal archives with them when embarking on extended travel, as it afforded them access to precedent, a record of those with whom they had corresponded, and essential ciphers to diplomatic codes.

Similar challenges were faced by ambassadors resident in Naples when the king was on campaign against his rebellious barons, as King Ferrante was from 1456 to 1464 and then again in 1486 to 1488. The ersatz Neapolitan court followed the king as he sought to subdue his nobility across the breadth of his sprawling kingdom.[44] Constrained by the exigencies of war, in such cases the king was unwilling or unable to supply any financial support or creature comforts to the ambassadors who travelled with him, and the instability and backwardness of the kingdom meant that locating suitable lodging was often very difficult. Giovanni Lanfredini stressed the dangers of having to travel with the king's court to Puglia. He complained of the physical dangers and the searing sun. Worst of all was the expense involved, because in the *osterie* in Puglia, one "could lose hide and skin".[45] But even this was probably not as bad as the fate of the Milanese ambassador Antonio da Trezzo, who, at the Battle of Sarno in 1460, lost his entire archive of letters during the rout of Ferranbte's army.[46]

Outbreaks of plague, too, forced ambassadors into unforeseen travel. Ambassadors begged for permission to leave their posts in such circumstances. When such permission was granted, the departure was hurried and even frantic. Antonio Bicchi, who was the Sienese ambassador in Naples, fled the city in June 1478 when the plague hit. Because of the threat of the disease, he was forced to take a route to Rome through the Maremma, "which was much longer and sad with little lodging available". Upon his arrival in Rome, he found the plague just as much of a threat there as it had been in Naples: merchants and courtiers had abandoned the city, as had all but four cardinals.[47] In 1458, Francesco Sforza ordered Antonio della Torre to join King Ferrante of Naples in Salerno, but he was unable to proceed to the Kingdom from Rome "because at Naples there is plague and also in numerous parts of the kingdom, so I will stay here for now and will take steps to fulfill my duty by third person, if I am able".[48]

Beyond Italy

It has been argued here that the confined environs and reliable communications of Italy were catalysts for the emergence of the new diplomacy.

The case becomes even stronger when we look at the challenges pertaining to ambassadorial travel beyond the Alps. Outside of Italy, there was a clear sense of difference, of separation. A widespread sentiment was evident that the French lands, even for those ambassadors from cities that had historically important ties with France (like Genoa, Milan, and Florence), were a far away, almost mythical place. As Michael Mallet has written, "the ambassador of an Italian state at the French or Imperial court was in a different world, a less intimate world, a world in which diplomacy was regarded with some suspicion, as a poor alternative to war".[49] France was on the periphery of the Italian political system; England and the Empire were further away still, geographically and culturally.

The prodigious distances involved in sending an ambassador abroad complicated nearly every aspect of the embassy. Once outside of Italy, the predictability of communication collapsed. Whereas ambassadors in Italy could rely on postal services and courier relays, there was no stable post in France. Letters travelling across France were thus far more likely to be intercepted and the relative reliability of communications seen in Italy completely nonexistent. The longer the journey, both for the ambassador and for the individual responsible for carrying any correspondence, the more likely a delay would occur. The requests made of ambassadors inside Italy that they write with great frequency (twice a week, or, in some cases, daily) were seldom made of diplomats at European postings. The ambassador serving outside Italy was necessarily afforded more autonomy; without detailed, regular instructions, he was forced to make decisions based on his own judgement. The added degree of uncertainty as to whether dispatches would reach their destination, occasioned by the distances and unfamiliar routes involved, meant that greater importance was placed on the final report – the *relazione* at mission's end – than on the regular flow of news customary of an Italian embassy.[50]

Concerns about finance and lodging abounded as well. Ambassadors generally had no ready access to capital abroad, and Italian states were understandably reluctant to send large amounts of cash over such long, unfamiliar distances – the result was that ambassadors were continuously running out of money while abroad. The funds afforded to them were invariably insufficient to cover the costs that they incurred while abroad, and they were forced to dip into their own pockets. This could be especially onerous for resident ambassadors drawn from the less wealthy rungs of society. The familiar *osterie* along well-travelled routes between Italian population centres were not available, and securing quality accommodation was difficult. Milanese ambassadors in France complained bitterly about their lodging while trying to keep up with the itinerant French court. The accommodations were poor; there was rarely enough room for everyone; and sometimes ambassadors were forced to eat, and even sleep, outside.[51]

The laments of Niccolò Roberti, sent by Duke Ercole d'Este to the French court in 1479, can be taken as representative. Niccolò complained about everything in France, including his sense of personal isolation – he described himself as being exiled *in finibus terre* – and the incessantly rainy climate, and he begged to be sent home. He was rarely certain whether his letters had reached their destination, nor did he receive regular news from Ferrara. He spent endless days in his room at the inn, where he "faded away and grew old".[52]

Land travel out of Italy also generally required passage through the Alps and all that crossing high mountain passes entailed.[53] The experiences could be harrowing. Florentine ambassadors, in November 1461, ascending the St Bernard Pass trudged through high winds and snow so thick that they could not see each other. They arrived at the top of the pass "more dead than alive".[54] It is scarcely surprising, then, that the ambassador Emanuele de Iacopo wrote to Francesco Sforza from Susa in November 1455 that "tomorrow, with God's grace, I will pass the mountains, which at the present time cannot be crossed without the greatest danger".[55]

In contrast to the Italian courts, which had largely established fixed locales, nearly all ultramontane European courts remained itinerant. The mobility of the French, Burgundian, and Imperial courts was a source of intense frustration for Italian ambassadors sent abroad, as they could not be sure where exactly they would be likely to find the princes in question.[56] Instructions issued to ambassadors headed to France demonstrated awareness of the uncertainty of the endeavour. "You will go to France, to whatever parts in which you are informed these Kings will be, with all possible haste", read the instructions issued to Angelo Accaiaoli from the Dieci of the Florentine *Signoria* in January 1453.[57] A year later, the same Accaiaoli, sent jointly by Duke Francesco Sforza of Milan and the *Signoria* of Florence to the king of France, spent weeks trying to find the monarch. To the duke he wrote, "I have had to go almost all around this Realm of France and have come to the remotest part of it (*infino allo extremo*). I do not know how otherwise to explain where we are, save that the place is called Saint-Maixent and lies on the borders of Anjou and Brittany. The King is lodged in the neighborhood, three miles away". On that same day, he wrote the Dieci of Florence, "If Your Lordships knew about the kind of journey I was required to make in order to find the King, I am certain that you would not consider that there has been any negligence on my part; but if I should describe that journey, you would conclude that I was pleasing myself rather than tending to business".[58] The familiar routes and reliable communications of Italy were far away, indeed.

Travel, distance, and the "new diplomacy"

The "new diplomacy" was the result of a range of factors. It has been suggested here that the relative reliability and celerity of travel and communications with the Italian peninsula should be considered keys among these. With the increased use of resident ambassadors, ambassadors themselves probably travelled less than they had previously; the real frequent flyers of this new diplomatic world were the couriers, those delivering the large volumes of correspondence sent by and to resident ambassadors. Whereas medieval diplomacy had been sporadic, focused on extraordinary missions, and often on face-to-face encounters between prominent political leaders, the new diplomacy was chiefly about political communication and the regular exchange of information, and couriers were its primary vectors.

By the beginning of the sixteenth century, the realities of the Italian political milieu had changed radically. The invasion by King Charles VIII of France in 1494 and the subsequent wars of Italy dramatically altered the political landscape. In the longer term, the true heirs to the new Italian diplomacy were the Spanish Habsburgs. Under Charles I and Philip II, the Habsburg diplomatic service was unrivalled in Europe. A great many of the features that distinguished the diplomatic service of the Habsburgs under Philip II were larger-scale versions of those already present in the fifteenth-century Italian network: these features included numerous and widely dispersed permanent, resident ambassadors who were expected to correspond on a regular basis, a highly developed postal service, and well-organized diplomatic chanceries and archives. The Habsburgs, however, with their far-flung empire, could not escape the repercussions of distance. For an early modern European and global empire, the dislocating effects of long-distance travel had inevitable policy implications. As Geoffrey Parker has shown, for Philip II distance was indeed "enemy number one". Transferred outside the confines of the Italian peninsula, the new diplomacy had to contend with great expanses of time and space. Diplomacy conducted on this scale necessarily lived under the tyranny of distance.

Notes

1 Isabella Lazzarini, *Communication and Conflict: Italian Diplomacy in the Early Renaissance, 1350–1520* (Oxford: Oxford University Press, 2015).

2 Until the appearance of Lazzarini's *Communication and Conflict*, the standard treatment of these changes was Garrett Mattingly, *Renaissance Diplomacy* (Boston: Houghton Mifflin, 1955). Good summaries are also Michael Mallett, "Diplomacy and War in Late Fifteenth-Century Italy," *Proceedings of the British Academy* 68 (1981): 267–88, and Riccardo Fubini, "Diplomacy and government in the Italian city-states of the fifteenth century," in *Politics and Diplomacy in Early Modern Italy*, ed. Daniela Frigo (Cambridge: Cambridge University Press, 2000), 25–48. For a focus on how these developments played out in Rome, see Catherine Fletcher, *Diplomacy in Renaissance Rome: The Rise of the Resident Ambassador* (Cambridge: Cambridge University Press, 2015).

3 Pier Candido Decembrio to Cicco Simonetta, January 13, 1457, Naples, letter 184 in *Dispacci sforzeschi da Napoli I (1444-2 luglio 1458)*, ed. Francesco Senatore (Naples: Carlone, 1997), 477–79.

4 Agostino Rossi to Galeazzo Maria Sforza, May 14, 1471, ASMi, AS, PE. Roma, Cart. 67.

5 Tommaso Rieti to Francesco Sforza, March 28, 1454 in *Dispatches with Related Documents of Milanese Ambassadors in France and Burgundy,*

1450–1483, ed. Vincent Ilardi and Paul M. Kendall, vol. 1 (Athens, OH: Ohio University Press, 1970), 1:148.

6 Daniela Frigo has pointed out that late medieval regimes were willing to spend a great deal of money to engage in pomp and bolster the reputation of a prince when sending out a short-term, extraordinary embassy—and a large number of horses was one means of achieving this. See "Corte, honore et ragion di stato: il ruolo dell'ambasciatore in età moderna" in *Ambasciatori e nunzi. Figure della diplomazia in età moderna. Chieron. Materiali e strumenti di aggiornamento storiografico*, ed. Daniela Frigo (Rome: Bulzoni, 1999), 13–55 (32).

7 Letter to Galeazzo Maria Sforza, September 7, 1471, Venice, ASMi, AS, PE – Venezia, Cart, 357.

8 Letter to Doge Pietro Mocenigo of March 18, 1475 in Ernesto Sestan, ed., *Carteggi diplomatici fra Milano sforzesca e la Borgogna*, vol. 1 (Rome: Istituto storico italiano per l'età moderna e contemporanea, 1985), 1:428–29.

9 Donald E. Queller, *Early Venetian Legislation on Ambassadors* (Geneva: Librairie Droz, 1966), 26.

10 Writing in 1550, Marino Cavalli, in his treatise *Informatione dell'Offitio dell'Ambasciatore*, emphasised the ordeal of travelling by carriage. See *Informatione dell'offitio dell'ambasciatore di Marino de Cavalli il Vecchio, MDL*, ed. Tommasso Bertele (Florence: Olschki, 1935), 40.

11 This topic has been the subject of a considerable amount of recent scholarship. See, for example, Joan-Pau Rubiés, "Late medieval ambassadors and the practice of cross-cultural encounters," in *The "Book" of Travels: Genre, Ethnology, Pilgrimage 1250–1700*, ed. Palmira Brumett (Leiden: Brill, 2009), 37–112.

12 One example was when the Duke of Milan in September 1478 asked the newly elected Venetian ambassador Bertuccio Gabriel to delay his entry into Milan so that he would arrive at the same time as the new Florentine ambassador. *Lorenzo de' Medici. Lettere. III (1478–1479)*, ed. Nicolai Rubinstein (Florence: Giunti, 1977), 180.

13 Catherine Fletcher, "The city of Rome as a space for diplomacy," in *Early Modern Rome 1341–1667*, ed. Portia Prebys (Ferrara: Edisai, 2012), 79–81. Such practices were also practiced outside of Italy. It was customary in France for the king to send a chamberlain or maître d'hôtel to accompany an ambassador to court. See René de Maulde de la Clavière, *La diplomatie au temps de Machiavel*, vol. 2 (Paris: E. Leroux, 1892), 2:170.

14 When the Venetian envoy Zaccaria Contarini stopped in Milan on his way to France in 1492, he was met two miles outside the city gates by the Venetian ambassador resident in Milan, along with two top Milanese councilors and forty men on horseback. This group then accompanied Contarini as far as the city gates, where they were greeted by members of the Milanese *Consiglio Segreto* and *Consiglio di Guistizia*, who then rode with the newly arrived ambassador as far as his *osteria*. Eugenio Albèri, ed., *Relazioni degli ambasciatori veneti al Senato*, series 1, vol. 4 (Florence: Tipografia all'insegna Di Clio, 1860), 4:4.

15 Donald E. Queller, *The Office of the Ambassador in the Middle Ages* (Princeton: Princeton University Press, 1967), 292.

16 On this, see "Tessa Beverly, Venetian ambassadors: An Italian Elite". Unpublished PhD dissertation, University of Warwick, 1993, 93–113; Donald E. Queller, "The civic irresponsibility of the Venetian nobility," in *Economy, Society and Government in Medieval Italy. Essays in memory of Robert L. Reynolds*, ed. David Herlihy (Kent, OH: Kent State University Press, 1969), 223–35. See also the request of Zaccaria Contarini to be allowed to return to Venice from the imperial court in "Dispacci al Senato Veneto di Francesco

Foscari ed di altri oratori presso l'imperatore Massimiliano I nel 1496." *Archivio storico italiano* 71 (1844): 723–948 (727).

17 Pandolfini's complaints went on for several months. See, especially, the letters of March 10, and April 6, 1487 (1486 in the Florentine calendar), found in BNCF, Manoscritti Ginori Conti, Carte Michelozzi, Ambascerie a Roma.

18 *Ibid.*, May 6, 1490.

19 In this letter, dated February 26, 1461, the Duke actually complained that, for all his expenditure on couriers, he had not heard from his ambassador in a month: letter 53 in *Dispacci sforzeschi da Napoli. IV (1 gennaio – 26 dicembre 1461),* ed. Francesco Storti (Naples: Carlone: 1998), 99–100.

20 Giovanni Ugolini, "Le communicazioni postali spagnole nell'Italia del XV secolo," *Ricerche storiche* 23 (1993): 283–373 (285).

21 Per a letter of Cicco of February 1454, published in Clemente Fedele and Mario Gallenga, *"Per servizio di Nostro Signore." Strade, corrieri e poste dei papi dal medievo al 1870* (Prato: Istituto di studi storico-postali, 1988), 209.

22 Eugenio Dupré Thesedier, *Niccolò Machiavelli diplomatico. I: L'arte della diplomazia del quattrocento* (Como: Carlo Marzorati, 1945), 192.

23 For example, Pier Filippo Pandolfini to Niccolò Michelozzi, August 2, 1484, Reggio, *Biblioteca Nazionale di Firenze, Manoscritti Ginori Conti, Carte Michelozzi,* 29.102.

24 Giovanni Lanfredini to Niccolò Michelozzi, June 20, 1487, Rome, BNCF, Manoscritti Ginori Conti, Ambasceria fiorentina a Roma. Similar sentiments were expressed by Battista Bendedei in a letter of May 8, 1481, to Duke Ercole d'Este. The ambassador wrote that although Ercole had written to him in code, he had chosen not to, because he was able to use a trusted courier named Giovanni Maria, the son of Brandelise Trotti, a scion of an influential Ferrarese family. ASMo, ASE, CA – Roma, B.2.

25 Among the names of the couriers used by the Marquis Ludovico Gonzaga for the route to Milan included Antonio, Ravanello, Giorgio, Morello, Biancho, Bianchino (presumably the son of Biancho), and Cipriano.

26 Letter of March 12, 1454 in Ilardi and Kendall, eds. *Dispatches,* 1:144–45.

27 Jacopo Trotti to Ercole d'Este, February 11, 1488, Milan, ASMo, ASE, CA – Milano, B.5.

28 Fernand Braudel, *The Mediterranean and the Mediterranean World in the Age of Philip II*, trans. Siân Reynolds, vol. 1 (New York: Collins, 1972), 1:362–66. Pierre Sardella, *Nouvelles et spéculations à Venise au début du XVIe siècle* (Paris: Armand Colin, 1948).

29 A term utilized by Isabella Lazzarini in *L'Italia degli stati territoriali: secoli XIII-XV* (Rome: Laterza, 2003), 142.

30 Luciana Frangioni, *Organizzazione e costi del servizio postale alla fine del Trecento*. Quaderni di storia postale, 3 (Prato: Instituto di Studi Storici Postali, 1983). For the later emergence of a papal postal system, see Fedele and Gallenga, *"Per servizio di Nostro Signore"*.

31 Federigo Melis, "Intensità e regolarità nella diffusione dell'informazione economica generale nel Mediterraneo e in occidente alla fine del medievo," in *I trasporti e le communicazioni nel medievo*, ed. Luciana Frangioni (Florence: Le Monnier, 1984), 179–223 (208).

32 *Ibid.*, 214–15.

33 *Ibid.*, 216.

34 A similar point is made by Francesco Senatore, *"Uno mundo di carta": Forme e struttura della diplomazia sforzesca.* (Naples: Liguori, 1998), 266.

35 Jacopo used this expression frequently in his years in Milan. He rarely suggested any concrete information on why a letter might have taken so long. On

exception was on March 15, 1486 when he remarked that a letter might have taken a full four days to reach him from Ferrara because of "the floods that are taking place here". Jacopo Trotti to Ercole d'Este, Milan, ASMo, ASE, CA – Milano, B.4.

36 The run of this correspondence has now been published as *Carteggio degli oratori mantovani alla corte sforzesca (1450–1500)*, ed. Franca Leverotti et al, 12 vols. (Rome: Ministero per i Beni e le attività culturali, 1999). See also Lazzarini, "L'informazione politico-diplomatica".

37 A numerical analysis of the time it took letters to and from the Mantuan ambassadors in Milan shows that between 1475-88, a letter took 3.43 days in the winter, 2.81 days in the spring, 2.80 days in the summer, and 3.46 days in the autumn. See Appendix I of Paul Dover, "Letters, Notes and Whispers: Diplomacy, Ambassadors and Information in the Italian Renaissance Princely State. Unpublished PhD dissertation, Yale University, 2003, 244–50.

38 Zaccaria Saggi to Federico Gonzaga, November 1, 1478, Cremona & November 5, 1478, Milan, ASMa, AG, AE. Milano, B. 1626.

39 Cited in Rodolfo Signorini, *Opus hoc tenue: La Camera dipinta di Andrea Mantegna* (Parma: Artegrafica Silva, 1985), 130.

40 On the assassination of Galeazzo Maria, see Bortolo Belotti, *Il drama di Gerolamo Olgiati* (Milan: G. Martinelli, 1929), 145–67; and Eugenio Casanova, "L'uccisione di Galeazzo Maria Sforza e alcuni documenti fiorentini." *Archivio storico lombardo*, 12 (1899): 299–332, which provides the observations of the Florentine ambassador, who also gave an account of the day's events.

41 Elisabetta Scarton, *Giovanni Lanfredini: uomo d'affari e diplomatico nell'Italia del Quattrocento* (Florence: Olschki, 2007), 201–52.

42 On this, see Paul Dover, "Saper la mente della soa beatitudine: Pope Paul II (1458-64) and the ambassadorial community," *Renaissance and Reformation* 31 (2008): 3–32.

43 Gregory Lubkin, *A Renaissance Court: Milan under Galeazzo Maria Sforza* (Berkeley: University of California Press, 1994), Appendix 3, 259–64.

44 See Francesco Senatore, *Spazi e tempi della guerra nel Mezzogiorno aragonese: l'itinerario militare di re Ferrante (1458-1465)* (Salerno: Carlone, 2002). For the latter conflict, see the dispatches in Giuseppe Paladino, ed. "Per la storia della congiura dei baroni: documenti inediti dall'archivio estense 1485-1487." *Archivio storico per le province napoletane*, new series, 5 (1919–1920, 44 (1919), 336–67; 45 (1920), 128–51, 325, 351; 46 (1921), 221–65; 48 (1923), 219–90) and Ernesto Pontieri, ed. *La politica mediceo-fiorentina nella congiura dei baroni napoletani contro ferrante d'Aragona, 1485-1492. Documenti inediti* (Naples: Società napoletana di storia patria, 1977).

45 Scarton, *Giovanni Lanfredini*, 212.

46 Senatore, "*Uno mundo di carta,*" 157. War-time travel was not always so bad. During the War of Ferrara, Jacopo Trotti, the ambassador of Duke Ercole d'Este, travelled for several weeks with the Milanese army as it sought to defend the *Ferrarese* – Jacopo stressed that the lodging provided him by Count Guilbert of San Vitale was better than he could expect were he at home in Ferrara. Jacopo Trotti to Ercole d'Este, May 14, 1483, *Ex salte penes felicia castra ducalis fellinum*, ASMo, ASE, CA – Milano.

47 Antonio Bicchi to the Consistory, June 28, 1478, Rome, ASS. Concistorio, N.2040.

48 Antonio della Torre to Francesco Sforza, June 9, 1458, Rome, in Senatore, ed. *Dispacci sforzeschi*, 647.

49 Michael Mallett, "Italian Renaissance Diplomacy," *Diplomacy and Statecraft* 12 (2001): 61–70 (68).

50 This is an observation made in Francesco Senatore, "I diplomatici e gli ambasciatori," in *Viaggiare nel medievo*, ed. Sergio Gensini (Rome: Ministero per i beni e le attività culturali, 2000), 267–98 (282).
51 See Gigliola Soldi Rondinini, "Le relazioni degli ambasciatori milanesi quali testimonianze della vita nelle corti di Francia e di Borgogna (seconda metà del secolo XV)," in *Saggi di storia e storiografia visconteo-sforzesca*, ed. Gigliola Soldi Rondinini (Bologna: Cappelli, 1984), 65–81 (74–6). Giovan Pietro Panigarola, the Milanese ambassador, in February 1466 just outside Pont-Andemur declared that the status of his lodging made him want nothing more than to go home (75).
52 Letter of February 24, 1479. Marco Folin, "Les ambassadeurs des Este à la cour de Valois (1470–1505)," in *Regards croisés: musiques, musiciens, artistes et voyageurs entre France et Italie au XVe siècle*, ed. Nicoletta Guidobaldi (Paris and Tours: Minerve, 2002), 135–70 (147).
53 For a discussion of the conflicts encountered by travellers through the Alps, see Joanne Anderson's article in this present volume.
54 Quoted in Gaetano Milanesi, "Il viaggio degli ambasciatori fiorentini nel 1461, descritto da Giovanni di Francesco di Neri Cecchi," *Archivio Storico Italiano* 37 (1865): 3–62 (45).
55 Ilardi and Kendall, eds. *Dispatches*, November 29, 1455, 1:186–89. In February 1475, the Milanese ambassador Giovanni Pietro Panigarola was forced to travel through the St. Bernard Pass on his way to the Burgundian court. Upon arriving on the other side, he wrote to the Duke of Milan that he was grateful to have survived the alpine passage: the weather had been awful (as bad as when he had travelled the same route twenty-five years earlier), so much so that the horses could not continue, and he was afraid that he had ruined his own mount. Giovanni Pietro Panigarola to Duke Galeazzo Maria Sforza, (18–25) February 1475 in Sestan, 1:406–07.
56 An observation of Donald Queller about medieval diplomacy remained true in almost all of non-Italian Europe in the fifteenth century, "Since medieval courts were peripatetic, ambassadors could not count upon finding the lord to whom they were sent at a given place". *The Office of the Ambassador,* 162.
57 *Ibid.*, January 26, 1453, 1:86–96.
58 *Ibid.*, November 18, 1451, 1:22, 28.

Bibliogaphy

Belotti, Bortolo. *Il drama di Gerolamo Olgiati.* Milan: G. Martinelli, 1929.

Bertele, Tommasso. *Informatione dell'offitio dell'ambasciatore di Marino de Cavalli il Vecchio, MDL.* Florence: Olschki, 1935.

Beverly, Tessa. *Venetian Ambassadors: An Italian Elite.* Unpublished PhD dissertation, University of Warwick, 1993.

Bicchi, Antonio. *"Antonio Bicchi to the Consistory."* June 28, 1478, Rome, ASS. Concistorio, N.2040.

Braudel, Fernand. *The Mediterranean and the Mediterranean World in the Age of Philip II.* Translated by Siân Reynolds, vol. 1. New York: Collins, 1972.

Casanova, Eugenio. "L'uccisione di Galeazzo Maria Sforza e alcuni documenti fiorentini." *Archivio storico lombardo 12* (1899): 299–332.

Clavière, René de Maulde de la. *La diplomatie au temps de Machiavel.* vol. 2. Paris: E. Leroux, 1892.

Contarini, Zaccaria. "Dispacci al Senato Veneto di Francesco Foscari ed di altri oratori presso l'imperatore Massimiliano I nel 1496." *Archivio storico italiano 71* (1844): 723–948.

de' Colli, Gerardo. "*Letter to Galeazzo Maria Sforza.*" September 7, 1471. Venice, ASMi, AS, PE – Venezia, Cart, 357.

Dover, Paul. *Letters, Notes and Whispers: Diplomacy, Ambassadors and Information in the Italian Renaissance Princely State.* Unpublished PhD dissertation, Yale University, 2003.

Dover, Paul. "Saper la mente della soa beatitudine: Pope Paul II (1458-64) and the ambassadorial community." *Renaissance and Reformation 31* (2008): 3–32.

Fedele, Clemente and Mario Gallenga. "*Per servizio di Nostro Signore.*" *Strade, corrieri e poste dei papi dal medievo al 1870.* Prato: Istituto di studi storico-postali, 1988.

Fletcher, Catherine. "The City of Rome as a Space for Diplomacy." In *Early Modern Rome 1341–1667.* Edited by Portia Prebys, 79–81. Ferrara: Edisai, 2012.

Fletcher, Catherine. *Diplomacy in Renaissance Rome: The Rise of the Resident Ambassador.* Cambridge: Cambridge University Press, 2015.

Folin, Marco "Les ambassadeurs des Este à la cour de Valois (1470–1505)." In *Regards croisés: musiques, musiciens, artistes et voyageurs entre France et Italie au XVe siècle.* Edited by Nicoletta Guidobaldi, 135–70. Minerve: Paris & Tours, 2002.

Frangioni, Luciana. *Organizzazione e costi del servizio postale alla fine del Trecento.* Quaderni di storia postale, 3. Prato: Instituto di Studi Storici Postali, 1983.

Frigo, Daniela, ed. *Ambasciatori e nunzi. Figure della diplomazia in età moderna. Chieron. Materiali e strumenti di aggiornamento storiografico.* Rome: Bulzoni, 1999.

Fubini, Riccardo "Diplomacy and government in the Italian city-states of the fifteenth century." *Politics and Diplomacy in Early Modern Italy.* Edited by Daniela Frigo, 25–48. Cambridge: Cambridge University Press, 2000.

Ilardi, Vincent and Paul M. Kendall, eds. *Dispatches with Related Documents of Milanese Ambassadors in France and Burgundy, 1450–1483.* vol. 1. Athens, OH: Ohio University Press, 1970.

Lanfredini, Giovanni. "*Giovanni Lanfredini to Niccolò Michelozzi.*" June 20, 1487. Rome, BNCF, Manoscritti Ginori Conti, Ambasceria fiorentina a Roma.

Lanfredini, Giovanni. *L'Italia degli stati territoriali: secoli XIII-XV.* Rome: Laterza, 2003.

Lazzarini, Isabella. *Communication and Conflict: Italian Diplomacy in the Early Renaissance, 1350–1520.* Oxford: Oxford University Press, 2015.

Leverotti, Franca et al. eds., *Carteggio degli oratori mantovani alla corte sforzesca (1450–1500).* 12 vols. Rome: Ministero per i Beni e le attività culturali, 1999.

Lubkin, Gregory. *A Renaissance Court: Milan under Galeazzo Maria Sforza.* Berkeley: University of California Press, 1994.

Mallett, Michael. "Diplomacy and War in Late Fifteenth-Century Italy." *Proceedings of the British Academy 68* (1981): 267–88.

Mallett, Michael. "Italian Renaissance Diplomacy." *Diplomacy and Statecraft 12* (2001): 61–70.

Mattingly, Garrett. *Renaissance Diplomacy.* Boston: Houghton Mifflin, 1955.

Melis, Federigo. "Intensità e regolarità nella diffusione dell'informazione economica generale nel Mediterraneo e in occidente alla fine del medievo." In *I trasporti e le communicazioni nel medievo.* Edited by Luciana Frangioni, 179–223. Florence: Le Monnier, 1984.

Milanesi, Gaetano. "Il viaggio degli ambasciatori fiorentini nel 1461, descritto da Giovanni di Francesco di Neri Cecchi," *Archivio Storico Italiano 37* (1865): 3–62.

Paladino, Giuseppe ed. "Per la storia della congiura dei baroni: documenti inediti dall'archivio estense 1485–1487." In *Archivio storico per le province napoletane*, new series, 5. 1919–1923.

Pandolfini, Per Filippo. *"Pier Filippo Pandolfini to Niccolò Michelozzi." August 2, 29, 1484. Reggio, Biblioteca Nazionale di Firenze, Manoscritti Ginori Conti, Carte Michelozzi*, 29.102.

Pontieri, Ernesto, ed. *La politica mediceo-fiorentina nella congiura dei baroni napoletani contro ferrante d'Aragona, 1485-1492. Documenti inedita.* Naples: Società napoletana di storia patria, 1977.

Queller, Donald E. *Early Venetian Legislation on Ambassadors.* Geneva: Librairie Droz, 1966.

Queller, Donald E. *The Office of the Ambassador in the Middle Ages.* Princeton: Princeton University Press, 1967.

Queller, Donald E. "The Civic Irresponsibility of the Venetian Nobility." In *Economy, Society and Government in Medieval Italy. Essays in memory of Robert L. Reynolds.* Edited by David Herlihy. Kent, OH: Kent State University Press, 1969.

Rondinini, Gigliola Soldi. "Le relazioni degli ambasciatori milanesi quali testimonianze della vita nelle corti di Francia e di Borgogna (seconda metà del secolo XV)." In *Saggi di storia e storiografia visconteo-sforzesca.* Edited by Gigliola Soldi Rondinini, 65–81. Bologna: Cappelli, 1984.

Rossi, Aostino. *"Agostino Rossi to Galeazzo Maria Sforza."* May 14, 1471. ASMi, AS, PE. Roma, Cart. 67.

Rubiés, Joan-Pau. "Late Medieval Ambassadors and the Practice of Cross-Cultural Encounters." In *The "Book" of Travels: Genre, Ethnology, Pilgrimage 1250–1700.* Edited by Palmira Brumett, 37–112. Leiden: Brill, 2009.

Rubinstein, Nicolai, ed. *Lorenzo de' Medici. Lettere. III (1478–1479).* Florence: Giunti, 1977.

Saggi, Zaccaria. *"Zaccaria Saggi to Federico Gonzaga."* November 1 & November 5, Cremona, 1478. Milan, ASMa, AG, AE. Milano, B. 1626.

Sardella, Pierre. *Nouvelles et spéculations à Venise au début du XVIe siècle.* Paris: Armand Colin, 1948.

Scarton, Elisabetta. *Giovanni Lanfredini: uomo d'affari e diplomatico nell'Italia del Quattrocento.* Florence: Olschki, 2007.

Senatore, Francesco, ed. *Dispacci sforzeschi da Napoli I (1444–2 luglio 1458).* Naples: Carlone, 1997.

Senatore, Francesco, ed. *"Uno mundo di carta": Forme e struttura della diplomazia sforzesca.* Naples: Liguori, 1998.

Senatore, Francesco, ed. "I diplomatici e gli ambasciatori." In *Viaggiare nel medievo.* Edited by Sergio Gensini, 267–98. Rome: Ministero per i beni e le attività culturali, 2000.

Senatore, Francesco, ed. *Spazi e tempi della guerra nel Mezzogiorno aragonese: l'itinerario militare di re Ferrante (1458–1465).* Salerno: Carlone, 2002.

Sestan, Ernesto. ed. *Carteggi diplomatici fra Milano sforzesca e la Borgogna.* vol. 1. Rome: Istituto storico italiano per l'età moderna e contemporanea, 1985.

Signorini, Rodolfo. *Opus hoc tenue: La Camera dipinta di Andrea Mantegna.* Parma: Artegrafica Silva, 1985.

Storti, Francesco, ed. *Dispacci sforzeschi da Napoli. IV (1 gennaio – 26 dicembre 1461)*. Naples: Carlone: 1998.

Thesedier, Eugenio Dupré. *Niccolò Machiavelli diplomatico. I: L'arte della diplomazia del quattrocento*. Como: Carlo Marzorati, 1945.

Trotti, Jacopo. "*Jacopo Trotti to Ercole d'Este.*" May 14, 1483. Ex salte penes felicia castra ducalis fellinum, ASMo, ASE, CA – Milano.

Trotti, Jacopo. "*Jacopo Trotti to Ercole d'Este.*" February 11, 1488. Milan, ASMo, ASE, CA – Milano, B.5.

Ugolini, Giovanni. "Le communicazioni postali spagnole nell'Italia del XV secolo." *Ricerche storiche 23* (1993): 283–373 (285).

8 Squadrons of inkpots

Pietro Aretino and the narrativity of conflict

William T. Rossiter

Francesco Guicciardini opens his *History of Italy* by equating the recent past of the peninsula which his account treats – 1490–1534 – with travel and conflict, propounding a national narrative of invited invasion whereby Italy became the battleground in which European rulers played out their warfare:

> I have determined to write about those events which have occurred in Italy within our memory, ever since French troops, summoned by our own princes, began to stir up very great dissension here.[1]

Guicciardini's approach to recording military mobility and conflict is ostensibly objective, yet such historiography is as personal as it is documentary, according to Hegel, who name-checked Guicciardini as representing what he termed "Original History":

> What is present and living in their environment is their proper material. The influences that have formed the writer are identical with those which have moulded the events that constitute the matter of his story. The author's spirit, and that of the actions he narrates, is one and the same. He describes scenes in which he himself has been an actor, or at any rate an interested spectator.[2]

The present chapter examines the continuum – really a dialectic – between personalized history and the grand narrative in a way that goes beyond the narrowly conceived Hegelian model and certainly beyond any sense of the personal (as distinct from the solipsistic) that could be claimed against Guicciardini. It does so by tracing how the Italian satirist Pietro Aretino inserted himself into the narrative of the Battle of Pavia and its political aftermath, which saw the French king Francis I captured by and transported to the court of the Holy Roman Emperor Charles V, an insertion that ultimately aligned Francis's defeat with the attempted assassination of Aretino.[3] The capture of the French king did not lead directly to the order for Aretino's assassination, of course, but was subsumed into the chain of events whereby Aretino constructed and represented his own history as being intimately bound up in the narrative of the Habsburg-Valois conflict and European history writ large.

Hegel's dramatic analogy of historian/material and actor/scene is apt in that rhetoric is the common factor, as the term history "denotes quite as much the *historia rerum gestarum*, as the *res gestae* themselves ... not less what has *happened*, than the *narration* of what has happened".[4] Indeed, this chapter concerns itself with narrative and rhetorical strategies, their risks and benefits, when they are played out in the public sphere. For Aretino, as for Francis, the private and public might be seen to constitute a false dichotomy; each is too intimately implicated in the other due to interception and censorship of letters on the one hand, and the immediacy of print culture with its seeming open access to the personal on the other. This dichotomy, in the case examined by this chapter, is glossed by an equally confected continuum between travel and conflict, a continuum that operates via the rhetorical figure of chiasmus, whereby conflict is the end of travel, and travel the end of conflict. However, Francis' capture and transit and Aretino's assassination and flight are not related via the false logical principle of *post hoc ergo propter hoc* so much as Aretino is exploiting and appending himself to the public account of the battle and its consequences in a way that fuels his existing personal conflicts, which are played out at the highest levels of power.

Of course one cannot expect the principles of nineteenth-century historiography to speak fluently to early modern history, but those principles operated upon early modern termini (such as the emergence of the state) and were enormously influential upon twentieth- and indeed twenty-first-century practices of reading and writing the early modern. As Maurizio Arfaioli argues in his discussion of how Italy became "a bloody laboratory of [military] experimentation", this "anachronistic view [that] military backwardness and political division opened the door to 'foreign' conquest...has been one of the unquestioned foundations of the Italian national narrative until very recent times".[5] Indeed, Arfaioli credits Guicciardini as being one of the *auctoritates* of this narrative.[6] Aretino appending his own history to the conflicts that took place around him seems strikingly modern by comparison with nineteenth-century Whig historicism, perhaps because early modern historiographers were more attuned to the narrativity of history.[7] His semi-dramatized records of the "public present" as providing materials for the future archive of the past, with himself at or near the centre, acknowledge the construction of historical narration as contemporary with its matter, although his mode is poetic-rhetorical rather than noetic-positivist.[8] And if, as Hegel argues, momentous conflicts between fixed systems and the contingencies adverse to them are the proper subject matter of historical study, then Aretino is certainly worthy of our study as both subject and practitioner, in a way that we do not read Guicciardini for himself as much as for his matter. As Brian Richardson summarizes, "Aretino was unique in his ability to use his writings in order to exert leverage on patrons...he could make opportunistic use of the press as an instant medium of diffusion for a range of works...and he flaunted the fact that this activity could be a powerful, if indirect, means of profit".[9] Neither an historian of conflict like Guicciardini nor an itinerant biographer like Cellini, Aretino is somewhere between the two personalized models, and distinct from each, as evinced by the range Richardson describes. Aretino could turn both fawning praise and pornographic outrage to the same ends as part of his construction. With these principles in situ we can return to Pavia, which itself forms part of another binary in Aretino's narrative, that of battlefield and court. It is the itineraries between these sites that this chapter traces.

The tragedy of Pavia

On 24 February 1525, after weeks of refusing to engage openly with Charles V's imperial army, headed by Charles de Lannoy, the French forces, led by Francis himself, were routed at the battle of Pavia.[10] It was the most extensive slaughter of French nobles since Agincourt, sending shockwaves across Europe that were felt along the lines of the ambassadorial information networks.[11] According to Ferdinand of Austria, writing to Henry VIII two days after the battle, Francis "was defeated and taken prisoner by De la Mota, maître d'hôtel of the duke of Bourbon".[12] This account is reinforced by the memoirs of Robert III de la Marck, marshal of Florange, who noted that Francis had been taken by "Le Mötte des Noyers" but would not formally surrender to him, either because he viewed him as a traitor – La Mothe had left the French king's service to join the Emperor – or because Francis was going to surrender only to a nobleman. In either case, he gave up his sword to Lannoy. The historical narrative of the conflict began on the battlefield, was developed in report, and continued by Francis himself, who constructed its history in accordance with poetic genres.[13]

After being captured Francis was transported to the Certosa di Pavia, a monastery six miles north of the city, where he composed two letters, one to the Emperor and one to his mother, Louise de Savoy, who subsequently ruled as regent during her son's captivity. In both letters Francis emphasizes the role that Fortune has played in his calamitous circumstances. He writes to his mother "so that you know how the rest of my misfortune proceeds: all is lost except honour and my life, which is saved".[14] Likewise, to Charles he writes that "I have no other comfort in my misfortune except for my esteem of your goodness". Charles, through Francis's rhetorical dexterity, becomes an instrument of divine grace: Francis, despite finding himself at the lowest point on Fortune's wheel, has firm hope that the emperor's virtue will not allow him to be treated in any way that is dishonourable, and commends himself to Charles' good grace.[15] As Francis writes to his mother, "I have hope that ultimately God will not abandon me here". Yet in order to verify this hope

Francis writes that Charles must also look into his own heart in order to determine how he should treat his prisoner, echoing his injunction to his mother that the messenger who bore her son's letter be granted safe passage to return to Spain "because he goes to the emperor, in order to know how he (Charles) wishes me to be treated".

The rhetorical strategies in operation here, echoing across the two letters, are derived from the late medieval conception of tragedy, Boethian philosophy, and the penitential psalms. The principles of late medieval tragedy are encapsulated by Chaucer's Monk:

> Tragedie is to seyn a certeyn storie,
> As olde bookes maken us memorie,
> Of hym that stood in greet prosperitee,
> And is yfallen out of heigh degree
> Into myserie, and endeth wrecchedly. ...
> For certein, whan that Fortune list to flee,
> Ther may no man the cours of hire withholde.
> Lat no man truste on blynd prosperitee.[16]

This model of tragedy is dependent upon the commonplace image of Fortune's wheel derived from Boethius's enormously influential *De consolatione Philosophiae* (*The Consolation of Philosophy*). That Francis knew this text cannot be doubted, nor indeed that he had it in mind during his imprisonment. Firstly, there is an extensive French tradition of translating Boethius's text into the vernacular, with perhaps the most famous being that of Jean de Meun, co-author of *Le Roman de la Rose*. Jean de Meun's translation was one of the earliest French texts to be printed (Lyon, 1483) and was reprinted in Paris in 1520.[17] Secondly, Francis's great-uncle, Charles, Duc d'Orléans (1394–1465), wrote a series of Boethian-inflected lyrics during his twenty-five years spent as a prisoner of war in England; he was captured at Agincourt.[18] Thirdly, Boethius himself wrote the *De consolatione* in Pavia in the sixth century whilst awaiting execution.[19] Francis simply could not have avoided Boethian echoes when considering his own position. The wheel usually shows a king coming undone, his crown toppling from his head as he reaches the nadir. It is at this nadir that the Biblical King David, according to the medieval and early modern traditions, accepted his abasement and impotence, reaching his lowest, most vulnerable point before he could accept and place himself at the will of divine grace. That these three narratives – medieval tragedy, Boethian Fortune, penitential humility – were still intertwined within the early modern imaginary is reinforced by Francis's own superimposition of them upon each other as part of a concerted rhetorical strategy for constructing his historical narrative. That strategy exculpates Francis from responsibility for the failure at Pavia; it was not his actions (mentioned below) but Fortune that caused the defeat. It also positions Charles as the instrument of divine grace, showing mercy to the abject sinner, and so enjoins him via implicit prospective public opinion to practice this virtue – he cannot be a truly Christian ruler otherwise.

Why employ such strategies in a letter to one's mother? The answer is that Francis must have known that his letter would be intercepted and so employed these strategies with a view to it being read by Charles himself and by a wider European audience. These pieces of seemingly private correspondence are *de facto* public statements. The letter to Louise de Savoy thus constitutes a form of textual diplomacy, the letter as envoy, as the king's orator and representative.[20] It was an analogue if not a model for the poetical-rhetorical form of historiography that Aretino practised. As with Francis's letter to his mother, Aretino frequently adopts in his correspondence a form of textual diplomacy that is privately addressed and composed, but public facing – a diplomacy entirely necessary to counterbalance Aretino's equally frequent lack of diplomatic tact.

As David Potter has shown, the letter to Louise "was obviously a text publicly disseminated (the Granvelle papers hold a faulty copy and a more accurate one was known to the authors of private journals of the period...)".[21] Potter shows how several of the king's letters were circulated thus, and it is no great stretch of the imagination to conjecture that the letter to Charles V was also spread aboard, given its addressee (who would ensure that this token of his victory would not remain invisible) and the fact that it was written in tandem with the letter to his mother, which we know was disseminated. It is my contention that Aretino's letters written to both Francis and Charles in the aftermath of Pavia are based upon his having read Francis's letters from Certosa di Pavia, and seeing in them the opportunity to establish or at least represent himself as a player in the major European drama of the day. It is to these letters we now turn.

Letters of recommendation

In the first instance, Aretino responds directly to Francis's repeated emphasis upon *mon infortune* in a letter of 24 April 1525:

> Francis, because of the trick that Fate played on him, has freed his mind from anxieties over whether it could make a King prisoner; and Charles, because of a gift of Fortune, has been enslaved by the thought that it could so to an Emperor. ...So do not complain about Fortune: she did all that she found it possible to do when she placed you in your present predicament. ...I make you not the equal but the superior of the victor, since by your prudence you overcome Fortune who has overcome you by Force.[22]

The injunction – "So do not complain about Fortune" – only makes sense if Aretino is responding directly to the letter Francis wrote to his mother. He also tempers his letter with contemporary accounts of the battle itself, appealing to Francis's conception of himself as a chivalric ideal:

> A man of brave heart should endure calamities and not flee from them, because their endurance reveals greatness of soul and their avoidance a cowardly heart. Whoever heard of a King such as you, who, in the sudden shifts of battle, and his sword still hot with enemy blood, made Fortune confess that the captive was he who fought and not he who made others fight, by declaring that human ways cannot be governed without reason but by an intricate network of causes totally hidden from us and predestined by immutable laws to their subsequent effects?[23]

Wraxall claimed that La Mothe recognized Francis despite "his face [being] covered with blood, caused by a deep wound which he had received across his forehead".[24] However, this account is based on earlier rumours of the king being seriously injured, which were contemporary with the battle, whereas Robert Knecht notes that such rumours "were dispelled by a medical examination, which revealed only a bruise on one leg and a scratch on one hand".[25] The image of Francis as the last man standing, his horse killed from under him, seriously wounded, "his sword still hot with enemy blood", is part of his public image as the last of the warrior kings: Francis fought; Charles made others fight.[26] The rumours became part of the historical narrative. Indeed, it was this chivalric fantasy of himself which, in part, led to the French catastrophe at Pavia, as Francis led the cavalry charge in front of his artillery, preventing his gunners from firing on the imperial forces for fear of hitting the king – a valiant image, but a poor strategy. It is this image which Aretino seeks to uphold in his letter to Francis, "I make you not the equal but the superior of the victor". Aretino knows the importance of the popular narrative of the battle; he spins the facts.

Aretino's rhetorical strategy here operates via a simple yet effective antonymy – he exalts Francis by denigrating Charles, albeit implicitly, through a kind of *praeteritio* – which inverts actuality. Thus, the vanquished is the true victor, "others lose by conquering and you conquer by losing"; Francis has been freed, Charles enslaved, for "he remains subjugated by your fortitude in suffering his inclemency". Ultimately, "victories are the ruin of him who wins them and the salvation of him who loses" through making the penitential counterturn of Fortune's wheel, "for victors, blinded by the insolence of pride, forget God and think only of themselves, while the vanquished, inspired by modest humility, forget themselves and remember God".[27] Aretino takes his cue from Francis's most famous statement, "all is lost except honour", expanding it into a panegyric through a deft use of *amplificatio,* which develops the defeated king's intermingling of humiliation and pride as a means of determining the history of the battle. The letter is, as Paul Larivaille notes, "a masterclass in malice and bad faith placed in the service of a militant politics...a string of facetious rationalisations punctuated by sentences invariably unfavourable to the emperor".[28] There are two fundamental questions raised by the letter: (i) what led Aretino to believe that he could and should write to Francis as an unbidden counsellor? And (ii) what are the militant politics of which Larivaille speaks?

In relation to the first question, Aretino had been introduced to Francis by his close friend Giovanni de' Medici, aka Giovanni dalle Bande Nere ("Giovanni of the Black Bands"), the infamous soldier of fortune who had very recently shifted paymasters from the emperor to the French king at the instigation and facilitation of the powerful papal datary Gianmatteo Giberti.[29] In the second week of January 1525, Aretino travelled from Rome to Pavia, where Francis had been quartered with his troops through the winter. Indeed, Aretino left the French camp shortly after 7 February 1525, around a fortnight before Francis was defeated and taken prisoner. Aretino's audience with Francis was a complete success, as Giovanni dalle Bande Nere records in an undated letter:

> I am forgetting to tell you that yesterday, with good reason, the King was saddened to find you were not with me as usual, whereupon I told him it was due to it being more pleasing to you to remain at court than at camp. His Majesty replied by saying that he would write to you to make you come here [to Pavia], and he swore that this would happen soon, if, when he writes to you, you obey him ... I know that you will come less for your own benefit than to see me, for I do not know how to live without Aretino.[30]

As Larivaille notes, the various encounters between Francis and Aretino during the two or three weeks spent at Pavia constitute a turning point in the writer's career. In addition to charming Giovanni de' Medici, dubbed the Great Devil, Aretino – renowned for his bonhomie and captivating conversation – also managed to enchant the Most Christian King.[31] No longer simply a satirist and would-be courtier, this "new victorious experiment in his powers of seduction" saw Aretino growing into his new persona: the celebrity whose company, endorsement, and opinions were sought by the rulers of Europe.[32] As such, it does not seem so untoward that Aretino should write to Francis following his defeat on 24 February as he had been laughing and joking with the king in the camp just a fortnight or so earlier.

However, just as Francis knew that his private letters to his mother and the emperor were also addressing a public audience, so did Aretino direct his letter to Francis to the hierarchy of the Roman Curia. The letter, as with the journey to the French camp, appeared to be Aretino's means of declaring his allegiance to the pro-French faction in the Curia, led by the papal datary Cardinal Gianmatteo Giberti.[33] What is curious about the letter to

Francis is its date, 24 April 1525, the day before the Feast of Pasquino in Rome, and the personal catastrophe that befell Aretino shortly after, almost in imitation of Francis falling foul of Fortune's mutability – but again, not in direct consequence, except via constructed history. Here we must address the second question asked earlier: what were the militant politics referred to by Larivaille? This refers to the partisan politics at the court of Pope Clement VII, Aretino's erstwhile friend and patron, and how these intersected with a personal conflict that developed between Aretino and Giberti, which compromised Aretino's adhesion to the French cause. At the heart of this conflict is a scandal that appears to be as far removed from the Pavia narrative as is possible.

The politics of pornography

1525 was a jubilee year organized under the auspices of Giberti, who was responsible for cleaning up the public image of the Church in response to the critiques of the reformers – early stirrings of the counter-Reformation proper.[34] Pilgrims were encouraged to travel to Rome where their sins would be cleansed and they would find themselves returned to a prelapsarian state, at least temporarily. Aretino chose to mark the jubilee so strategically curated by Giberti in his characteristic way, by penning a series of pornographic poems: the *Sonetti lussuriosi*.[35]

Aretino's sequence of sixteen *sonetti caudati* (tailed sonnets, appropriately) were written to accompany the sexually explicit engravings made by Marcantonio Raimondi in 1524 of sixteen drawings completed by Raphael's successor, Giulio Romano. The images, known as *I modi* (*The Positions*), were immediately suppressed, with only a few later copies and fragments remaining today following the Pope's (or Giberti's) order for their mass destruction. The images accompanying Aretino's *sonetti lussuriosi* of 1525 were inferior reproductions of certain of the original series. As Bette Talvacchia has argued, the traditional narrative of the scandal is flawed. In this version of events, Romano painted the images in Rome then fled to Mantua to avoid repercussions, unlike Raimondi, who was imprisoned after reproducing them as engravings. Aretino was so outraged at Raimondi's imprisonment – due to what he saw as the hypocrisy of a papal court that practised every kind of sexual vice – that he added a sonnet to each one and had them republished.[36] In fact, Romano was already planning to leave Rome for a position at the Mantuan court and had been waiting to do so for some time; Romano's transfer to Mantua was being arranged by Baldessare Castiglione before he left Mantua to take up his new position as papal nuncio to the imperial court. It was thus a kind of knowledge transfer: the courtier would depart Mantua to work for the Pope, whilst the Pope would send the artist to Mantua. Talvacchia suggests that the paintings were given to Raimondi as a parting gift in order to provide income for the Roman workshop that he took over upon the death of his *maestro* Raphael. It was Raphael who had instituted the lucrative relationship with Raimondi; with Romano gone the workshop would become defunct. As such, Romano did not leave Rome under a cloud, which leads to the conclusion that "the drawings were not criminal; it was only their appearance in prints that was culpable".[37] There was an anxiety over the technology and its control, a technology that Aretino was one of the first to exploit.

The view that Romano was at risk of the same fate as Raimondi comes from a statement made by Vasari over fifty years later in his *Vite* (1568), "if, when [*I modi*] was published, Giulio had not already left for Mantua, he would have been harshly punished by the pope for it in his anger".[38] Vasari's account is, when taken in chronological order, the third competing historical narrative we have of the fallout from *I modi*, following Ludovico Dolce's 1557 dialogue on art entitled *Aretino*, and a letter of Aretino's from 1537(?) to a certain Battista Zatti of Brescia. Each narrative comes with its own hypothesis of when Aretino produced the sonnets. Vasari implies that the sonnets were produced simultaneously with Raimondi's engravings, "so that I do not know which was worse: the vision of Giulio's drawings to the eye, or the sound of Pietro's words to the ear". Vasari also adds that "it would have been a catastrophe for [Raimondi] if Cardinal [Ippolito] de' Medici and Baccio Bandinelli, who served the pope in Rome, had not rescued him".[39] This account runs contrary to that provided by Aretino ten years after the event:

> When I obtained from Pope Clement the liberty of Marcantonio Bolognese, who was in prison for having engraved on copper plates the Sixteen Positions et cetera, I felt a desire to see the figures that were the source of Giberti's complaints, who demanded that such a fine virtuoso be crucified. And having seen them, I was touched by the spirit that moved Giuliano Romano to design them … I exhibit them above the Sonnets that stand below, whose lewd memory I dedicate to you, pace all hypocrites. I despair of the bad judgement and damnable habits that forbids the eyes what delights them most.[40]

Unsurprisingly, Aretino claims that he was Raimondi's salvation, which does not entirely go against Vasari's account as Aretino did have the ear of Clement and may well have used his favour with the Pope to obtain the engraver's release – no doubt much to the chagrin of Cardinal Giberti, who already was unhappy with Aretino's influence.[41] However, Aretino makes it very clear that his sonnets were *a posteriori* to the engravings and not coincident with them. The third account of the events, the dialogue by Dolce – the humanist scholar who previously had been Aretino's secretary – is slightly confused, partly on purpose. Aretino's interlocutor Fabrini asks if "Raphael showed a sense of decency when he designed those cartoons of men and women embracing one another lasciviously and also indecently and had Marcantonio make copperplate engravings from them?", to which Aretino replies "it was not Raphael who designed them, but Giulio Romano, his pupil and successor... This same Marc'Antonio, but for the action I took, would have been deservedly punished for his temerity by Pope Leo".[42] It is not evident as to why Dolce, who knew Aretino well, clarifies that it was Romano and not Raphael who designed the images but then claims that it happened during the papacy of Leo X. Admittedly, it does sound like the kind of scandal that *would* have happened under Leo, a reputed bon viveur, but the error is curious.[43] It is possible that Vasari (who also knew Aretino personally) was drawing on and correcting Dolce's account as there is a clear echo of Dolce's phrase "sarebbe stato da Papa Leone della sua temerità degnamente punito" in Vasari's "sarebbe stato dallo sdegno del papa aspramento castigato". Dolce also reinforces Aretino's claim that it was he who liberated Raimondi. Nevertheless, it would appear to be that the scandal broke after 6 October 1524, when Romano departed Rome for Mantua and that Aretino produced his sonnets, as he says, following Raimondi's release from incarceration in 1525.

To return to the point of departure, and the question of constructed causality between the personal and the grand narrative, it would also appear to be the case that the sonnets were written in the aftermath of the Battle of Pavia as "the final verse of one of two sonnets that appear on a folio missing from the unique sixteenth-century edition [of the *Sonetti lussuriosi*] known to all today alludes clearly to the imprisonment of Francis I", and as such cannot have been composed prior to 24 February 1525:

> I would not depart,
> Dear lady, from such sweet chatter,
> If I knew well that the King of France were alive.[44]

Reports of the battle and Francis's capture were sent on the day itself but took a couple of days to reach their respective destinations, prompting rumours, as one reporter wrote to Cardinal Wolsey on 28 February: "For two days there has been a rumour of the defeat of the French, and capture of the King, with great effusion of blood on either side; but...accounts differ".[45] If in the immediate aftermath of Pavia before the reports filtered through it was not known if Francis had survived, then we can assume a date very close to 24 February for this sonnet at least; given what we know of Aretino's habit of writing in ferocious bursts of activity, it might be assumed that the rest were written within a similar timeframe. These missing verses provide a political direction to the pornographic sonnets, which subsumes them into the Pavia narrative that both Francis and Aretino were separately composing. It allows Aretino to outrage the leading member of the French faction at the curia whilst outwardly signalling his own continued allegiance to Francis. Whether this allegiance was sincere remains to be seen; what matters is the constructed historical narrative.

This reference to Francis's imprisonment in the *Sonetti lussuriosi* confirms Larivaille's claim that a letter sent to Aretino from Giovanni dalle Bande Nere, which is essential for tracing the consequences of the scandal of *I modi*, has been given the wrong date:

> I beg you, upon receiving this letter, to leave Arezzo and to come stay with me. This is something I desire warmly, even though I should not, for I am upset with you for having placed yourself in line with Fra Nicolo [Schomberg] and Vaison [Girolamo Schio]. Because in having lost Gian Matteo [Giberti], you have also lost the pope. So that you who could rule the world have ruined yourself; and not without damage to me, since while you were at the Roman court, I had one who would defend, regardless, the rightness of my position, of doing what I have done, and what I might well do again. Now, I await you. Because certainly, you went out of bounds for good intentions, and for no other reason. And I want to praise you for this: that everyone can cause unhappiness sometimes, but you, never yet.[46]

The letter is dated 3 August 1524, but this appears an erratum for 3 August 1525. The letter is sent from Fano, which the Pope conceded to Giovanni *after* the Battle of Pavia and where he did not take up residence until the middle of July 1525.[47] In the summer of 1524, Aretino was still on good terms with Giberti, and allied with the French faction – as opposed to the imperial faction of Schomberg and the Bishop of Vaison. And in any case, relationships between the two factions were not especially tense during that summer, unlike in 1525, again *after* Pavia. In the autumn of 1524, Aretino was at the peak of his career as a roman courtier so is unlikely to have fallen from grace dramatically in the preceding summer. Aretino's conflict with Giberti belongs to 1525, following the Festival of Pasquino (detailed below) and the publication of the *Sonetti lussuriosi*, which is surely what Giovanni is referring to when he says that Aretino has "lost" Giberti. Crucially, Giovanni upbraids Aretino because he was the person at the Roman curia who defended Bande Nere's actions – notably, his defection from serving the emperor to serving the French king, which took place in late 1524. However, Aretino could not reply to de' Medici's letter: on 28 July 1525, Aretino had been left fighting for his life after Giberti ordered his assassination.

Character assassination

The identification of the content of the missing *sonetto lussurioso* and the correction of the date of Giovanni de' Medici's letter to Aretino from August 1524 to August 1525 embed Aretino's conflict with Giberti securely within the aftermath of the Battle of Pavia. But what caused this conflict? And when did it begin? After all, in February 1525 Aretino was sharing jokes around the campfire with Francis I and wrote a laudatory poem in praise of Giberti (*Canzone in laude del datario*), who led the French faction at Clement's court. On 24 April 1525, the day before the Festival of Pasquino, Aretino wrote his sympathetic, anti-Caroline letter to the French king.

Pasquino was the name of a dilapidated statue in Rome that became a forum for public satire. Originally used by humanist scholars to criticize the clergy, when Aretino arrived in Rome, the satires suddenly became far more vicious and ad hominem. It must be noted that vituperation had been a characteristic of the tradition, but it had previously stopped short of slander, as Anne Reynolds has shown.[48] Aretino makes a point of not stopping short. Indeed, Aretino used Pasquino to great effect in the papal conclave that resulted in the election of Adrian V, whom Aretino had attacked repeatedly. In consequence, Aretino had to flee Rome in 1522, resulting in his first sojourn at the camp of Giovanni de' Medici.

In many ways this festival was designed for Aretino – it was licensed attack upon both the clergy and the pedants, Aretino's *bêtes-noires*. Each year, a *magister ludi* was appointed to ventriloquize the voice of Pasquino, who became Aretino's alter ego. However, as Reynolds notes, "censorship clearly increases as the century progresses", and the satires came to be vetted before being submitted to the festival.[49] In 1525 Aretino's poems were censored, at least indirectly, by Giberti, who was their direct target.[50] Again, on the micro level, Aretino's adoption of the organized disorder licensed by the fixed system of the papal state effects a collision between himself and the "existing, acknowledged duties, laws, and rights" represented by Giberti. According to Hegel's account with which we began, such collisions provide the coordinates for historical narratives that are enabled by the state, the sanctioned and constructed past of the "public present", and ritualized feast days are a kind of shorthand for such a past. Indeed, one might have anticipated the conflict between Giberti's role as major-domo of the jubilee and his attempts to ameliorate the church's public image, and Aretino's role as the festival's Lord of Misrule, relentlessly exposing curial hypocrisy and vice.

Aretino had threatened to expose this vice in his comedy *La Cortigiana* (1525). As Aretino's Histrion I says to his Roman audience, "I'm warning you – I wouldn't hesitate to expose every last one of you [*io vi vitupererò tutti*] ... I'll tell everyone which of you are doing it, and which of you are having it done. ... I'd make all of your sins a matter of public record. I know all about them".[51] Aretino's later incarnation as the self-styled "secretary of the world" had begun in Rome.[52] Whilst salacious sex stories damaged the church's credibility, they were not as dangerous as revealing state secrets. On 1 April 1525, a treaty was drawn up with the imperial lieutenant Lannoy – to whom, as we recall, Francis had surrendered at Pavia – yet within two months, Giberti was engaged in secret talks with France via the Bishop of Bayeux.[53] However, despite Giberti being only one of two contemporary figures who receive honourable mentions in the play, which suggests it was composed certainly prior to July, its relentless ridicule of the ethical vacuum of the curia cannot but have irked Giberti by undermining his and Clement's ongoing PR job.[54]

Can we now trace the development of the conflict between Aretino and Giberti? It escalated quickly. Aretino had remained supportive of Giberti at least until February 1525, when the *Canzone in laude del datario* was published. He also exonerates Giberti from the taint of curial vice in *La Cortigiana*. However, Aretino's praise fell on deaf ears: Giberti did not rise to the bait Aretino usually set for prospective patrons. Aretino's newfound posture of silence and respect towards the papal court – prior to the publication of *La Cortigiana* – was viewed with suspicion or disbelief by its members, and Aretino's fellow *pasquinisti* saw it as political and material opportunism.[55] Without recompense for his support, Aretino no longer had a reason to keep Giberti onside. If, as Aretino claimed, the engraver Raimondi had been imprisoned at the instigation of Giberti – who either superseded or exerted pressure on Tomasso del Badio, the *Maestro del sacro palazzo*, responsible for censorship and the imprimatur in Rome since 1515 – then Aretino's intervention to obtain his friend's release would have been an affront to the datary.[56] Despite the play's exoneration of Giberti, the savage indictment of the Roman court in *La Cortigiana* would have only deepened the rupture between the two, which was exacerbated further by Giberti's censoring of the personal attacks upon him penned for the Festival of Pasquino on 25 April, whereby Aretino redressed his previous decision to spare the datary the lashing tongue of the Scourge of Princes – the title conferred on Aretino by Ariosto in *Orlando furioso* (1532 ed.). Seeing the potential in the Aretino–Giberti conflict, the imperialist faction at Clement's court – as recorded in the letter from Giovanni de' Medici of 3 August 1525 (not 1524) – perhaps incited him to turn more vehemently against Giberti, which he did by publishing the *Sonetti lussuriosi* alongside Raimondi's engravings. Through this final outrage Aretino, to use de' Medici's phrase, had "lost" Giberti and in doing so had lost the Pope. Giberti responded to Aretino's ad hominem attacks with an ad hominem attack of his own: he arranged Aretino's assassination.

On the night of 28 July, Aretino was repeatedly stabbed by a member of Giberti's household named Achille della Volta. Tellingly, the first account of the assassination is sent on 30 July by Girolamo da Schio, Bishop of Vaison, to Aretino's long-term patron, Federico II Gonzaga, Marquis of Mantua, a fellow imperialist. Vaison refers to "the strange fate that befell our Maestro Pietro Aretino the other night...he received two stab-wounds in the chest, one of which is mortal".[57] However, the Marquis' ambassador, Francesco Gonzaga, reported back to Federico that "notwithstanding the wounds are in his chest, and noticeably very deep, nonetheless by the virtue of a medic here ... he is out of danger".[58] A cover story was circulated that Aretino and della Volta fought over a woman who worked as Giberti's cook, but this attempt to spin the narrative was not credited by the Roman populace, as evinced by a characteristically malicious poem written by one of Aretino's enemies, Francesco Berni, to mark the occasion.[59] Indeed, Clement VII admitted the following year that he had sacrificed justice for his friend Aretino for the sake of his minister of secrets, Giberti.[60] Specifically, della Volta was later made an apostolic nuncio despite all of Rome knowing what he had done and on whose orders.[61] The court had become too deadly. Aretino had to travel to a safe haven – the site of conflict.

From court to camp

Aretino took two months to recover from his injuries, and as soon as he was able, he departed Rome for good. He travelled initially to a place of comfort and familiarity – the battlecamp of Giovanni de' Medici, whose invitation Aretino received whilst in extremis. The camp was a setting that fostered liberty and amity; it was the curia that was more dangerous for Aretino. As he inverts private and public, so does he invert the locales of conflict, as Larivaille notes, "In the company of Giovanni de' Medici [Aretino] recovers above all that freedom of language of which Rome had tragically deprived him, that faculty to express his own opinions on facts and people without reservation".[62] By way of inverse analogue, Francis was removed from the battlefield at Pavia and made a guest-prisoner at the emperor's court. The site of open conflict is privileged; Aretino travelled from court to regain freedom at the camp, whilst Francis lost his liberty in transit to court. De' Medici had warned Aretino of the covert conflict of court compared to the honest conflict of the field, as he related a comment he made to Francis, "it is 'more pleasing to you [Aretino] to remain at court than at camp [*in corte che in campo*]'".[63] The camp brought Aretino peace after the conflict with Giberti – the battle was over; the war far from it, as far as Aretino was concerned – and marked the end of Aretino's life as a courtier. He had tasted censorship, self-censorship, state censorship, and violently imposed silence, and he did not savour it. Coincident with the end of his life as a *cortigiano* was the end of his travels. His Roman years had been peripatetic, spent fleeing the consequences of courtly conflicts: within the space of three years, he had travelled from Rome to Arezzo, Mantua, Pavia, and Fano, amongst others. In 1527, he arrived penniless in Venice but soon became La Serenissima's most celebrated adopted son. He repaid the republic's indulgence of his freedom to express himself as himself with unstinting loyalty until his death in 1556.

However, we are in danger of over-romanticizing: battlefields are still battlefields. On 24 November, 1526, Giovanni de' Medici was shot in the leg by a falconet – the same leg in which he had been injured by an arquebus shot prior to Pavia – and died four days later. Aretino was by his friend's side throughout, as the surgeons amputated his blasted limb and as he dictated his will and epitaph, and recorded the death and life of the great *condottiere* in vivid, touching detail that illustrates perfectly his form of personalized historiography:

> "Love me when I shall be dead", he [de' Medici] replied.
>
> "The virtues which have won you such glory", said the Marquis [of Urbino] "will make you worshipped, let alone loved, both by me and others".
>
> At the end he turned to me and commanded that Madonna Maria should send Cosimo to him...his household, no longer observing their respectful reticence, surged about him...Such was the passing of the great Giovanni de' Medici, who possessed from the time he was a baby in arms the most noble character imaginable.[64]

The fate of Giovanni de' Medici is a reminder of what a perilous position Francis had been in at Pavia. De' Medici died the soldier's death Francis would have wanted. But Francis, after Pavia, was embarking on his own travels. At first he was taken to the castle of Pizzighettona midway between Cremona and Piacenza. On 31 May 1525, Francis boarded a ship at Genoa bound for Spain. On 17 June, the ship reached Palamos, and Francis was royally received at Barcelona two days later. Shortly after he was taken to the castle of Tarragon, where he was nearly killed by a stray bullet from an arquebus, again an uncomfortable echo of Giovanni de' Medici's death.[65]

In Aretino's account of the great soldier's death, we read the kind of history that was condemned by later historians, such as the *Annales* School in the twentieth century, which included Fernand Braudel and Jacques Le Goff.[66] As Hayden White asks, "How can one condemn narrative on grounds of its 'novelizing' effects?"[67] The answer he provides speaks directly to the culture of *individualismus* that underpinned early modern historiography:

> One suspects that it is not the "dramatic" nature of novels that is at issue but a distaste for a genre of literature that puts human agents rather than impersonal processes at the center of interest and suggests that such agents have some significant control over their own destinies.[68]

It is this agency that Aretino ascribes to Giovanni de' Medici and which, by virtue of the epideictic tradition, he arrogated to himself. By contrast, in the immediate aftermath of Pavia, Francis appeared bereft of agency, having conceded control of his destiny in accordance with a Boethian poesis, which also runs contrary to the state-sanctioned historical narrative.

On 27 March, Francis was visited by a papal nuncio, of whom he "asked after John de Medicis" and asked that he himself be recommended to

Fortune. The nuncio "[t]old the King that all was lost; at which he was much moved, saying nothing else was to be expected" and reported to Clement that "[i]t was a pitiable sight for he does not appear to be a prisoner of the emperor and his captains, so much as of his guards, who all claim a right in him. He behaves courteously and liberally and jokes with them, thinking it no less virtue to accommodate himself to Fortune than to command a kingdom. [Anne, Duc de] Montmorency is his only comfort".[69] The nuncio's report would no doubt have circulated at the Roman court – whether in copy or by rumour – and would have reached Aretino's ears. He knew the French king was still alive, he knew he was despondent, and he knew he blamed Fortune. He took up his pen. Indeed, he did not forgo the French king until much later; the argument that at this point Aretino had switched sides in the Franco–Imperial conflict separates the private and public, the personal and the political, in a way that Aretino does not allow.[70] I believe that Aretino turned against Giberti not because of a political ideology but for more selfish reasons. I also do not believe that turning against Giberti entailed burning bridges with Francis. Indeed, in 1533 Aretino was still being wooed and having gifts bestowed on him by the French king, who sent Aretino an engraved orator's chain. However, Aretino did not appreciate the delay between the pledge and the gift proper, as he wrote later (8 June 1537) to the same Montmôrency who had been Francis's only comfort during his captivity:

> I too am a captain, but my soldiering does not steal its soldiers' pay, nor does it incite to mutiny or surrender fortresses. On the contrary, with its squads of inkpots, and with Truth painted on every banner, it acquires more glory for the prince it serves than all the new lands that his armed men could conquer. And my pen pays in the ready money of honour and shame.[71]

Like the king's letter to his mother, like Aretino's letter to Francis, this private letter was understood and intended as a public document. It speaks to and develops Aretino's European-wide reputation and the rhetorical strategies upon which it is predicated. In this sense the letter is performative: it is a European document, proof that Aretino's armies of inkpots are advancing upon the historical narrative.

Conclusion: the itinerary of conflict

The anomalous presence of a reference to the French king's captivity in a sequence of otherwise relentlessly pornographic sonnets – whilst one might fancifully argue that it is consonant with Francis's amorous proclivities – confirms that Pavia was an opportunity for Aretino, just as it was for the imperial faction at the papal court. The fact that Aretino's letter of counsel to Francis was dated 24 April, the day before the Feast of Pasquino, offers a number of conclusions. One might at first be tempted to interpret it as signifying a means of amending his anti-Gibertine invective, writing an anti-Caroline letter of support to Francis to offset his pillorying the leader of the pro-French curial faction. However, it seems more likely that we can in part divorce Aretino's actions from Franco–Imperial factionalism by this point. The letter shows Aretino going beyond the patronage system of the curia; a system that, as he came to realize, had not really favoured him. Why attempt to ingratiate himself with the leader of the French faction when he could befriend the French monarch, one captain to another, around the battlefield campfire? Due to his developing enmity with Giberti, it suited Aretino to seek support from the Roman imperialists, but this was due to a personal rather than a political conflict. He still sought Francis's favour. And more importantly, Francis had sought Aretino. Indeed, Pavia and its aftermath signalled a sea-change in Aretino's attainment and use of patronage, which would come to full fruition in his Venetian years (1527–1556). From being the Voice of Pasquino, focussing on localized satire, and a courtier struggling to find support in Clementine Rome (when he was not fleeing the Eternal City for one scandal or threat or another), he became the Scourge of Princes, the Secretary of the World, who was courted by powerful would-be patrons, who served no ruler but the liberty expressed and defended by the Venetian Republic. As such, he was willing to risk personal conflict in pursuit of the greater goal of complete independence. He just had to stay alive long enough to achieve it.

Notes

1 Francesco Guicciardini, *The History of Italy*, trans. Sidney Alexander (1561; New York: MacMillan, 1969), 3.
2 Georg Wilhelm Friedrich Hegel, *The Philosophy of History*, trans. J. Sibree (New York: Dover, 1956; repr. 2004), 30–31.
3 Satirist does not explain Aretino adequately but is one of the many roles he performed.
4 Historiography thus becomes the *tertium quid*. I am indebted to Hayden White's discussion of Hegel's model in "The Question of Narrative in Contemporary Historical Theory," *History and Theory* 23 (1984): 1–33.
5 See Maurizio Arfaioli, *The Black Bands of Giovanni: Infantry and Diplomacy During the Italian Wars (1526–1528)* (Pisa: Pisa University Press, 2005), 2–4.
6 *Ibid.*
7 See for example Benjamin G. Kohl, "Petrarch's Prefaces to *De viris illustribus*," *History and Theory* 13 (1974): 132–44, representing (early) humanist historiography.
8 See White, 19.
9 Brian Richardson, *Printing, Writers and Readers in Renaissance Italy* (Cambridge: Cambridge University Press, 1999), 95.
10 See Robert J. Knecht's analysis of Pavia in *Francis I: Renaissance Warrior and Patron* (Cambridge: Cambridge University Press, 1996), chapter 11.
11 When Francis had first invaded Milan in 1524, he had with him 28,000 troops, 5000 more than the Imperial forces.

12 J. S. Brewer, ed., *Letters and Papers, Foreign and Domestic, Henry VIII, Volume 4, 1524–1530*, (London: 1875) 4, no. 1127, hereafter cited as *LP*.
13 See the *Mémoires du maréchal de Florange dit le jeune adventereux*, ed. Robert Goubaux and P. André Lemoisne, 2 vols (Paris: H. Laurens, 1913–1924), 2:232. For the argument that Francis did not surrender to La Mothe because he considered him a traitor, see, for example, Nathanial William Wraxall, *The History of France, Under the Kings of the Race of Valois*, 2 vols (London: McMillan, 1807), 1:330. La Mothe later challenged Lannoy to a duel over who had taken Francis prisoner (see Knecht, 224).
14 For the original letter and that to Charles V see Aimé Champollion-Figeac, *Captivité du roi François Ier* (Paris: Imprimerie Royale, 1847), 129–31. David Potter notes, with appropriate incredulity, "there is as yet no project for a systematic edition of [Francis's] letters and how this might be achieved", hence the recourse of commentators to nineteenth-century partial collections. See "The Letters of Francis I: Culture, Style of Rule, and the Written Word," *French History* 29 (2015): 445–68.
15 Champollion-Figeac, 130–31.
16 See "The Monk's Prologue and Tale" (7:1973–1977, 1991–1997) in *The Riverside Chaucer*, gen. ed. Larry D. Benson, 3rd ed. (Oxford: Oxford University Press, 1988).
17 See Glynnis M. Cropp, "Boethius in Medieval France: Translations of the *De consolatione philosophiae* and Literary Influence", in *A Companion to Boethius in the Middle Ages*, ed. Noel Harold Kaylor Jr and Philip Edwards Phillips (Leiden: Brill, 2012), 312–55. For early print editions, see Andrew Pettegree, Malcolm Walsby and Alexander Wilkinson, ed., *French Vernacular Books: Books Printed in the French Language before 1601*, 2 vols (Leiden: Brill, 2007), 1:182.
18 See Mary-Jo Arn, ed., *Fortunes Stabilnes: Charles of Orlean's English Book of Love—A Critical Edition* (Binghamton, NY: Medieval Texts and Studies, 1994).
19 He is buried in Pavia in the Augustinian Basilica di San Pietro in Ciel d'Oro.
20 The "hard" diplomacy took place later with the discussions that led to the Treaty of Madrid (January 14, 1526), whereby Francis ceded Burgundy and all claims to Italy. Yet literary scholars, historians, and International Relations researchers involved in the recent "diplomatic turn" in early modern studies have been at pains to emphasize the importance of the soft powers or rhetorical modes of diplomacy to a fuller understanding of how early modern diplomacy was organized and implemented. See, for example, Tracey A. Sowerby and Jan Hennings, ed., *Practices of Diplomacy in the Early Modern World c. 1410–1800* (London: Routledge, 2017).
21 Potter, 445.
22 *Lettere*, 1:3. I have used Paolo Procaccioli's six-volume edition of Aretino's *Lettere* for the Edizione Nazionale throughout (Rome: Salerno, 1997–2002), hereafter cited by volume and letter number. I have also used Procaccioli's two-volume edition of the letters written to Aretino, *Lettere scritte a Pietro Aretino* (Rome: Salerno, 2003–2004), hereafter *Lettere a P.A.*, following the same method of citation. For the English translation see Pietro Aretino, *Selected Letters*, trans. George Bull (Harmondsworth: Penguin, 1976), 51–52.
23 *Ibid.*; Bull trans. 52.
24 Wraxall, 1:330. Cf. *LP*, 4, no. 1124, "he received two wounds in the hand and the face".
25 Knecht, 226.
26 This is not to deny that Francis acquitted himself bravely – he continued to fight on foot until he was entirely surrounded, as the accounts agree. In addition to

Florange, see, for example, the *Mémoires de Martin et Guillaume Du Bellay*, ed. Victor-Louis Bourrilly and Fleury Vindry, 4 vols (Paris: Renouard, 1908), 1:356–57.

27 *Lettere*, 1:3; Bull trans., 52–53.

28 Paul Larivaille, *Pietro Aretino* (Rome: Salerno, 1997), 99–100.

29 *Ibid.*, 408 n. 40. Larivaille suggests the transfer of allegiance began in December 1524 and was completed by January 1525.

30 *Lettere a P.A.*, 1:2. It is very similar to a letter sent on February 18 and must have been composed between February, 7 shortly after which Aretino left the camp, and February 24, when Pavìa was taken.

31 Francis was addressed by Aretino with the honorific *Cristianissimo Sire* in the letter of April 24.

32 Larivaille, 97.

33 On the office of datary and Giberti's expansion of it see Bette Talvacchia, *Taking Positions: On the Erotic in Renaissance Culture* (Princeton, NJ: Princeton University Press, 1999), 13.

34 On Giberti's career, see Adriano Prosperi, *Tra evangelismo e controriforma: Gian Matteo Giberti (1495–1543)* (Rome: Edizioni di storia e letteratura, 1969; repr. 2011).

35 Talvacchia includes the entire sequence of *sonetti lussiorosi* as Appendix B in *Taking Positions*.

36 I disagree respectfully with Talvacchia's position that the sonnets were not published until 1527. See *Taking Positions*, 82–84.

37 *Ibid.*, 9.

38 *Ibid.*, 6.

39 *Ibid.*

40 *Lettere*, 1:308; trans. Talvacchia, 6. Aretino worked for the legendary banker Agostino Chigi when he first arrived in Rome.

41 Aretino campaigned for de' Medici during the conclave of 1522, unsuccessfully, and remained in public support of him until the events of 1525.

42 See Mark W. Roskill, trans., *Dolce's* Aretino *and Venetian Art Theory of the Cinquecento* (New York: New York University Press, 1968; repr. Toronto: University of Toronto Press, 2000), 162–65.

43 Dolce might simply have misremembered after more than thirty years – Leo died in December 1521 and the scandal broke in late 1524 – or he might have been rewriting history so as to exonerate Clement's papacy of such scandals (the Sack of Rome in 1527 was more than enough for Clement's legacy). However, this is entirely conjectural.

44 See Larivaille, 92–93. Positions 5 and 6 are missing from the remaining sixteenth-century edition but can be found in later editions, such as Alcide Bonneu's *Les sonnets luxurieux du divin Pietro Aretino* (Paris: Lisseux, 1882), 12. Bonneau lists the sonnet erroneously as number IV. For the reconstruction of the original content, see Lynn Lawner, *I modi: The Sixteen Pleasures: An Erotic Album of the Italian Renaissance* (Evanston, IL: Northwestern University Press, 1989), 24–26, 72–73, 109.

45 *LP*, 4, no. 1131 (Clerk to Wolsey). See also *LP*, 4, no. 1121 (Lannoy to Henry VIII), 4, no. 1123 (Freundsburg to Archduke Ferdinand), and 4, no. 1130 (Archduke Ferdinand to Margaret of Savoy).

46 *Lettere a P.A.*, 1:3; Talvacchia, 16. Many letters in the first volume of *Lettere* were dated wrongly.

47 Larivaille, 109.

48 Anne Reynolds notes that in the early years of the Festival, from its inception ca. 1501(?) to around 1520, "Vituperation is designated as the keynote of the

celebration since 'nemo tam sanctus atque innocens est, quem calumniari non possis'; it is, however, vituperation well within the bounds of slander (maledicentia)". See "Cardinal Oliviero Carafa and the Early Cinquecento Tradition of the Feast of Pasquino," *Humanistica Lovaniensia* 34 (1985): 178–208 (196). Note Aretino's use of "vitupererò" in the Histrion's speech from *La Cortigiana*, below.

49 Reynolds (1985), 207.

50 Talvacchia, 17.

51 Pietro Aretino, *Cortigiana*, trans. J. D. Campbell and Leonard G. Sbrocchi (Ottawa: Dovehouse, 2003), 52. For the Italian, see *Cortigiana (1525 e 1534)*, ed. Paolo Trovato and Federico delle Corte (Rome: Salerno, 2009).

52 He describes himself thus in a letter of November, 28 1537 to the scholar Francesco Alunno (*Lettere*, 1:257).

53 Larivaille, 110.

54 I agree with Larivaille (108) that even if this play were composed and produced within the ambit of a courtly coterie or Aretine *accademia*, it still would have reached the ears of Giberti.

55 *Ibid.*, 108.

56 Raimondi had produced one of Aretino's earliest portraits.

57 Quoted in Larivaille, 116.

58 *Ibid.*

59 See Giovanni Mazzuchelli, *La vita di Pietro Aretino* (Padua: Comino, 1741), 26–27.

60 See the letter from Girolamo Montaguto, dated December 5, 1537, in *Lettere a P.A.*, and Larivaille, 119.

61 Giberti himself, unsurprisingly, always maintained his non-involvement, as he wrote to Federico Gonzaga on February 9, 1530, "if anyone ever did anything to Master Pietro it was without my order, my consent, or my knowledge". See Alessandro Luzio, *Pietro Aretino primi suoi anni a Venezia* (Turin: Loescher, 1888), 91.

62 Larivaille, 120.

63 *Lettere a P.A.*, 1:2. It is very similar to a letter sent on February 18 and must have been composed between February 7, shortly after which Aretino left the camp, and February 24, when Pavia was taken.

64 Bull trans., 57–58.

65 For Francis's itinerary see Knecht, chapter 11, *passim*.

66 See White, 8–10.

67 *Ibid.*, 10.

68 *Ibid.*

69 *LP*, 4, no. 1219.

70 See Talvacchia, 15, for example, who divides Giberti and Aretino according to faction.

71 *Lettere*, 1:144. Bull trans., 93–94. For Aretino's discussion of the inscription see his letter of November 10, 1533 (*Lettere*, 1:36).

Bibliography

Aretino, Pietro. *(Les) sonnets luxurieux du divin Pietro Aretino*. Edited by Alcide Bonneu. Paris: Lisseux, 1882.

Aretino, Pietro. *Selected Letters*. Translated by George Bull. Harmondsworth: Penguin, 1976.

Aretino, Pietro. *Lettere*. Edited by Paolo Procaccioli. Rome: Salerno, 1997–2002.

Aretino, Pietro. *Cortigiana*. Translated by J. D. Campbell and Leonard G. Sbrocchi. Ottawa: Dovehouse, 2003.

Aretino, Pietro. *Cortigiana (1525 e 1534)*. Edited by Paolo Trovato and Federico delle Corte. Rome: Salerno, 2009.

Arfaioli, Maurizio. *The Black Bands of Giovanni: Infantry and Diplomacy during the Italian Wars (1526–1528)*. Pisa: Pisa University Press, 2005.

Arn, Mary-Jo ed. *Fortunes Stabilnes: Charles of Orlean's English Book of Love—A Critical Edition*. Binghamton, NY: Medieval Texts & Studies, 1994.

Bourrilly, Victor-Louis and Fleury Vindry, eds. *Mémoires de Martin et Guillaume Du Bellay*. 4 vols. Paris: Renouard, 1908.

Brewer, J. S. ed. *Letters and Papers, Foreign and Domestic, Henry VIII, Volume 4, 1524–1530*. London: 1875.

Champollion-Figeac, Aimé *Captivité du roi François Ier*. Paris: Imprimerie Royale, 1847.

Chaucer, Geoffrey. *The Riverside Chaucer*. Edited by Larry D Benson, et al. 3rd ed. Oxford: Oxford University Press, 1988.

Cropp, Glynnis M. "Boethius in Medieval France: Translations of the *De consolatione philosophiae* and Literary Influence." In *A Companion to Boethius in the Middle Ages*. Edited by Noel Harold Kaylor Jr and Philip Edwards Phillips, 312–55. Leiden: Brill, 2012.

Goubaux, Robert and P. André Lemoisne, eds. *Mémoires du maréchal de Florange dit le jeune adventereux*. 2 vols. Paris: Renouard & H. Laurens, 1913–1924.

Guicciardini, Francesco, *The History of Italy*. Translated by Sidney Alexander. New York: MacMillan, 1969.

Hegel, Georg Wilhelm Friedrich. *The Philosophy of History*. Translated by J. Sibree. New York: Dover, 1956.

Knecht, Robert J. *Francis I: Renaissance Warrior and Patron*. Cambridge: Cambridge University Press, 1996.

Kohl, Benjamin G. "Petrarch's Prefaces to *De viris illustribus*." *History and Theory 13* (1974): 132–44.

Larivaille, Paul. *Pietro Aretino*. Rome: Salerno, 1997.

Lawner, Lynn. *I modi: The Sixteen Pleasures: An Erotic Album of the Italian Renaissance*. Evanston, IL: Northwestern University Press, 1989.

Luzio, Alessandro. *Pietro Aretino primi suoi anni a Venezia*. Turin: Loescher, 1888.

Mazzuchelli, Giovanni. *La vita di Pietro Aretino*. Padua: Comino, 1741.

Pettegree, Andrew, Malcolm Walsby and Alexander Wilkinson, eds. *French Vernacular Books: Books Printed in the French Language before 1601*. 2 vols. Leiden: Brill, 2007.

Potter, David. "The Letters of Francis I: Culture, Style of Rule, and the Written Word." *French History 29* (2015): 445–68.

Procaccioli, Paolo. *Lettere scritte a Pietro Aretino*. Rome: Salerno, 2003–2004.

Prosperi, Adriano. *Tra evangelismo e controriforma: Gian Matteo Giberti (1495–1543)*. Rome: Edizioni di storia e letteratura, 1969.

Reynolds, Anne. "Cardinal Oliviero Carafa and the Early Cinquecento Tradition of the Feast of Pasquino." *Humanistica Lovaniensia 34* (1985): 178–208.

Richardson, Brian. *Printing, Writers and Readers in Renaissance Italy*. Cambridge: Cambridge University Press, 1999.

Roskill, Mark W., trans. *Dolce's Aretino and Venetian Art Theory of the Cinquecento*. New York: New York University Press, 1968.
Sowerby, Tracey A. and Jan Hennings, eds. *Practices of Diplomacy in the Early Modern World c. 1410–1800*. London: Routledge, 2017.
Talvacchia, Bette. *Taking Positions: On the Erotic in Renaissance Culture*. Princeton, NJ: Princeton University Press, 1999.
White, Hayden. "The Question of Narrative in Contemporary Historical Theory." *History and Theory 23* (1984): 1–33.
Wraxall, Nathanial William. *The History of France, Under the Kings of the Race of Valois*. 2 vols. London: McMillan, 1807.

9 The wars in Europe and the journeying play

Thomas Drue's *The Duchess of Suffolk* (1624)

David Nicol

In Thomas Drue's play, *The Duchess of Suffolk* (1624), the eponymous heroine flees England in search of safety from persecuting Catholics during the reign of Queen Mary. After a flight through the streets of London and a narrow escape in Essex, she arrives at the seaport of Emden in East Frisia, accompanied by her husband Richard Bertie, their children, and an entourage of servants.[1] But the town is full of enemies, so the Duchess must immediately move on and Bertie must pretend not to know her. Before she leaves, the Duchess utters a speech that seems directed more toward the audience than to anyone onstage:

> You people happy in a land of peace,
> That joy your consciences with the world's increase,
> Look with indifference into my sad life.
> Here my poor husband dares not know his wife,
> And I a princess, to avoid like danger,
> Must use my own dear husband as a stranger.
>
> (3.2.93–98)

Touched by her fortitude, the Duchess's servant Cranwell asks "what pale trembling coward's heart would faint, / To wade through danger with so pure a saint?" (3.2.106–07). On the surface, this sequence simply asks its 1624 audience to admire and feel sympathy for an exiled Protestant heroine of the 1550s. However, its disdain for those who would not "wade through danger" for such a lady links the play to debates surrounding more contemporary conflicts in Europe. Studies of *The Duchess of Suffolk* have identified a network of allusions designed to ensure that the plight of the Duchess and Bertie would have recalled another noble couple: Elizabeth Stuart, daughter of King James, and her husband Frederick, Elector of the Palatinate and King of Bohemia.[2] The couple had fled Prague in 1620 when the Holy Roman Empire invaded both Frederick's homeland of the Palatinate and his recently acquired kingdom of Bohemia; by the time of the play's licensing on 2 January 1624, the exiled pair were living in The Hague under the protection of Prince Maurice of Orange, and Frederick was fighting a losing war to reclaim his lands. All this time, James had refused Frederick any significant military assistance.[3] If the character of the Duchess is indeed designed to evoke Elizabeth of Bohemia, the latter would thus appear in this speech to be accusing the English audience of looking upon her with "indifference" from the safety of their "land of peace".[4] And since in practice there was no such indifference in London – outrage at James's pacific response to the "Bohemian Crisis" was frequently expressed in pamphlets, in the pulpit, and in parliament – the speech is by implication directed more toward King James himself.[5] Does that mean, then, that this speech about journeying through a land of peace is in fact a call to arms?

Most commentators on the play seem to think so, as *The Duchess of Suffolk* is often discussed alongside the writings of the militant pamphleteers of the 1620s. But demands for war were not the only way of contesting King James's response to the Bohemian crisis, and *The Duchess of Suffolk* is not so bellicose as it may seem. Although it evokes great sympathy for Elizabeth of Bohemia and arouses hopes for her eventual rescue, it nonetheless resists the opportunity to depict heroic Protestant militarism and instead expresses a remarkably hesitant and doubtful attitude toward violence. This pacific side to the play has been overlooked because critics have been more interested in its brief topical allusions than in its overall structure. Marsha S. Robinson rightly observes that the play's content is "the stuff of romance", being "a series of adventures" in which our heroes "are repeatedly rescued from the jaws of death".[6] However, the tendency of the few scholarly studies of *The Duchess of Suffolk* to categorize it as a "history play" reflects a general indifference to the invented scenes of action and adventure that make up most of its length.[7] Jerzy Limon's important study, for example, dismisses the bulk of the play as repetitive melodrama: "having escaped ... from one source of oppression the fugitives immediately fall prey to another, equally perilous. And this continues, practically without interval, for the duration of the middle three acts".[8] I propose that to ignore these scenes of adventure is to overlook one of the play's genres, which is that of the episodic "journeying play".[9] In his study of early modern plays about voyages, Peter Holland brings together examples such as Shakespeare's *Pericles* (1606–1608), John Day, William Rowley and George Wilkins's *The Travels of the Three English Brothers* (1607), and Thomas Heywood's *The Fair Maid of the West Part 1* (*c.* 1603), to examine the "dramaturgical difficulty" inherent in staging journeys through multiple locations: their repetitive structure of travel to place after place, with an adventure in each, risks "shapeless indeterminacy".[10] Holland proposes that the more successful of these plays use the repetitive structure as a tool for exploring multiple dimensions of a theme. Thus, *The Travels of the Three English Brothers* is structured as a series of encounters between different cultural groups who hold different attitudes toward the taking of prisoners, and *The Fair Maid of the West* is structured around a series of tests of its heroine's chastity.[11] Similarly, Suzanne Gossett observes that, in *Pericles*, each location dramatizes a different kind of father-daughter relationship.[12] That there could be a political dimension to this genre is evident in Claire Jowitt's work on *The Fair Maid of the West Part 1*, which sees the stages of the heroine's journey as allegorical representations of the problems and limits of Elizabeth I's rule.[13]

The Duchess of Suffolk follows a similar format as its heroine travels through London and thence to Leigh, over the sea to Emden, inland to Wesel, and ultimately to sanctuary in the Palatinate. Fights, deaths, and other such incidents occur in each episode, but although the play is clearly written for an audience who enjoyed onstage violence, it is remarkably sceptical of the *need* for the violence that it depicts. Viewing *The Duchess of Suffolk* as a journeying play reveals that, as in the plays described above, there is an underlying structure to the repeated rescues that the Duchess undergoes. That structure is a pacifist one, for, as the Duchess's apparent need for violent aid builds in intensity the closer she gets to the Palatinate, each rescue questions the value of violence in different ways. As we will see, Drue's use of this structure illustrates that support for Elizabeth of Bohemia did not inherently require a bellicose stance in the 1620s, and it links the play instead to the more pacific forms of support that existed at the time.

Drue's play tells the story of Katherine Willoughby, Duchess of Suffolk, ana historical figure whose story was familiar to English audiences thanks to its retelling in John Foxe's Acts and Monuments and in a ballad by Thomas Deloney.[14] According to these tales, the Duchess travelled with her husband Bertie across Europe in search of refuge from the persecutions of Protestants during the reign of Queen Mary. She endured a series of incidents, including extremities of weather and giving birth in a church porch, before ultimately accepting an offer of sanctuary from the King of Poland. She then returned to England upon the ascension of Queen Elizabeth. In his play, Drue invents additional adventures for the Duchess but follows the essence of Foxe's and Deloney's narratives. On the surface, then, the story is a straightforward tale of Protestant virtue rewarded and would seem unlikely to offend political sensibilities in Jacobean England. However, a note in the office-book of Sir Henry Herbert, Master of the Revels, reveals that it in fact required censorship before it could be licensed for performance:

> *For the Palsgraves company. January*
>
> The history of the Dutchess of Suffolk by M^{r}. <Drewe being full of dange>rous matter was much reformed & for my <paines this 2^{d} Jan. 1623 [*i.e.,* 1624, N.S.] I had 2li.>[15]

Herbert "much reformed" the "dangerous matter" that he found in the play; he did it to such an extent that he charged twice his usual fee.[16] The extant text presumably represents the aftermath of Herbert's reformations, since it was he who licensed it for printing a few months later, although it ultimately remained unpublished until 1631.[17] The specifics of what Herbert disliked and the nature of his alterations are unknown, but scholars have detected within the extant text clues to a likely explanation: Herbert recognized that Drue had retold the Duchess's story in such a way as to draw parallels with the present-day situation of Elizabeth of Bohemia on the continent.[18] Why we can still recognize these parallels despite Herbert's "reformations" is unclear, but one explanation is that he removed only explicit statements and was less concerned about mere inferences.[19]

Parallels between the Duchess and Elizabeth could be "dangerous" because to draw them was to intervene in ongoing debates about King James's foreign policy. Elizabeth had married Frederick, the Elector Palatine, in 1613, leaving England for his seat at Heidelberg. In 1618–19, Frederick accepted the offer of the Bohemian crown in the aftermath of a Protestant revolution in that country, and he and Elizabeth were duly crowned King and Queen in Prague. The decision to offer Frederick the crown was a momentous one, as it marked a significant shift in the balance between Protestant and Catholic power in Europe.[20] The ultimate result was the invasion of both Bohemia and the Palatinate in 1620 by armies of the Holy Roman Empire, forcing Frederick and Elizabeth into exile.[21] This "Bohemian Crisis" spawned a network of conflicts that would together become known as the Thirty Years' War.[22] Frederick gathered supporters in Germany and fought a gruelling, losing battle against imperial forces to try to regain the Palatinate.[23] His efforts had a wider significance: in England, Frederick's struggle was seen by militant Protestants of all ranks as a crisis point in a contest between Catholicism and Protestantism on the continent that would need to be resolved with a holy war.[24] King James, however, had never approved of Frederick's acceptance of the Bohemian crown, disliking not only its unsettling of the balance of power but also the very idea of an elective monarchy.[25] He thus resisted significant military engagement and pursued diplomatic solutions instead, most notably the project of achieving rapprochement with the Catholic powers by marrying his son Charles to the Infanta of Spain.[26]

The plight of Elizabeth became a rallying point for those advocating English participation in the war. One MP told Parliament in 1621, "our King's only daughter hath nowhere to rest her head of her own ... Who can listen to this and not be ready to give [to the war effort]?"[27] The anonymous author of the 1622 pamphlet *Tom Tell-Troath* claimed to be relaying the typical opinions of the common people when he warned King James that they "will hardlie be perswaded that you are their Father; because, they see, the lamentable Estate, whereto you suffer Thinges to run, comes nearer to Destruction than the Nature of fatherly Correction"; if James wished to avoid the "Scorn and Opprobry of the Nations of the Earth" he must "send over an Armye of good *English*" who would adhere to their mission "as much out of Affection, as Obligation to the Cause, and Persons of your Children".[28] In a series of proclamations from 1620 to 1622, James responded to such criticism of his policies with demands that commoners refrain from engaging in discussion of "matters of state", but covert discussions and publications continued nonetheless.[29]

Drue's play was written in late 1623, a time when the topic of war could be addressed more openly due to shifts in the mood of the court. On October 5, King James's plan for a Spanish match collapsed when Prince Charles returned from Madrid without a bride and was greeted by mass public celebrations. As Thomas Cogswell has described, a sea-change took place in the government: Charles and Buckingham began actively to advocate for war with Spain, and even James reluctantly considered it.[30] In this new context, a flurry of pro-war pamphlets appeared in 1624.[31] *The Duchess of Suffolk* is one of a group of plays that appear to echo this revitalization of the pro-war movement, the others being Philip Massinger's *The Bondman* (December 1623) and Thomas Middleton's *A Game at Chess* (August 1624). Specifically, Drue appears to echo those pamphleteers who continued to use Elizabeth's plight as a motivating factor. In one such pamphlet, John Reynolds directly addressed the King:

> Let your Courage but animate your Designes, ... and your Nobilitie and Gentrie (out of theyr true zeale and innate affection, to the famous Princesse your Daughter, her Husband and Posteritie) will flye from *Thames* to *Rhyne*, as to a Fayre, or Wedding ... Soe your Majestie shall undoubtedlie finde, that the rightest and shortest way for you to recover the Pallatynat will bee by your Sword. ... And noe sooner shall your Drummes beate, and your Coullers bee displayed uppon the banckes of *Rhyne*; but your Majesties sword shall put a newe face uppon *Germanie*, and make *England* (consequentlie) assume her olde one, which was ever woont to looke more Martiall, and lesse Effemynate.[32]

In creating a dramatic allegory of Elizabeth's struggles, the journeying play was the natural genre for Drue to use because it enabled him to utilize broad parallels between the journeys of the Duchess and Bertie and those of Elizabeth and Frederick. Both suffered calamity at the hands of Catholics opposed to them because of their Protestant beliefs; both went on the run in central Europe; both suffered humiliation and the loss of their riches and were at the mercy of strangers; both women were pregnant and gave birth in unfavourable conditions; and both eventually found refuge thanks to a generous nobleman who had been a former suitor: the King of Poland in the Duchess's case, Prince Maurice in Elizabeth's.[33] Scholars of the play have identified several ways in which Drue alters the historical accounts of the Duchess in order to include allusive words and images, such as the resonant word "Palsgrave", which encourage his audience to view the play with current events in mind and thus make the parallels between the Duchess and Elizabeth more likely to be noticed.[34]

Drue's broad purpose in drawing these parallels is clear: he evokes sympathy for Elizabeth via the tear-jerking story of the Duchess and then portrays her rescue by an English monarch. Pity is aroused most strongly during Bertie's lament when the Duchess is forced to give birth in a church porch during a rainstorm:

Alas, alas, this is a homely place,
To bring a princess of such state to bed.
A wide church-porch is made her bedchamber,
And the cold stones her couch. Here are no curtains,
But the bleak winds, cold clouds, and storms of hail,
And they begirt her round. Heaven, for thy mercy,
This poor distressed princess shield and save,
Whose cold head lies upon some dead man's grave.

(3.3.47–54)

The audience members who perceived the parallels, perhaps encouraged by the repeated reference to the Duchess as a "princess", might have applied their emotional response to Elizabeth, intensifying it with the knowledge that the living woman had suffered similar things only recently and could still be restored to her former position. Her possible restoration would have then been imagined when, in the last act, a messenger announces that Queen Mary is dead and that her successor, Queen Elizabeth I, will welcome the Duchess home:

My sovereign, hearing that the Lady Katherine,
The Suffolk duchess, her ally in blood,
Did live obscurely in these provinces,
In want, in misery, and great distress,
Sends to repeal both her, and all her friends.

(5.1.97–101)

Elizabeth I's swift aid to her "ally in blood" implicitly contrasts with King James's indecision over aiding his own daughter. In the political climate of the 1620s, this ending, though happy, had the potential to provoke anger in the audience watching the play as they registered the parallels between support for the Duchess in comparison to the lack of support for Elizabeth of Bohemia.

Given this context, it is easy to understand why scholars tend to describe *The Duchess of Suffolk* in the same breath as militant Protestant demands for war on the continent. Limon calls it a "severe criticism of James I's foreign policy and one of the voices in the widespread campaign for the salvation of endangered Protestantism and its new martyrs".[35] This leads him to group the play with others that he regards as part of "a propaganda campaign" and with belligerent works such as *Tom Tell-Troath*.[36] Similarly, Julia Gasper discusses the play alongside "earlier militant Protestant plays" that also draw on Foxe, and Marsha S. Robinson relates it to Thomas Scott's *Vox Coeli*, another work that uses historical English Protestants to urge James to "intervene in the Palatine cause".[37] For Hans Werner, the play makes Elizabeth an "emblem of an apocalyptic struggle against the forces of international catholicism", while Tristan Marshall sees it as part of a trend in the drama toward "favouring war with Spain" and

Dutton and Galbraith see it as "little short of an open invitation to the English to fight on behalf of their exiled princess".[38]

However, while the play's sympathy for Elizabeth is undeniable, to see it as advocating a military response is to misread its tone, as the structure of Drue's travel romance questions the value of violence at every stage of the journey. The Duchess and her household experience many adventures on their travels, and they are chased by persecutors who want to burn the Duchess at the stake. Yet the only unquestionably useful acts of violence are performed as slapstick comedy: in Leigh, the Duchess's servant Fox pushes the Bishop Bonner down a well, which results in the pyromaniacal persecutor crying as he is rescued, "I am almost drowned. Oh lead me to some fire" (3.1.148). Fox later defeats another pursuer by sawing off the tree branch on which his adversary is sitting (4.4.55 71). In most of the play's scenes, though, violence is taken far more seriously: swords are drawn and death is a possibility, and the threats mount in intensity as the characters draw closer to the Palatinate. Throughout these scenes, a sceptical approach to violence emerges, in which qualms and doubts about its value are expressed.

The first serious encounter with violence occurs when, as they leave Emden and travel through wilderness toward the town of Xanten, the Duchess's entourage is set upon by thieves who attempt to rob and rape her. Bertie saves the day by disguising as a thief, luring most of the bandits away with a story about treasure, and then fighting the one who remains. Violence thus saves the Duchess, but it is surprising how restrained Bertie is: despite the vicious attacks on the Duchess, he holds back, refusing to "stain [his] hands" with the thieves' blood, and instead merely ties him up and rolls him into a ditch (3.2.150–52). The honourable behaviour of Bertie is contrasted with the dishonourable behaviour of the bandits, emphasizing the desirability of resolving conflict through nonviolent means.

The value of restraint in violence is then demonstrated in the next sequence. Arriving in the Rhine port of Wesel during a storm, the Duchess and Bertie can find no shelter, which results in the Duchess giving birth in a church porch. A ballad by Thomas Deloney is the source for what happens next: a sexton arrives and tries to evict the couple, but, instead, Bertie beats him. Drue follows this sequence but makes significant changes and expansions. In the ballad, the sexton is drunk and, "void of shame" starts "spuming forth" at the Duchess, and thus is apparently deserving of his beating.[39] In the play, the incident arises instead out of miscommunication: the sexton (speaking stage Dutch that is easily understandable to the audience but not to Bertie) thinks they are peddlers, but Bertie angrily tells him "what thou speakest I know not" and beats him for being an "uncivil fellow" (3.3.60). The sexton runs away and raises the alarm, so that Bertie's impulsive violence, which contravenes his earlier qualms, has worsened the Duchess's plight. The situation is saved only by the arrival of none other than Erasmus, who just happens to be passing by. This brief appearance by the famous scholar is, perhaps, one of Drue's more eccentric elaborations on his sources, but it has some thematic relevance. Although he remained a Catholic, Erasmus's translation of the New Testament proved influential to Protestants and he held friendships across national and confessional divides.[40] Erasmus and Bertie communicate in Latin and arrange help for the Duchess. The salvation of the Duchess is thus not Bertie's violent defence of her, but rather a humanist internationalism based upon reason that permits rescue to be found across religious and national divides.

The peaceful town of Wesel gives the Duchess "time / To breath a space" in "a quiet port" (4.1.10–11), but Drue quickly shatters this idyll by introducing a new and greater threat: the Duke of Brunswick and his army. The threat of soldiers introduces for the first time the imagery of European war, not only raising the stakes but also alluding to the concerns of 1623–1624. Yet in the ensuing sequence of the Duchess's escape from Wesel (an addition to his sources), Drue dwells on European conflict more in sadness than in belligerence. The Duchess and her entourage escape by disguising as a funeral procession, with the Duchess and her newborn baby hiding inside a coffin. Perecell, a helpful Walloon who plans the escape, explains that they will be imitating an actual funeral that is about to take place for a man who "late returned from the wars, / And falling sick, within the city here, / Died some few days since", and whose brother wants him "buried / Amongst his ancestors" (4.1.99–103). Even though the dead man's backstory is irrelevant to the plot, Drue explores it further when the Duke of Brunswick and his soldiers watch the false funeral go by: the Portreeve of the town identifies the dead man as "young Vandermast":

BRUNSWICK: What was this Vandermast? A citizen?
PORTREEVE: No, my good Lord, he was a soldier,

A proper gentleman, and one had served
The emperor and others in their wars,
Yet was ordained to end his days at home.

(4.2.45–49)

In this reference to a soldier of the Holy Roman Empire, Drue alludes once again to the contemporary wars in Europe. But even though Vandermast was a warrior for the Catholic enemy, Drue does not celebrate his death; instead, the death of this "proper gentleman" (4.2.47) whose family only want him buried "whereas his ancestors are all entombed" (4.2.40) is described with solemnity, implying that ending one's days at home, rather than on foreign fields, is preferable.

The image of armies marching across Europe intensifies as Brunswick and his men go in pursuit of the Duchess upon realizing she has escaped. She and Bertie arrive near Weinheim in the Palatinate, home to the Duchess's former suitor, the Palsgrave (who is also King of Poland in Drue's re-working of history).[41] But the Duchess and Bertie do not know where they are, and the soldiers of the Palsgrave do not know who they are, so violence breaks out again, this time to a much greater degree than previously. Bertie angrily strikes a soldier and later kills one offstage (4.4.47–49, 70), and, remarkably, the Duchess herself joins in: "farewell, woman-weakness, welcome sword", she announces, "For once I'll play the man, to save my lord", and "*She fights, beats them off*" (4.4.51–52). In this sequence, Drue appears finally to evoke unequivocally the fantasies of the militant Protestants: the Duchess's wielding of a sword within the Palatinate is an obvious allusion to Elizabeth of Bohemia's status as rallying point for pro-war voices.[42] It also gestures to the popular notion that Elizabeth of Bohemia bore the qualities of her namesake Elizabeth I; as Reynolds put it, the Queen of Bohemia "inheriteth the Name and Vertues, the Majestie and generositie of our Immortall Queen *Elizabeth*".[43] Yet even as he conjures this English Amazon, Drue continues to resist glamorization of the battle. Bertie's reaction to having killed an enemy is not triumph but paralyzing horror: he fears for his soul, saying "If he be slain, require not heaven his blood / Of miserable Bertie", and when the Duchess tries to get him to move, he declares he is "so amazed [he knows] not where to go". Bertie is not physically wounded; rather, it is his conscience that immobilizes him. The Duchess has to help him hide in a hayloft as he complains that "[his] grief's so great / It weighs [him] down, and [he] must needs retreat" (4.4.76–77, 85–86, 96–97). And when the soldiers and local justices arrive, and the Duchess draws her sword again (4.4.104–06), Bertie defuses the situation by acknowledging his guilt and requesting only the "due process of the law" (4.4.134). As he does so, the Duchess realizes that while they were distracted, a "savage hand hath taken [her] children hence, / Whilst [they] were busied in this luckless brawl" (4.4.151–53); her taking up the sword has distracted her from maternal care. All through this sequence, then, violence produces guilt, remorse, and disaster even after it apparently saves the couple.

The play does not, of course, end in catastrophe for the protagonists; when the Palsgrave recognizes the Duchess, he forgives the killing of the soldier and the couple's children are then found safe. But when the Duke of Brunswick arrives, the spectre of European war reappears. The Palsgrave warns Brunswick that he is prepared to "sacrifice ten thousand lives" before he will allow the "religious souls" of Bertie and the Duchess to die, warning him to leave or be "withstood with fire and sword" (5.1.64–65, 75). Brunswick accuses him of "disobedience" and warns that the Church will be unhappy, but the Palsgrave ignores him and ennobles those whom he has rescued (5.1.78–86). In a militant pamphlet, imagery such as this – a Protestant Palsgrave threatening fire and sword to a Catholic prince – might have looked triumphalist, but given Drue's handling of the previous moments, this sequence should perhaps be read instead as a terrifying moment at which

Europe teeters on the brink of a conflict that will involve countless deaths. It is a joyous moment then when, just as war appears inevitable, the news arrives that Elizabeth I is now queen and wishes the Duchess and Bertie to come home. The Duchess attributes this final rescue not to warriors but to "the director of all human lives" and his "gracious providence" (5.1.108, 118).

The tone of the Duchess's "dangerous pilgrimage" (4.4.12) to salvation is thus very different from that of bellicose writers such as Reynolds. Every moment of serious violence is hedged with doubt, caution, regret, and guilt, and the rescues of the Duchess come more from the kindness of strangers of many faiths than they do from battle. Even when Elizabeth I brings home the Duchess at the end, she does do so not by the threat of war but rather by a simple exertion of her will, as if the playwright's wish is not so much for James to take up the sword as for him to somehow acquire the effortless authority for which Queen Elizabeth was nostalgically famed.[44]

To link *The Duchess of Suffolk* purely to the militant wing of English Protestantism is, then, to misrepresent it and, indeed, the multiplicity of responses to the Bohemian Crisis. The play offers its own distinctive take on the situation, one that is different again from the anti-war voices that did exist in print from 1621–3. Some strategies adopted by those writers involve repeating James's use of the maxim *dulce bellum inexpertis* ("war is sweet to those who have never experienced it") to criticize those who demanded war without ever having experienced it, stressing the horrific effects of war upon civilians, and simply criticizing any questioning of royal policy.[45] The play achieves its effect quite differently: it instead evokes disquiet about violence by stressing its inadequacy in aiding the protagonists and its negative impact on the consciences of its perpetrators. Drue's combination of evoking sympathy for Elizabeth and Frederick while also expressing reluctance and distaste for violence demonstrates that there were positions available beyond the divide between fulminating militancy or obedient pacifism. *The Duchess of Suffolk* needs to be recognised as a play that is not a call for war; it is rather a fantasy of rescuing the princess without it.

The Duchess of Suffolk's travels end in a hopeful peace, and Drue's attitude toward the conflicts that his play alludes to is communicated through his use of the "journeying play" structure. The Duchess's rescues may be repetitive, but each rescue questions in different ways the necessity for violent solutions. The episodes of violence express the importance of restraint; the value of international and interdenominational cooperation; the sadness and the horrors of inter-faith war in Europe; and the trauma of taking life, even of one's enemies. Just as England itself was considering a journey toward conflict in the Palatinate, the play dramatizes that movement but makes both travel and conflict seem far less appealing than peace at home in England. Meanwhile, the journey that Elizabeth of Bohemia herself took was a far more complex and indeterminate one, and the horrors that it was associated with lasted longer than any writer of fiction could have imagined.

Notes

1 See Thomas Drue, *The Duchess of Suffolk*, ed. Richard Dutton and Steven K. Galbraith (Columbus: Ohio State University Press, 2015), 3.2.77. Act, scene and line references refer to this edition.
2 For the most detailed analysis of these allusions, see Jerzy Limon, *Dangerous Matter: English Drama and Politics 1623–1624* (Cambridge: Cambridge University Press, 1986), 40–61; for a summary, see Dutton and Galbraith, 9–13.
3 For a summary of these events, see Nadine Akkerman, ed., *The Correspondence of Elizabeth Stuart, Queen of Bohemia, Volume 1: 1603–1631* (Oxford: Oxford University Press, 2015), 27–44.
4 Dutton and Galbraith, 99 n. 40.
5 For an overview, see Thomas Cogswell, *The Blessed Revolution: English Politics and the Coming of War, 1621-1624* (Cambridge: Cambridge University Press, 1989), 12–20.
6 Marsha S. Robinson, *Writing the Reformation: Actes and Monuments and the Jacobean History Play* (Aldershot: Ashgate, 2002), 45–46.
7 Irving Ribner, *The English History Play in the Age of Shakespeare* (Princeton: Princeton University Press, 1957), 293–95; Judith Doolin Spikes, "The Jacobean History Play and the Myth of the Elect Nation," *Renaissance Drama* 8 (1977): 117–49 (143–44); Limon, 59; G. K., Hunter, *English Drama, 1586-1642: The Age of Shakespeare* (Oxford: Clarendon Press, 1997), 273; Robinson, 1.
8 Limon, 44–45; see also Hunter 277–8. An exception is Robinson 45–46, who studies the Providential language in the Duchess's escapes.
9 This term is used by Claire Jowitt and David McInnis to encompass not only "travel plays", which are "based on a particular documented voyage", but also the "voyage plays" that depict fictional journeys of all kinds. As Jowitt and McInnis note, the term is useful because "it is perfectly possible for a play to be simultaneously both a travel and a voyage drama" if it combines historical accounts of a voyage with fictional material; this is exactly what Drue does in *The Duchess of Suffolk*. See "Introduction: Understanding the Early Modern Journeying Play," in *Travel and Drama in Early Modern England: The Journeying Play*, ed. Claire Jowitt and David McInnis (Cambridge: Cambridge University Press, 2018), 1 20 (2–3).
10 Peter Holland, "'Travelling Hopefully': the Dramatic Form of Journeys in English Renaissance Drama," in *Travel and Drama in Shakespeare's Time*, eds. Jean-Pierre Maquerlot and Michèle Willems (Cambridge: Cambridge University Press, 1996), 160–78 (160–61, 168).
11 *Ibid.*, 168–69, 171.
12 See William Shakespeare, *Pericles*, ed. Suzanne Gossett (London: Arden Shakespeare, 2004), 84–85.
13 Claire Jowitt, *Voyage Drama and Gender Politics, 1589–1642: Real and Imagined Worlds* (Manchester: Manchester University Press, 2003), 40–47.
14 See Dutton and Galbraith's introduction, 5–9.
15 N. W. Bawcutt, *The Control and Censorship of Caroline Drama: The Records of Sir Henry Herbert, Master of the Revels 1623–73* (Oxford: Clarendon Press, 1996), 148.
16 Dutton and Galbraith, 13.
17 Janet Clare,*"Art Made Tongue-Tied by Authority": Elizabethan and Jacobean Dramatic Censorship* (Manchester: Manchester University Press, 1990), 189.
18 This idea was first proposed by Judith Doolin Spikes in 1977 (143, n. 38) and elaborated on by Margot Heinemann in *Puritanism and Theatre: Thomas Middleton and Opposition Drama under the Early Stuarts* (Cambridge: Cambridge University Press, 1980), 205–06. It was then developed in far greater

detail by Jerzy Limon (40–61). Dutton and Galbraith agree with Limon's interpretation (9–13).

19 Clare, 190; Bawcutt, 71; Dutton and Galbraith, 12–13.

20 Geoffrey Parker, *The Thirty Years' War* (London: Routledge & Kegan Paul, 1984), 52–55; Geoff Mortimer, *The Origins of the Thirty Years War and the Revolt in Bohemia, 1618* (Basingstoke: Palgrave Macmillan, 2015), 171–72.

21 Akkerman, 27–36.

22 Peter H. Wilson, *Europe's Tragedy: A New History of the Thirty Years War* (London: Penguin, 2010), 8–9.

23 Wilson, 314–47; Akkerman 38–43.

24 Marvin A. Breslow, *A Mirror of England: English Puritan Views of Foreign Nations, 1618–1640* (Cambridge, Mass: Harvard University Press, 1970), 12–15; Cogswell, 16–17.

25 W. B. Patterson, *King James VI and I and the Reunion of Christendom* (Cambridge: Cambridge University Press, 1997), 303; Brennan C. Pursell, *The Winter King: Frederick V of the Palatinate and the Coming of the Thirty Years' War* (Aldershot: Ashgate, 2003), 80–81.

26 Cogswell, 16–20; Patterson, 305–14.

27 Wallace Notestein, Frances Helen Relf and Hartley Simpson, eds. *Commons Debates 1621*, 7 vols. (New Haven: Yale University Press, 1935), 2:86; see also Robert Zaller, *The Parliament of 1621: A Study in Constitutional Conflict* (Berkeley: University of California Press, 1971), 47.

28 Anon, "Tom Tell-Troath: Or, a Free Discourse Touching the Manners of the Time. Directed to His Majestie by Way of Humble Advertisement," in *The Harleian Miscellany*, ed. Thomas Osborne, 5 vols. (London: Thomas Osborne, 1744), 2:400–18 (2:401, 406 and 408).

29 Cogswell 20–21, 32–34; Cyndia Susan Clegg, *Press Censorship in Jacobean England* (Cambridge: Cambridge University Press, 20015), 7–9 and 185–87.

30 Cogswell, *passim*; Patterson, 345–48.

31 Cogswell. 289–90, 294–99.

32 John Reynolds, *Votivae Angliae* (Utrecht, 1624), D4r-E1r.

33 Limon, 53–54.

34 On "Palsgrave", see Limon, 60; Clare, 190; Dutton and Galbraith 11–12. For other allusions, see Limon, 5–5.

35 Limon, 41.

36 *Ibid.*, 131; 51 and 55.

37 Julia Gasper, *The Dragon and the Dove: The Plays of Thomas Dekker* (Oxford: Clarendon Press, 1990), 143; Robinson, 158.

38 Hans Werner, "*The Hector of Germanie, or The Palsgrave, Prime Elector* and Anglo-German Relations of Early Stuart England: The View from the Popular Stage", in *The Stuart Court and Europe: Essays in Politics and Political Culture*, ed. Malcolm Smuts (Cambridge: Cambridge University Press, 1996), 113–32 (115); Tristan Marshall, *Theatre and Empire: Great Britain on the London Stages Under James VI and I* (Manchester: Manchester University Press, 2000), 178; Dutton and Galbraith, 11.

39 Thomas Deloney, "The Duchess of Suffolk's Calamity", lines 117–19. Line references are to the edited text of the ballad in Dutton and Galbraith, Appendix B.

40 Diarmaid MacCullouch, *Reformation: Europe's House Divided, 1490–1700* (London: Penguin, 2004), esp. 33–34, 83, 99–101, 151.

41 Limon, 60.

42 Dutton and Galbraith. 11, 133 n. 72.

43 Reynolds, D3r. For other examples, see Breslow, 37–38 and Cogswell, 95–96.

44 For a brief discussion of Elizabethanism on the Jacobean and Caroline stage, see Martin Butler, *Theatre and Crisis, 1632–1642* (Cambridge: Cambridge University Press, 1984), 198–210. On the positive and negative afterlives of Elizabeth, see John Watkins, *Representing Elizabeth in Stuart England: Literature, History, Sovereignty* (Cambridge: Cambridge University Press, 2002) and Julia Walker "Bones of Contention: Posthumous Images of Elizabeth and Stuart Politics", in *Dissing Elizabeth: Negative Images of Gloriana*, ed. Julia Walker (Durham and London: Duke University Press, 1998), 252–76.

45 Cogswell, 27–28, 30–31.

Bibliography

Akkerman, Nadine, ed. *The Correspondence of Elizabeth Stuart, Queen of Bohemia, Volume 1: 1603–1631*. Oxford: Oxford University Press, 2015.

Anon. "Tom Tell-Troath: Or, a Free Discourse Touching the Manners of the Time. Directed to His Majestie by Way of Humble Advertisement." In *The Harleian Miscellany*. Edited by Thomas Osborne, 2:400–418. London: Thomas Osborne, 1744.

Bawcutt, N. W. *The Control and Censorship of Caroline Drama: The Records of Sir Henry Herbert, Master of the Revels 1623–73*. Oxford: Clarendon Press, 1996.

Breslow, Marvin A. *A Mirror of England: English Puritan Views of Foreign Nations, 1618–1640*. Cambridge, MA: Harvard University Press, 1970.

Butler, Martin. *Theatre and Crisis, 1632–1642*. Cambridge: Cambridge University Press, 1984.

Clare, Janet. *"Art Made Tongue-Tied by Authority": Elizabethan and Jacobean Dramatic Censorship*. Manchester: Manchester University Press, 1990.

Clegg, Cyndia Susan. *Press Censorship in Jacobean England*. Cambridge: Cambridge University Press, 2001.

Cogswell, Thomas. *The Blessed Revolution: English Politics and the Coming of War, 1621–1624*. Cambridge: Cambridge University Press, 1989.

Dutton, Richard and Steven K. Galbraith, eds. *The Duchess of Suffolk*. Columbus: Ohio State University Press, 2015.

Gasper, Julia. *The Dragon and the Dove: The Plays of Thomas Dekker*. Oxford: Clarendon Press, 1990.

Gossett, Suzanne, ed. *Pericles*. London: Arden Shakespeare, 2004.

Heinemann, Margot. *Puritanism and Theatre: Thomas Middleton and Opposition Drama under the Early Stuarts*. Cambridge: Cambridge University Press, 1980.

Holland, Peter. "'Travelling Hopefully': the Dramatic Form of Journeys in English Renaissance Drama." In *Travel and Drama in Shakespeare's Time*. Edited by Jean-Pierre Maquerlot and Michèle Willems, 160–178. Cambridge: Cambridge University Press, 1996.

Hunter, G.K. *English Drama, 1586–1642: The Age of Shakespeare*. Oxford: Clarendon Press, 1997.

Jowitt, Claire. *Voyage Drama and Gender Politics, 1589–1642: Real and Imagined Worlds*. Manchester: Manchester University Press, 2003.

Jowitt, Claire and David McInnis, "Introduction: Understanding the Early Modern Journeying Play." In *Travel and Drama in Early Modern England: The Journeying Play*. Edited by Claire Jowitt and David McInnis, 1–20. Cambridge: Cambridge University Press, 2018.

Limon, Jerzy. *Dangerous Matter: English Drama and Politics 1623–1624*. Cambridge: Cambridge University Press, 1986.
Marshall, Tristan. *Theatre and Empire: Great Britain on the London Stages Under James VI and I*. Manchester: Manchester University Press, 2000.
Mortimer, Geoff. *The Origins of the Thirty Years War and the Revolt in Bohemia, 1618*. Basingstoke: Palgrave Macmillan, 2015.
Notestein, Wallace, Frances Helen Relf and Hartley Simpson, eds. *Commons Debates 1621*. 7 vols. New Haven: Yale University Press, 1935.
Parker, Geoffrey. *The Thirty Years' War*. London: Routledge and Kegan Paul, 1984.
Patterson, W.B. *King James VI and I and the Reunion of Christendom*. Cambridge: Cambridge University Press, 1997.
Pursell, Brennan C. *The Winter King: Frederick V of the Palatinate and the Coming of the Thirty Years' War*. Aldershot: Ashgate, 2003.
Reynolds, John. *Votivae Angliae*. Utrecht, 1624.
Ribner, Irving. *The English History Play in the Age of Shakespeare*. Princeton: Princeton University Press, 1957.
Robinson, Marsha S. *Writing the Reformation: Actes and Monuments and the Jacobean History Play*. Aldershot: Ashgate, 2002.
Spikes, Judith Doolin. "The Jacobean History Play and the Myth of the Elect Nation." *Renaissance Drama 8* (1977): 117–149.
Walker, Julia. "Bones of Contention: Posthumous Images of Elizabeth and Stuart Politics." In *Dissing Elizabeth: Negative Images of Gloriana*. Edited by Julia Walker, 252–276. Durham and London: Duke University Press, 1998.
Watkins, John. *Representing Elizabeth in Stuart England: Literature, History, Sovereignty*. Cambridge: Cambridge University Press, 2002.
Werner, Hans. "*The Hector of Germanie, or The Palsgrave, Prime Elector* and Anglo-German Relations of Early Stuart England: The View from the Popular Stage." In *The Stuart Court and Europe: Essays in Politics and Political Culture*. Edited by Malcolm Smuts, 113–132. Cambridge: Cambridge University Press, 1996.
Wilson, Peter H. *Europe's Tragedy: A New History of the Thirty Years War*. London: Penguin, 2010.
Zaller, Robert. *The Parliament of 1621: A Study in Constitutional Conflict*. Berkeley: University of California Press, 1971.

Part IV

The art of travel and imaginary journeys

10 *Ars apodemica* gendered

Female advice on travel

Gábor Gelléri

In early modern European culture, debates on the place of women are extensive; however, it was widely agreed that this place is not on the road. From biblical exegesis through mythological examples to physiological considerations, many fields of knowledge were mobilized to argue this point. The story of Dinah, Jacob's daughter, was regularly used as a justification to restrict women's access to the public sphere.[1] Genesis 34 tells us the story very briefly: Dinah went out of her father's camp to visit the women of Shechem and was subsequently raped; this led ultimately to the massacre of the town's menfolk by Jacob's sons. The biblical story itself doesn't comment on Dinah's behaviour any further – in fact, she doesn't even speak. It was possibly this minimal context that allowed a range of Bible commentators, starting with Saint Jerome and including major figures such as Luther and Calvin, to use this story as evidence of why women should not go abroad. There were slight differences in the interpretations (to what extent was Dinah to blame for her rape?), but they all rebuked women who "walked out to see" – a double condemnation of mobility and curiosity. The preferred counter-example was the Virgin Mary, who never left her house unaccompanied except when going to church.

The important proto-feminist figure, Arcangela Tarabotti (1604–1652), who was forced into a convent and powerfully protested against it, commented on the spectacular male misappropriation of a rape story to justify women's restriction to the domestic or monastic sphere.[2] However, this had little impact on the prevailing patriarchal interpretation. While we know that a fair number of women did roam early modern roads, these were usually cases of professional or forced mobility. It was particularly "useful travel" that was inaccessible to women, "the inter- and extra-continental travel associated with education...was ordinarily the privilege of a male elite".[3] As Erica Harth argues, well into the late early modern period, even the most educated women were excluded from a range of elements of the "educational apparatus", including formal schooling and educational voyages.[4]

On this basis, scholarship has challenged the very notion of female "travel". In her survey of female travels in Renaissance France, Collette H. Winn reminded us that what she discusses is not the same notion as the case of male forms of "voyage", and we do not see the usual connotations of conquest, exploration, or professional and intellectual training.[5] More recently, Bernadette Andrea questions the widespread use of the dichotomy between "empty" spatial *movement* and "meaningful" *travel*, the latter being almost exclusively male.[6] She suggests that this approach helps to perpetuate one of the most classic oppositions between the resident female and the itinerant male, that of Penelope vs. Ulysses – a conception in which any woman not conforming to this schema will inevitably be likened to the sorceress Circe. Her suggestion is to consider that a lack of "volition", that is, lack of a positive desire to travel, doesn't preclude travelling; thus, involuntary forms of mobility, such as exile or captivity, could and should also be considered "travels".

In this essay, I will study this conflict of interpretations about female mobility (both as movement and as discourse) from a hitherto understudied perspective. The focus will be on the role played by women within the culture of *ars apodemica*, the early modern genre of the "art of travel". *Ars apodemica* weaves together education and travel, two elements of the "educational apparatus" listed by Harth, but questions of mobility without volition are also present. Recognized first as a significant body of texts by Clare Howard, "apodemic" literature has witnessed increasing scholarly interest over the last decades, from early works by Justin Stagl and Normand Doiron, through a range of major studies by authors such as Joan-Pau Rubiés and Daniel Carey, to the launch of the "Art of Travel 1500–1850" database.[7] Its role not only in shaping the mindset of potential and actual travellers but also in the formation of scientific methodology is now widely acknowledged.[8] However, its "gendered" aspect is mostly ignored.

On the rare occasions when the question of women in *ars apodemica* is tackled in scholarship, it is to highlight that most authors of this genre were against the idea of women being allowed to travel. Occasionally, women appear in *ars apodemica* texts as objects: usually, they feature as sirens of sorts, exposing the traveller to idle curiosity and to the moral and physical danger of seeking dubious pleasures on the road (as represented by the "rich embroidery of Signora Livia's knickers" in Montaigne's ironic comments on the travelling noblemen's curiosities).[9] But one aspect that has been systematically ignored is that of women participating in apodemic culture as authors: women writing instructions for travelling, for both men and women.

Carl Thompson indicates a turn within scholarship on female travel writing, moving from the quest for "female" stylistic features towards questions of female agency mediated through travel writing.[10] I intend to tackle female apodemic production from a similar angle: to read it as an opportunity for female expression in this apparently unlikely format. Women penetrating these two related and almost exclusively male universes, travel (understood here as mobility for curiosity and education) and rhetoric (*ars apodemica* being a rhetorical exercise, performed by the educated, normally male, elite[11]), was bound to create conflicts on many fronts.

Gendered exclusions?

Ars apodemica is a form of normative discourse aimed at creating a secular, regulated form of travel; particularly during its early period in the sixteenth century, it was defined in direct opposition to pilgrimage. Within this corpus, the condemnation of female travel appeared more or less immediately. This fact prompted Carey to suggest that the elimination of female travel, made possible with the decline of pilgrimage, could have been a major goal for, and a component in, the rise of the genre.[12] Already the presence of women among the pilgrims had often been met with disapproval: the condemnation of female pilgrimage became one of the numerous possible uses of Dinah's story and, as early as 747, Saint Boniface urged Rome to forbid female pilgrimage on the basis that, during their trip, "they perish, few remaining pure".[13] Female presence among the pilgrims sometimes led to the condemnation of pilgrimage as such: Sara Warneke cites both Wyclif who suggested the separation of genders in pilgrimage and a fifteenth-century manuscript that condemns pilgrimage "for comunely...pilgrimagis ben mayntenyg of lecherie, of gloterie, of drunkenesse, of extersiouns, and wordly vanytes".[14]

Carey has provided a thorough survey of objections to female mobility in *ars apodemica* essays.[15] Among the better known examples he cites are Hieronymus Turler, who noted that "wéemen are forbidden, as it were of honestie and womanhoode, not to take long or often iourneis in hand",[16] and Thomas Palmer, who – partly inspired by Turler – lists women next to "decrepite persons", "fooles", and "lunaticks" among those who should not travel.[17] Similarly to the treatment that Dinah's story received, St. Peter's comment on women being "weaker vessels"[18] (1 Peter 3:7), initially an invitation for respect and protection, was misappropriated by commentators to argue for men's superiority over women.[19] Based on this, male authors were happy to argue that women were unable to face physical hardship or to perform the sort of dynamic physical activity that travel required. This "proof" that the biological, God-ordained nature of women prevents them from travelling is present from the earliest apodemic texts[20] and survives well into the eighteenth century, when Jean-Jacques Rousseau condemns women travelling "in softness and repose", and such level of comfort by definition precludes worthwhile forms of travel.[21] But it is moral and physical danger, as illustrated by the story of Dinah, that prevails. Various authors observed that in some countries, female mobility was less frowned upon than in others, and even Palmer could imagine exceptions to the rule of excluding women from travelling (although he never clearly defined what they were).[22] However, most authors, both literary and apodemic, agreed that the travelling woman is of dubious reputation, and Turler even suggested that they are usually promiscuous. This trope is reflected even in female writing: a character bearing the meaningful name of Constantia in Hannah Woolley's *Gentlewomen's Companion* (1655), protests strongly against being called a female traveller, "I hope, Sir, you do not take me for a Lady-errant".[23] More than a century later, the theme is still alive, with only a slight modification: in one of the discourses anonymously submitted to the 1785–1787 prize contest of the Lyon Academy on pedagogical travel, the author admits that some women might be fit for travel but that they would find it very difficult to avoid gaining the reputation of, or even becoming, "adventuresses".[24]

Woolley's character continues thus: "[H]owever, Sir, I shall acknowledg I have travel'd through the Universe, and yet was never out of my own Country". Travelling through reading is widely considered as a desirable alternative for women. Women are "fit for home", as Samuel Purchas indicates in this *Microcosmus*.[25] If any female curiosity is allowed, it has to be pursued within a domestic context.[26] But if an author deems women unfit for travel, we should also consider whether he praises, or even suggests, travel for men. In Woolley's dialogue, the lady Constantia might not wish to be known as a "Lady-errant", but the male traveller (the aptly-named Erraticus) is also presented as "impertinent and lying".[27] Richard Mulcaster, a supporter of women's education, was against the idea of female travel but was not particularly keen on male educational mobility either.[28] And the same can be said about the text, which features the most elaborate and spectacular opposition of male mobility and female immobility, Rousseau's *Émile* (1762).

The opposition between the two halves of book 5 of *Émile* – Sophie's education and considerations on Émile's travel – is striking. Not only does Émile travel while Sophie remains with her family, but this opposition is also furthered by Sophie's bookish fascination for Telemachus, the hero of bishop Fénelon's famous education novel presented through a travel setting, and by the Ulysses and Circe illustration that opens book 5.[29] As Yaël Schlick argues, the word "travel" hides an entire range of things that are unavailable to women according to Rousseau: physical freedom, logical reasoning, and citizenship (along with its advantages and duties).[30] However, while this looks like a simple exclusion of women from travelling, travel as such is restricted to an elite in Rousseau and possibly limited to a single (fictional) child. Émile, the exceptional traveller, is definitely male, and so is his tutor; however, the exclusion touches not only womankind, but also most of mankind.

The political and the maternal

Female-authored advice for travel appears both in private and public forms. The emphasis in this article is on women's participation in the public discussion on travel; however, private instructions provide important insights. The women who write these private instructions are generally mothers, who might be called upon to take a more important role in their son's education in the absence of the father.[31] A particularly intriguing case comes from Transylvania: Judit Vér, widow of Mihály Teleki, member of one of the most powerful houses of Transylvania, sent her third son on a trip to Western Europe in 1695.[32] Although most of the instruction for the young man's tour was probably written by professors from the college of nearby Enyed (today Aiud), we can spot his mother's intervention in more than one instance.[33] This is particularly visible in the instructions against lustful temptations, "If he wishes to remain my child, and not fall out of the House and Fortune of his father, and particularly of God's good will, he should not bring home, instead of Wisdom, the French sickness".[34] The success of the young man's travel answers several sets of expectations: that of the mother; that of the father's family; that of God; and that of his homeland, Transylvania. Several interesting questions arise here: what is the exact articulation of roles, how much is this text "parental", and how much is it "gendered"? Should we consider "motherly instructions" as a separate category at the meeting point of the two? The involvement of Frances Cromwell in her son's education through the means of travelling, discussed in the *Introduction*, is another example when gendered parental roles are explicitly articulated.

Female participation in the genre of *ars apodemica* pre-dates the appearance of instructions for female mobility. The fact that several female authors seem to be content to provide advice for male travelling practices, would seem to indicate that they are, implicitly, accepting educational travel to be a male prerogative. But women seeking to participate in *ars apodemica* as an intellectual and textual tradition (a field even more male-dominated than contemporary travelling practices) is in itself meaningful. Furthermore, a closer study of the authors and of their texts indicates that even the female author–male traveller binary opposition requires further thought: while the travel instructions are written for men, they could be accompanied by references to, or calls for, female mobility.

It is probably no surprise that the first author we encounter here is Margaret Cavendish, who also features as a starting point in Karen Lawrence's book on the inversion of ideas relating to female travel.[35] As is well established, Cavendish did not travel herself in the strictest sense, but her exile and related movements could be considered as such under the definition suggested by Andrea and discussed above. It is in Cavendish's writings that travel is most often considered – and, as several scholars argue, that we see her "travelling". "Assaulted and Pursued Chastity", written during her exile in 1656, is the story of Miseria, also known under the name Travelia. In her name and in the story, the travel/travail duality and conflict is accompanied by the feminization of these experiences: not only is she forced into mobility, but she also needs to repeatedly cross-dress for her safety. As the preface explicitly states, "Young, Beautiful and Virtuous Women, if they wander alone, find but very often rude entertainment from the Masculine Sex".[36] Lawrence argues that writing fictions of mobility was a form of "errancy" through imagination, and, as such, a revolt against social constraints, at a time when physical mobility with volition was not available to Cavendish and her female contemporaries.[37] In her famous letter "To the two Universities", one of several cases where she complains about women's powerlessness, Cavendish states with despair that women are "kept like birds in cages, to hop up and down in our houses, not suffered to fly abroad to see the several changes of fortune and the various humours ordained and created by nature".[38] These opinions have to be taken into consideration when studying Cavendish's apodemic essay.

With the Stuart Restoration, not only do the Cavendishes return to England, but the practice of educational travel abroad comes back to fashion in England. This allows Margaret Cavendish to discuss travel, the possibility of its control, and its relationship to statecraft. She does so in her brief but powerfully phrased "Oration concerning the Foreign Travels of Young Gentlemen".[39] The text, for the most part, doesn't add much novelty to the already century-long debate on educational travel: following the lead of a range of authors before her, Cavendish suggests that unregulated travel is far from being beneficial and is, indeed, dangerous. From the author of *The Blazing World*, a "science fiction" story featuring various types of animal-men, we are not surprised to see this idea being developed through vivid animal imagery:

> for our Young men in this Age get nothing by their Travels, but Vanity and Vice, which makes them Fools; for they gain not any Profitable Understanding, or Knowledge, to make them Wise men; the truth is, they go forth of their Own Country, Civil Men, but return Brute Beasts, as Apes, Goats, and Swine, and some few return Foxes, so that their Travels Metamorphose them from Men to Beasts.[40]

Cavendish develops on notions of state-controlled mobility already present in Plato's *Republic*, but the proposed solution is rather original. She suggests "that none should Travel without Leave of the State or Publick Counsel, and at their Return should be Accountable to the State and Publick Counsel of their Travels".[41] While getting an authorization for the foreign travel was received practice at various times (see the letter of Frances Cromwell quoted in the *Introduction*), in most cases it did not carry the obligation to report afterwards.[42] The idea was indeed appreciated to some extent: the only known reaction to Cavendish's "art of travel" within the apodemic corpus, the *Three Diatribes* by prolific writer and theologian Edward Leigh, quotes this sentence approvingly.[43]

Her *Oration* is an interestingly phrased variation on the common theme of criticism of travelling; it is written at a time when, with the restoration of monarchy in England, she returns to a position in which she and her husband are hoping – mostly in vain, as it turns out – to increase their importance, and when, following the normalization of political relations with France, foreign travel for young English gentlemen is likely to become more widespread again. This critical approach to male travel could be read, somewhat speculatively, as a translation of Cavendish's frustration in the face of female immobility, but – as suggested by the fact that Leigh quotes it without reflecting on its gendered nature – it can also be understood without consideration of gender and more from the point of view of aspirations for power and influence. This first female-authored apodemica is meaningful in another respect: as indicated by the genre she chooses to write in (*Orations*), a woman takes part in public discourse. Katie Whitaker's biography of Cavendish indicates that the duchess was fascinated by the "strange hidden mystery" of rhetoric as a skill reflected in certain genres like orations but was also afraid of engaging with this field due to her lack of formal training.[44] Trying her hand at rhetoric and normative discourse was as much of a transgressive move as setting out on a journey.

The concluding words of Cavendish's oration compare travel, when performed well, to a "nursery and school" for young men.[45] This would be the only instance in this text when Cavendish uses vocabulary that can seem somewhat maternal in tone. Another female-authored apodemic text, while political in its main focus and ambition, makes a more direct use of the maternal angle. Unlike Margaret Cavendish and Madame de Genlis (the other two authors of female apodemic texts), Julie (or Juliane) de Giovane is yet to receive the scholarly attention she deserves.[46] Born in Würzburg in 1766 as Juliane von Mudersbach, she was involved from her teens in German intellectual life, writing for the first German ladies' magazine. At the age of 20, she married an Italian prince and moved to the Bourbon court of Naples. It seems that the idea of taking up writing (again) might have been suggested to her by Goethe, who met her during his Italian trip, just like Herder did two years later. She first studied her surroundings: witnessed the eruption of Mount Vesuvius, collected minerals on the mountain, and wrote a dissertation on these. In 1790 she left behind not only an unhappy marriage but also her son and moved to Vienna, where she gained a reputation for her writings and became *Hofmeisterin* to the princess Marie Louise, granddaughter of her patron in Naples, and future second wife to Napoleon. She wrote in a variety of genres and in three languages, became the second female honorary member of the Berlin Academy of Sciences (after Catherine the Great – the first female full membership wasn't awarded until 1964), and became an honorary member of the Swedish Academy.

Her life is not devoid of mobility, but – like Cavendish – she was no traveller in the conventional sense, nor does female travel appear in her work. In her handbook for the education of princesses, she suggests that a princess should speak four modern languages and have an active and useful knowledge, not only of history, geography, and literature but also of law, trade, and "state history". One could argue that every element necessary for an "art of travel" is present in her work, as is the notion of "active education", but the idea of travel isn't mentioned as such.[47] Of course, the education of a princess is different from that of women in general, and princesses could be forbidden to leave the country. Giovane's educational programme provides princesses with the type of knowledge they could gain from travel but does not actively challenge the restriction of female travel.

She is, however, the author of a book suggesting rules for a particular form of male travel: travel in service of the state. The political and the maternal are closely intertwined right from the opening of her *Plan pour faire servir les voyages à la culture des jeunes gens qui se vouent au service de l'état dans la carrière politique* (1797).[48] In a key paratextual position, opposite the title page is an engraving of the author that comprises both her scholarly credentials (membership of academies) and a passage from the book "Virtue, who has been my guide on the arduous path that I have had to follow: be that of my Son, too!" (Fig. 10.1). Her difficult life experiences, personal qualities, and her status as a mother who had to leave her son behind are thus immediately associated to her work. The book is dedicated to her son: she hopes that one day he will be among those young men who will benefit from these instructions.

Giovane's *Plan pour faire servir les voyages* is an elaborate work, offering a survey of the apodemic tradition; general discussion on travelling; instructions for domains of observation (such as political situation and institutions, geography, and trade); and advice for taking notes and structuring the information gathered into tables. Her contribution was nevertheless mostly overlooked because, in many respects, it compares poorly to a major piece in the apodemic tradition that was published only a handful of years earlier: the *Essay to direct and extend the Inquiries of Patriotic Travellers* by the Moravian aristocrat Leopold Berchtold, published in London for the author in 1789. Berchtold's text is a compendium of possibly all fields that could be discussed in relation to travel. It is probable that, although she elected not to refer to it, Giovane was familiar with Berchtold's work, which was translated from the English and published in French in the same year she published her book on travel. Both in Berchtold and in Giovane, all activity is directed towards the idea of improvement of the state, hence the term "patriotic" in Berchtold's title. But the contributions of individual travellers are key not only for their own state. Both Berchtold, the Moravian aristocrat publishing his book in English in London and then cooperating with the revolutionary government in Paris, and Giovane, moving between Germany, Naples, and Vienna, were cosmopolitan figures. The information gathering they wish

Figure 10.1 Julie de Giovane, *Plan pour faire servir les voyages à la culture des jeunes gens qui se vouent au service de l'Etat* (Vienna, 1797), frontispiece. Utaz Q136, Library and Information Centre of the Hungarian Academy of Sciences (MTAKIK).

to improve and regulate is not that of classified state information; beyond the benefit of specific states lies the wider ambition of their texts being of benefit to mankind. For Berchtold, a philanthropist, the priority is the issue of public health.[49] Giovane, who participated earlier in an academic contest on the theme of "the public good", speaks generally of "good government".

Although less elaborate than Berchtold when discussing travellers' note-taking and observation practices, Giovane came up with one important original idea regarding the regulation of travel. She was in contact with scholars from Göttingen and was aware of the classes on the method of travel taught there by Ludwig von Schlözer.[50] But she considered that university education, limited to smaller numbers, might not be the most appropriate method for teaching travel, particularly if it is limited to one university; also, in her view, this teaching remained too theoretical. Instead, she suggested introducing travel as a topic into the curriculum taught in schools. While this appears as a democratizing gesture, it is also restrictive: according to her, only those who completed this education should be allowed to travel. This "class on travel" would essentially cover all fields discussed in an apodemic treatise, from moral preparation through the use of tools such as maps, all the way to what to do upon one's return.[51] Later, she defined "the travel of young men as an intermediary cultural instruction, between their early education, and their establishment as citizens in their society".[52] This idea of introducing travel as an element of public education and as a key step in the education of any professional elite, was never implemented – we can only speculate whether her untimely death in 1801, when she was only age 39, had a role in this. Her proposal, although clearly written with practical application in mind, remains an abstract speculation – it is an unusual combination of a proposal for perfect travel; a suggestion for a slight democratization and wider scope of "good" travel practices; and, ultimately, a maternal *Wunschtraum*.

Both Cavendish and Giovane successfully penetrated the apodemic genre; as shown by Leigh quoting Cavendish's orations and the numerous praises of Giovane, their works were recognized as worthwhile contributions to the debate. Both of them lived through experiences of mobility that were far from pleasurable travel (Cavendish's exile, Giovane fleeing an unhappy marriage), and this might play a role in the fact that, in their writings, female travel appears only briefly (in Cavendish's letter to the universities) or not at all (in Giovane's works). Both authors came up with innovative ideas when it came to the question of who has the right to travel: it is as if their own frustrated mobilities invited them to suggest a stronger regulation of male mobility. A fully female apodemica appeared, however, in the works of an author who did have the opportunity to travel with volition.

A travelling Sophie?

Daniel Roche concludes his survey of apodemic literature with words of praise for an item written by a woman: Madame de Genlis's *Le Voyageur* (1800). He appreciates it as an "ambulant school of economy" and as a meeting point between abstract apodemic instructions and practical advice for the road (itineraries and even recipes easy to prepare on the road).[53] Genlis's *Le Voyageur* accompanies a different publication, the *Manuel du Voyageur*, a series of language dialogues she published, both translating her personal experience as an aristocratic refugee and making this experience slightly less uncomfortable through the income thus generated.[54] The dialogues and letter schemas provided are directed towards both genders, some of them clearly feminine (such as a format to follow when writing a letter requesting to be admitted to a convent), others clearly male; most of them are useful for both sexes.

The 1800 text *Le Voyageur* contains a section entitled "Advice for travellers and fugitives who want to undertake long trips and have no experience in such matters".[55] The dual audience, "travellers and fugitives", could be another proof supporting Bernadette Andrea's invitation to understand "travel" as a wider category that includes cases where volition is less clear: the experience of travellers and fugitives is similar enough that it can be addressed in a single volume. It is a gender-neutral text, providing practical advice to both sexes – and Genlis, formerly an avid traveller, did provide a limited amount of (rather traditional and derivative) advice for observation and note-taking as well. With its double scope, combining practical advice with these limited considerations on observation and note-taking, Genlis's *Le Voyageur* is situated on the periphery of the apodemic genre.[56]

However, scholarship fails to acknowledge that Madame de Genlis's activity within the apodemic field starts more than a decade earlier and from a radically different position. While still in a position of respectability and even relative power, both as a popular author of novels and books on education and as the governess of the children of the duke of Orléans, she published in 1781 a pedagogical novel, *Adèle et Théodore*.[57] The text presents no less than three programmes for educational travel, including possibly the only programme featuring a female educational traveller. Although the text is a novel, there is no reason for it not to be included in the apodemic corpus; as it has been pointed out many times, Rousseau's *Émile*, containing one of the best known eighteenth-century apodemic discussions, is itself a novel in many respects.[58] A novel might even be a particularly appropriate space to locate a female contribution to the debate on travelling: not only was the novel generally considered a female form of expression in eighteenth-century France, it was also the dominant form of female reading. As such, for an author who had among her ambitions to inspire real and prospective mothers, communicating via a novel was appropriate.

The relationship between *Adèle et Théodore* and *Émile* is well documented and clear.[59] References to Rousseau in Genlis' text are numerous; leaving aside the similarity of the initial situation, these references are more often critical than praising.[60] Madame de Genlis develops Rousseau's core idea of a "natural education", carried out away from the negative influence of the metropolis. The setting of Genlis' model education is Languedoc, and, contrary to Rousseau's, it takes place in a family context and with clear aristocratic overtones. A father and a mother, Monsieur and Madame d'Almane, with the help of a handful of servants and various acquaintances, take care of the education of their two children. It is in the educational travel programme developed by Genlis that we see the most radical break with Rousseau – and indeed, with the whole of apodemic tradition – because Genlis is the first writer to challenge female immobility.

Three pedagogical travel experiences are mentioned in the text. The first is for a boy, an acquaintance of the d'Almanes in Languedoc, who travels with his grandfather.[61] As the young man in question will be seeking to marry young Adèle (Adelaide in the English translation), there is a parallel here with the travel-cum-marriage setup that we encounter in *Émile*, inspired probably by John Locke, but with the major difference that, in this instance, the prospective bride shall travel as well. A second travelling programme is developed briefly through the letters of a friend of the d'Almanes who is in charge of the education of a prince.

The emphasis is on the travel of the d'Almane family – the programme for the trip is presented in letter 19 of the second volume;[62] the travel itself lasts well into the middle of volume 3. Conceived to last for eighteen months, it is later prolonged to two full years. It might be a unique case in the apodemic literature because the whole family is travelling; another unusual aspect, emphasized by Madame de Genlis, is the age of the siblings: Théodore is 13, Adélaïde is 12 at departure, and Madame d'Almane proudly points out that her children are clever enough to travel at such a young age. Rousseau's project in which Émile travels as a near-adult is in opposition to many other apodemic ideas regarding the age of the traveller; for instance, Montaigne recommends that the child travel as early as possible.[63]

Genlis's pedagogical goals are appropriate to the age of Théodore and Adélaïde, and they seem to be gender-specific as well, as the children are supposed to each follow a different programme:

> it is neither men nor laws that we are to study in Italy: my children will acquire a taste for polite arts, and perfect themselves in drawing. Whilst they amuse themselves in admiring the monuments and ruins of Roman grandeur, they will gain a complete knowledge of that interesting history. In short, my son, conducted by a father,…will learn to write a good journal, and in it nothing frivolous; in a word, reap all the fruits of travelling. I shall bring back Adelaide at fourteen, an excellent musician, an adept in drawing, talking and singing like a native Italian;

> and entirely divested of all those little feminine delicacies which nothing but travelling can radically cure. She will neither fear the sea, nor bad roads; sleep as well in an alehouse, as in her own apartment. She will learn to be contented with a bad supper. (224)

During the trip, both children will not only sketch but also write a travel journal. Additionally, both children become excellent linguists: they have been educated in French and English so far, and they pick up other languages during the Italian trip. The above programme is sketched by Madame d'Almane before the travel takes place. The modifications to this programme that occur systematically reduce the differences between the gendered experiences. Although this is not phrased explicitly in the text, it could be argued that these discrepancies between programme and experience indicate that even the d'Almanes realize that the children's behaviour overwrites the initial intentions of the parents.

We have seen that apodemic literature excludes women, among other things, because of their perceived weakness and inability to face physical hardship. Well before enduring exile, Madame de Genlis discussed this aspect in the educational travel programme: on several occasions when passing through mountains, her characters have to alight from the carriage and walk. Physical difficulties during travel, including an episode of seasickness (alluded to in the brief programme quoted above as fear of the sea), are incorporated into the programme to form a part of moral education.[64] During a difficult sea passage, first Theodore demonstrates his qualities by maintaining pleasant conversation despite regularly vomiting; daughter and mother then decide to follow his example, and they perform a duet despite experiencing seasickness, which leads the father and his son to praise their resilience. The idea of making the young travellers accustomed to all sorts of discomfort and to have them learn to resist fear, a regular feature of apodemic texts, is extended here to both sexes; a clearly aristocratic touch is added to it by the insistence on not losing dignity. Other moral lessons also become age and gender-specific. The recurring apodemic theme of overcoming national prejudice and accepting practices that are outside of one's established ways is here brought down to the level of an anecdote: the twelve-year-old Adelaide laughs at local women and their clothes and then realizes she is being laughed at as well by these women.[65]

Daniel Roche praises Madame de Genlis's *Le Voyageur* as a case where the abstract prescriptions of the art of travel are not only brought down to the level of everyday practicalities but also to applicability.[66] The same can be said of the idea of travel in *Adèle et Théodore*, particularly in opposition to *Émile*. Although this is a distinctly elite programme of education that is free from any constraints as regards to time, space, or financial means, the children Adelaide and Theodore are not as hypothetical as Émile. Madame de Genlis's programme of education is considerably closer to reality than Rousseau's. Although it is still an elite form of travel and of education, it would not be theoretically – or practically – impossible for higher levels of society to achieve. And it is at this level, the higher echelons of society, that the ultimate barrier, women's entrance into the field of travel for curiosity and education, can be overcome.

However, it would be an error for us to limit discussions of female considerations around this trip to the person of young Adelaide. Not one, but two women are present. Madame d'Almane states in one of her letters that her identity as a traveller is almost nonexistent: when in Rome, she is so absorbed in overseeing the children's learning that she barely sees the city herself.[67] But this role as a female "bear-leader", as the tutors of travelling young aristocrats are jokingly known at the time, is a crucial component of the entire programme. Referencing Fénelon's *Télémaque*, a classic of educational travel fiction, Madame de Genlis invents not only the idea of a female Telemachus but also that of a female Mentor – reflecting an often neglected gendered aspect, namely that Mentor is, in fact, the goddess Minerva in disguise. Daniel Roche called Madame de Genlis "a skirted mentor" when some observational training is combined with kitchen recipes and sanitary advice in her later works.[68] We see in this novel that the female, maternal mentor role predates her exile and initially covers the full range of apodemic considerations. And in this case fiction will become reality: some years later, Madame de Genlis will accompany one of her pupils, the daughter of the Duke of Orléans, to England, on what was, for that period, a rare case of female-only travel experience.[69]

Conclusion

Besides these three apodemic texts by Cavendish, Genlis, and Giovane, we could cite a poem on travel – more a praise of the heroic traveller than suggestions for travelling – by the princess of Salm, Paris salonnière of the Napoleon era, and some brief, mostly critical comments on the incorrect uses of travel by established female authors such as English historian Catherine Macaulay and the celebrated French translator of Homer, Anne Dacier.[70] This form of reflection remains rare – the double privilege men hold over travel and over rhetoric and normative discourse explains their rarity, and also their slow emergence. The apodemic debate is already over a hundred years old by the time the first female contribution, Cavendish's oration, appears.

This leads us to the question whether women breaking into the domain of *ars apodemica* is a case of what Carl Thompson calls, for travel writing, a "journey into authority". Despite Mme de Genlis's unquestionable popularity and impact on her contemporaries, opinions around female travel are very slow to change all over Europe. An Italian apodemic poem from 1793, although critical of all travel, particularly ridicules female travellers, "is there really any need to see how States are governed throughout Europe, in order to better govern the home?"[71] As late as 1840, the writer Paolina

Leopardi is complaining that her family won't let her travel;[72] the heritage of the male interpretations of Dinah's story seems to be very much alive. The apodemic contributions by Cavendish and Giovane could be appreciated without questioning whether Penelope should accompany or even replace Ulysses, because they didn't mention female travel as such. Even scholarship seems to perpetuate this image to some extent. Both the bibliography of apodemic texts established by Justin Stagl and his collaborators, and Daniel Roche's essay, acknowledge Madame de Genlis's work only from the time of her exile and not the text in which the ideal travelling female child travels alongside a male sibling as an equal partner.

Harriet Guest highlights a moment in Mary Wollstonecraft's *Vindication of the Rights of Women* (1792) where a discussion of divergent travelling attitudes is used to highlight female and male preoccupations.[73] This comment appears in chapter 4, which addresses "the state of degradation to which woman is reduced by various causes"; the observation presents the male traveller as always oriented towards the outcome and benefits of the trip, while the female traveller is more concerned by the impression she makes on others while travelling, blocking her from being able to reap the full benefits of the experience of travel. We can detect in this image of the self-conscious and anxious female traveller the long tradition of the dubious reputation of travelling women; any non-immobile female can be likened to Circe, be perceived as a lady-errant, or face extreme dangers such as Dinah.

On the path to building a consciously gendered conception of mobility and recognizing the conflicts inherent in this distinction, female *ars apodemica* texts are important milestones. Cavendish first ventured into the field of rhetoric and, to some extent, politics; Giovane's essay showcased that women can participate in ongoing scientific debates in the fields of travel, and also politics, and even develop radically new pedagogical ideas. More directly, Madame de Genlis's reinterpretation of Émile's travels played a role in turning travel writing into one of the genres in which women of later periods were able to make claims for intellectual independence; more generally, these texts turned travel into an activity that women would be allowed to perform, enjoy, and exploit on their own and for their own purposes.

Notes

1 The following remarks are based on Joy Schroeder's survey of the uses of this story in *Dinah's Lament: The Biblical Legacy of Sexual Violence in Christian Interpretation* (Minneapolis: Fortress Press, 2007), chapter 1.

2 Arcangela Tarabotti, *Paternal Tyranny*, ed. and trans. Letizia Panizza (Chicago: University of Chicago Press, 2004), 115.

3 Karen Newman, "Armchair Travel," in *Crossing Boundaries: Attending to Early Modern Women*, eds. Jane Donawerth and Adele Seefe (Newark: University of Delaware Press, 2000), 211–26 (211).

4 Erica Harth, *Cartesian Women: Versions and Subversions of Rational Discourse in the Old Regime* (Ithaca: Cornell University Press, 1992), 62.

5 Colette H. Winn, "Des femmes en mouvement," *Romanic Review*, 94 (2003): 115–51 (150). She lists cases of forced mobility, health and business trips, but also some cases of "touristic" visits.

6 Bernadette Diane Andrea, "'Travelling Bodyes': Native Women of the Northeast and Northwest Passage Ventures and English Discourses of Empire," in *Rethinking Feminism in Early Modern Studies: Gender, Race, and Sexuality*, ed. Ania Loomba and Melissa E. Sanchez (London: Routledge, 2016), 135–48.

7 Clare Howard, *English Travellers of the Renaissance* (London: John Lane, 1914); George B. Parks, "Travel as Education," in *The Seventeenth Century: Studies in the History of English Thought and Literature from Bacon to Pope*, ed. Richard Foster Jones (Stanford: Stanford University Press, 1951), 264–90; Justin Stagl, *A History of Curiosity: The Theory of Travel 1550–1800* (Chur: Harwood Academic, 1995); Normand Doiron, *L'Art de voyager: le déplacement à l'époque classique* (Sainte-Foy and Paris: Presses de l'Université Laval/ Klincksieck, 1995); Joan-Pau Rubiés, "Instructions for Travellers: Teaching the Eye to See," *History and Anthropology* 9 (1996): 139–90; for summary see Daniel Carey, "Advice on the Art of Travel," in *The Cambridge History of Travel Writing*, ed. Nandini Das and Tim Youngs (Cambridge: Cambridge University Press, 2019), 391–407.

8 Although, as Stagl has pointed out, the "direct application" of *ars apodemica* is always a question. See the "Postscriptum" to chapter 1 of *A History of Curiosity*.

9 Michel de Montaigne, *The Complete Essays*, trans. Michael Andrew Screech (London: Penguin, 1991), 172.

10 Carl Thompson, "Journeys into Authority: Reassessing Women's Early Travel Writing, 1763–*1862," *Women's Writing* 24 (2017): 131–50.

11 In my monograph *Lessons of Travel in Eighteenth-Century France: From Grand Tour to School Trips* (Martlesham: Boydell and Brewer, 2020), I call for *ars apodemica* to be considered not only a "discourse on practice", but also a "practice of discourse": all reading of texts of this corpus should acknowledge its rhetorical heritage and properties (7).

12 Carey, "Advice on the Art of Travel", 403.

13 Sara Warneke, *Images of the Educational Traveller in Early Modern England* (Leyden: Brill, 1995), 21.

14 *Ibid.*, 22.

15 See Carey, "Advice on the Art of Travel", 403.

16 Hieronymus Turler, *The Traveiler of Ierome Turler* (London, 1575), 96. This is the English translation of his *De peregrinatione et agro Neapolitano* (Strasbourg, 1574), cited in Carey, "Advice on the Art of Travel," 403.

17 Thomas Palmer, *An Essay of the Meanes how Make our Travailes, into Forraine Countries, the More Profitable and Honourable* (London, 1606), B1r. Turler and Palmer are the two male authorities against female travel quoted in Bernadette Andrea's "Travelling Bodyes" (140). On the connections between travel literature and images of the moon, including lunacy, see Rachel Willie's chapter in the present volume.

18 The expression features for example in Palmer, *An Essay on the Meanes*, 7.

19 For a longer history of this expression and English attitudes towards women, see Antonia Fraser, *The Weaker Vessel: Women's Lot in Seventeenth-Century England* (London: Mandarin, 1989), 1.

20 See, for example, one of the most elaborate apodemic dissertations, Theodor Zwinger's *Methodus apodemica in eorum gratiam, qui cum fructu in quoqunque tandem vitae genere peregrinari cumpiunt* (Basel, 1577), 56.

21 Jean-Jacques Rousseau, *Émile, or on Education*, trans. Alan Bloom (New York: Basic Books, 1979), 411. The importance of the conception that women are morphologically unable to perform any dynamic movement is highlighted by Yaël Schlick, *Feminism and the Politics of Travel* (Lewisburg: Bucknell University Press, 2012), 31–32.

22 Although, as Alison Games points out, Fynes Morrison's *An Itinerary* clearly stated that women shouldn't travel under any circumstances, even when the laws allow it. See Alison Games, *The Web of Empire: English Cosmopolitans in an Age of Expansion 1560–1660* (Oxford: Oxford University Press, 2008), 22.

23 Hannah Woolley, *The Gentlewoman's Companion* (London, 1670), 258.

24 Académie de Lyon, Ms 237-3, submission no. 2, f. 54v-55. It would appear to be the only mention of female travel in the contest. See a detailed analysis of the contest in Daniel Roche, *Humeurs vagabondes: de la circulation des hommes et de l'utilité des voyages* (Paris: Fayard, 2003), 77–87 and a new reading in chapter 5 of my monograph *Lessons of Travel.*

25 Samuel Purchas, *Purchas his Pilgrim, Microcosmus or the historie of man* (London:, 1619), 488. As Carey highlights, this text indicates travel as especially dangerous for men because they might become effeminate (403–04).

26 On the undesirability of female curiosity and the associated negative images of Pandora and Eve, see Barbara M. Benedict: *Curiosity: A Cultural History of Early Modern Inquiry* (Chicago: University of Chicago Press, 2001), esp. chapter 3.

27 Woolley, 258–61 – Erraticus speaks in nonsensical terms of every country he visited.

28 Richard Mulcaster, *Positions Concerning the Training Up of Children*, ed. William Barker (Toronto: University of Toronto Press, 1994), 212.

29 Schlick, 26.

30 *Ibid.*, 20.

31 Several British examples are mentioned by Warneke, 34–36.

32 Zsuzsa Font, ed., *Teleki László külföldi tanulmányútja: Levelek, számadások, iratok 1695–1700*, (Szeged: Szegedi Oktatástörténeti Munkaközösség, 1989).

33 *Ibid.*, 416–17.

34 *Ibid.*, 7.

35 Karen R. Lawrence, *Penelope Voyages: Women and Travel in the British Literary Tradition* (Ithaca: Cornell University Press, 1994), chapter 1.

36 Margaret Cavendish, "Assaulted and Pursued Chastity" (1656), in *Nature's Picture Drawn by Fancies Pencil to the Life*, 2nd ed. (London, 1671), 394–514 (394). After multiple misadventures, Miseria/Travelia ultimately becomes ruler of the land.

37 Lawrence, 29–30.

38 Cavendish, "To the two Universities," in *Philosophical and Physical Opinions* (London, 1655), B2v. For further considerations on this and other texts on women's powerlessness, see Jacqueline Broad and Karen Green, *A History of Women's Political Thought in Europe 1400–1700* (Cambridge: Cambridge University Press, 2009), 221.

39 Cavendish, *Orations of Divers Sorts, Accommodated to Divers Places* (London, 1662), 71–73.

40 *Ibid.*, 74.

41 *Ibid.*, 74–75.

42 If a report, oral and written, is to be prepared for the authorities, the travel experience enters into the realm of "intelligencing". On the interface of *ars apodemica* and intelligencing, see Elizabeth Williamson: '"Fishing after News" and the Ars Apodemica: The Intellingencing Role of the Educational Traveller in

the Late Sixteenth Century", in Joad Raymond and Noah Moxham eds., *News Networks in Early Modern Europe* (Brill e-books series, 2016), 542–62.

43 Edward Leigh, *Three Diatribes or Discourses First of Travel, or a Guide for Travellers into Forein Parts...* (London, 1671), 2.

44 Katie Whitaker, *Mad Madge: Cavendish, Duchess of Newcastle, Royalist, Writer and Romantic* (London: Vintage, 2004), 251–53. On Cavendish's frustrations and ambitions regarding rhetoric art, see Andrew Black, "'Persuade us out of our selves': Margaret Cavendish's regulation of rhetorics," *Seventeenth-Century Studies* 33 (2018): 303–22. For the relationship between travel and rhetoric, see Natalya Din-Kariuki's chapter in this volume.

45 Cavendish, "Oration concerning the Foreign Travel of Young Gentlemen," 75.

46 See Benedetto Croce, "La Duchessa Giovane," *Rassegna Pugliese* 30 (1887): 275–77; Heiner Reitberger, "Juliane Herzogin Giovane di Girasole (1766–1805)," in *Frauengestalten in Franken. Eine Sammlung von Lebensbildern.* ed. Inge Meidinger-Gerse (Würzburg: Weidlich, 1985), 124–29.

47 Julie de Giovane, *Lettres sur l'éducation des princesses* (Vienna: Joseph Stahel, 1791).

48 "Plan to make travelling useful for the education of young people who plan to enter the service of the state".

49 After extensive work to improve public health in various states, Berchtold died in 1809 when, having transformed his own palace into a hospital, he contracted a fever. On Berchtold, see Stagl, *A History of Curiosity*, chapter 5.

50 Giovane, 15, footnote. On Schlözer's role in the development of ethnography and science of the state, see Stagl, *A History of Curiosity*, chapter 6; Han F. Vermeulen, *Before Boas. The genesis of ethnography and ethnology in the German Enlightenment* (Lincoln: University of Nebraska Press, 2015), chapter 6.

51 Giovane, 14–25.

52 *Ibid.*, 40, footnote.

53 Roche, 168–69.

54 Madame de Genlis (Stéphanie-Félicité Crest), *Manuel du Voyageur* (Berlin: F. T. de Lagarde, 1799, numerous editions during the following decades).

55 Genlis, *Le Voyageur, ouvrage utile à la jeunesse et aux étrangers* (Berlin: J. F. de La Garde, 1800), XIII–XLII.

56 An item with practical instructions for the travelling women features in the bibliography of apodemic literature (Justin Stagl, Klaus Orda and Christine Kämpfer, *Apodemiken: eine räsonnierte Bibliographie der reisetheoretischen Literatur des 16., 17. und 18. Jahrhunderts*, Paderborn: Schöning, 1983): Sidonia Hedwig Zäunemannin's *Curieuser und immer währender Astronomisch-Meteorologisch-Oeconomischer-Frauenzimmer-Reise und Händ-Kalender* (Erfurt, 1737). This work features no considerations on travel beyond practicalities, thus it cannot be considered a real *ars apodemica*.

57 Genlis, *Adèle et Théodore, ou Lettres sur l'éducation, contenant tous les principes relatifs aux trois différents plans d'éducation des Princes, des jeunes personnes, et des Hommes*, 3 vols (Paris: Lambert & Baudouin, 1781); critical edition by Isabelle Brouard-Ardens (Rennes: Presses Universitaires de Rennes, 2006). I am quoting from the English edition by Gillian Dow, *Adelaide and Theodore, or Letters on Education* (London: Pickering & Chatto, 2007). I develop here ideas discussed in chapter 4 of my *Lessons of Travel.*

58 On fiction in Émile, see Laurence Mall, *Émile ou les figures de la fiction* (Oxford: Voltaire Foundation, 2002).

59 See a summary in Gillian Dow's introduction to *Adelaide and Theodore*, XIV.

60 Mary Seidman Trouille in her *Sexual Politics in the Enlightenment: Women*

Writers Read Rousseau (Albany: State University of New York Press, 1997), points at Madame de Genlis' ambivalence between the strong influence of Rousseau and her criticism of his model (238).
61 Genlis, *Adelaide and Theodore*, 124; this trip is mentioned briefly on a number of occasions.
62 *Ibid.*, 224–25.
63 Highlighted by Juliette Morice, "Voyage et anthropologie dans l'*Émile* de Rousseau," *Revue de métaphysique et de morale* 1 (2013): 127–42 (138).
64 Genlis, *Adelaide and Theodore*, vol. 2, letter 34.
65 Genlis, *Adelaide and Theodore*, vol. 2, letter 39.
66 Roche, *Humeurs vagabondes*, 169.
67 Genlis, *Adelaide and Theodore*, vol. 2, letter 47.
68 Roche, *Humeurs vagabondes*, 170.
69 See a discussion of this trip in my *Philosophies du voyage: visiter l'Angleterre aux 17e-18e siècles* (Oxford: Voltaire Foundation, 2016), pp. 235–41.
70 Constance de Salm, "Sur les voyages," in *Poésies de Madame la Princesse Constance de Salm* (Paris: Firmin Didot, 1811), 115–18; Catherine Macaulay, *Letters on Education, with Observations on Religious and Metaphyical Subjects* (London: C. Dilly, 1715), 140; *L'Odyssée d'Homère, traduite en François, avec des remarques, Par Madame Dacier* (Amsterdam: Aux Dépens de la Compagnie, 1717), 7.
71 Ippolito Pindamonte's *I Viaggi* (Venice, 1793), cited in Ricciarda Ricorda, "Travel Writing," in *A History of Women's Writing in Italy,* ed. Letizia Panizza and Sharon Wood (Cambridge: Cambridge University Press, 2000), 107–21 (108).
72 *Ibid.*, 107.
73 Harriet Guest, "Travel Writing," in *Cambridge Companion to Women's Writing in Britain, 1660–1789,* ed. Catherine Ingrassia (Cambridge: Cambridge University Press, 2015), 196–209 (197).

Bibliography

Andrea, Bernadette Diane. "'Travelling Bodyes': Native Women of the Northeast and Northwest Passage Ventures and English Discourses of Empire." In *Rethinking Feminism in Early Modern Studies: Gender, Race, and Sexuality*. Edited by Ania Loomba and Melissa E. Sanchez, 135–48. London: Routledge, 2016.

Benedict, Barbara M. *Curiosity: A Cultural History of Early Modern Inquiry.* Chicago: University of Chicago Press, 2001.

Black, Andrew. "'Persuade Us Out of Our Selves': Margaret Cavendish's Regulation of Rhetorics." *Seventeenth-Century Studies 33* (2018): 303–22.

Broad, Jacqueline and Karen Green. *A History of Women's Political Thought in Europe 1400–1700* Cambridge: Cambridge University Press, 2009.

Carey, Daniel. "Advice on the Art of Travel." In *The Cambridge History of Travel Writing*. Edited by Nandini Das and Tim Youngs, 391–407. Cambridge: Cambridge University Press, 2019.

Cavendish, Margaret. *Philosophical and Physical Opinions*. London, 1655.

Cavendish, Margaret. *Orations of Divers Sorts, Accommodated to Divers Places.* London, 1662.

Cavendish, Margaret. *Nature's Picture Drawn by Fancies Pencil to the Life*, 2nd ed. London, 1671.

Croce, Benedetto. "La Duchessa Giovane." *Rassegna Pugliese, 30* (1887): 275–77.

Doiron, Normand, *L'Art de voyager: le déplacement à l'époque classique*. Sainte-Foy and Paris: Presses de l'Université Laval & Klincksieck, 1995.

Font, Zsuzsa, ed. *Teleki László külföldi tanulmányútja: Levelek, számadások, iratok 1695–1700*. Szeged: Szegedi Oktatástörténeti Munkaközösség, 1989.

Fraser, Antonia. *The Weaker Vessel: Women's Lot in Seventeenth-Century England*. London: Mandarin, 1989.

Games, Alison. *The Web of Empire: English Cosmopolitans in an Age of Expansion 1560–1660*. Oxford: Oxford University Press, 2008.

Gelléri, Gábor. *Lessons of Travel in Eighteenth-Century France: From Grand Tour to School Trips*. Martlesham: Boydell and Brewer, 2000.

Gelléri, Gábor. *Philosophies du voyage: visiter l'Angleterre aux 17e-18e siècles*. Oxford: Voltaire Foundation, 2016.

Genlis, Stéphanie-Félicité Crest (Madame de). *Manuel du Voyageur*. Berlin: F. T. de La Garde, 1799.

Genlis, Stéphanie-Félicité Crest (Madame de). *Le Voyageur, ouvrage utile à la jeunesse et aux étrangers*. Berlin: J. F. de La Garde, 1800.

Genlis, Stéphanie-Félicité Crest. *Adelaide and Theodore, or Letters on Education*. London: Pickering and Chatto, 2007.

Giovane, Julie de. *Lettres sur l'éducation des princesses*. Vienna: Joseph Stahel, 1791.

Guest, Harriet. "Travel Writing." In *Cambridge Companion to Women's Writing in Britain, 1660–1789*. Edited by Catherine Ingrassia, 196–209. Cambridge: Cambridge University Press, 2015.

Harth, Erica. *Cartesian Women: Versions and Subversions of Rational Discourse in the Old Regime*. Ithaca: Cornell University Press, 1992.

Homer. *L' Odyssée d'Homère, traduite en François, avec des remarques, Par Madame Dacier*. Amsterdam: Aux Dépens de la Compagnie, 1717.

Howard, Clare. *English Travellers of the Renaissance*. London: John Lane, 1914.

Lawrence, Karen R. *Penelope Voyages: Women and Travel in the British Literary Tradition*. Ithaca: Cornell University Press, 1994.

Leigh, Edward. *Three diatribes or discourses first of travel, or a guide for travellers into forein parts*. London, 1671.

Macaulay, Catherine. *Letters on Education, with Observations on Religious and Metaphyical Subjects*. London: C. Dilly,, 1715.

Mall, Laurence. *Émile ou les figures de la fiction*. Oxford: Voltaire Foundation, 2002.

Montaigne, Michel de. *The Complete Essays*. Translated by Michael Andrew Screech. London: Penguin, 1991.

Morice, Juliette. "Voyage et anthropologie dans l'*Émile* de Rousseau." *Revue de métaphysique et de morale 1* (2013): 127–42.

Mulcaster, Richard. *Positions Concerning the Training Up of Children*. Edited by William Barker. Toronto: University of Toronto Press, 1994.

Newman, Karen. "Armchair Travel." In *Crossing Boundaries: Attending to Early Modern Women*. Edited by Jane Donawerth and Adele Seefe, 211–26. Newark: University of Delaware Press, 2000.

Palmer, Thomas. *An Essay of the Meanes how Make our Travailes, into Forraine Countries, the More Profitable and Honourable*. London, 1606.

Parks, George B. "Travel as Education." In *The Seventeenth Century: Studies in the History of English Thought and Literature from Bacon to Pope*. Edited by Richard Foster Jones, 264–90. Stanford: Stanford University Press, 1951.

Purchas, Samuel. *Purchas his Pilgrim, Microcosmus or the historie of man.* London, 1619.

Reitberger, Heiner. “Juliane Herzogin Giovane di Girasole (1766–1805).” In *Frauengestalten in Franken. Eine Sammlung von Lebensbildern.* Edited by Inge Meidinger-Gerse, 124–29. Würzburg: Weidlich, 1985.

Ricorda, Ricciarda. “Travel Writing.” In *A History of Women's Writing in Italy.* Edited by Letizia Panizza and Sharon Wood, 107–21. Cambridge: Cambridge University Press, 2000.

Roche, Daniel. *Humeurs vagabondes: de la circulation des hommes et de l'utilité des voyages.* Paris: Fayard, 2003.

Rousseau, Jean-Jacques. *Émile, or on Education.* Translated by Alan Bloom. New York: Basic Books, 1979.

Rubiés, Joan-Pau. “Instructions for Travellers: Teaching the Eye to See.” *History and Anthropology 9* (1996): 139–90.

Salm, Constance de. “Sur les voyages.” In *Poésies de Madame la Princesse Constance de Salm,* 115–18. Paris: Firmin Didot, 1811.

Schlick, Yaël. *Feminism and the Politics of Travel.* Lewisburg: Bucknell University Press, 2012.

Schroeder, Joy. *Dinah's Lament: The Biblical Legacy of Sexual Violence in Christian Interpretation.* Minneapolis: Fortress Press, 2007.

Stagl, Justin, Klaus Orda and Christine Kämpfer. *Apodemiken: eine räsonnierte Bibliographie der reisetheoretischen Literatur des 16., 17. und 18. Jahrhunderts.* Paderborn: Schöning, 1983.

Tarabotti, Arcangela. *Paternal Tyranny.* Edited and translated by Letizia Panizza. Chicago: University of Chicago Press, 2004.

Thompson, Carl. “Journeys into Authority: Reassessing Women's Early Travel Writing, 1763–1862.” *Women's Writing 24* (2017): 131–50.

Trouille, Mary Seidman. *Sexual Politics in the Enlightenment: Women writers read Rousseau.* Albany: State University of New York Press, 1997.

Turler, Hieronymus. *The Traveiler of Ierome Turler.* London, 1575.

Vermeulen, Han F. *Before Boas. The Genesis of Ethnography and Ethnology in the German Enlightenment.* Lincoln: University of Nebraska Press, 2015.

Warneke, Sara. *Images of the Educational Traveller in Early Modern England.* Leyden: Brill, 1995.

Whitaker, Katie. *Mad Madge: Cavendish, Duchess of Newcastle, Royalist, Writer and Romantic.* London: Vintage, 2004.

Williamson, Elizabeth. “‘Fishing after News’ and the Ars Apodemica: The Intellingencing Role of the Educational Traveller in the Late Sixteenth Century”, In *News Networks in Early Modern Europe.* Edited by Joad Raymond and Noah Moxham, 542–62. Leiden: Brill e-books series, 2016.

Winn, Colette H. “Des femmes en mouvement.” *Romanic Review 94* (2003): 115–51.

Woolley, Hannah. *The Gentlewoman's Companion.* London, 1670.

Zäunemannin, Sidonia Hedwig. *Curieuser und immer währender Astronomisch-Meteorologisch-Oeconomischer-Frauenzimmer-Reise und Händ-Kalender.* Erfurt: Joh. Michael Funcke, 1737.

11 Travel, utopia, and conflict

Patterns of irony in early modern utopian narratives

Daniel Carey

From the inception of the genre of utopian writing by Thomas More in 1516 through a series of instances across the early modern period, the relationship between travel and utopia remained deeply entwined. Although travel is not a necessary condition for utopia – after all, authors could simply dream up alternative social and political systems without mapping them in space or fabricating journeys to reach them[1] – the narrative form depends on displacement, the encounter with something deeply unfamiliar that invites, even if it does not require, the change of location made possible by travel. The separation between home and elsewhere, self and other, the known and unknown, becomes more immediate and pronounced when it is made integral to the storytelling element of utopia, whether realist or fantastical in execution. At the same time, this separation initiates a key dimension of utopian fictions: the process of translation between settings (the near and the remote). Such a process is productive of meaning, but it is never "innocent": the reflexive form of reading it implies depends on critique and negation of traditions adopted at home.

In the critical practices encouraged by split locations, we find the space of conflict within the utopian argument, both hidden and overt. Utopianism is intrinsically conflictual because the ideal it conjures must remain discordant with "reality". Narratives of utopia thrive by exposing home itself as the locus of conflict, and the alternative to home as a place of resolution and harmony, realizing an element of wish-fulfilment while censuring those held responsible for abridging it.

In this essay, I explore how travel enables the reimagining essential to utopian thinking, the potential for irony, and conditions of conflict out of which it emerges. Rather than beginning at the beginning, so to speak, with Thomas More, I open the discussion by considering Shakespeare's *The Tempest* (1612), composed nearly a century after More's *Utopia* (1516), as it recognizes the complexities of utopianism with particular richness, showing the depth of reflection made possible by More's intervention. This account will lead us back to More and then onward to consider the durability of travel and conflict as principles in utopian writing.

Utopian irony

In Act 2 of *The Tempest*, a brief but salient description of what might be called an Ovidian utopia occurs, in the midst of a sequence of potential political transformations and reprisals. Following shipwreck during a storm blown by Prospero, a distracted group of Italian noblemen wanders the island in a state of fatigue. The Neapolitan lord, Gonzalo, takes the occasion to imagine a different way of configuring human affairs. He addresses himself to Alonso, King of Naples, whom he serves as counsellor, in part to distract him from the loss of his son Ferdinand, presumed to have perished in the sea. He introduces the reflection in these terms:

> Had I plantation of this isle, my lord,—
>
> ...
>
> I'th' commonwealth I would by contraries
> Execute all things...[2]

The invitation to remake the island begins with a colonial concept of planting, but the model is premised on reversal. Everything would be done differently. But before Gonzalo can begin his account in earnest, his remarks are interrupted (indicated by the first ellipsis above).[3] No sooner does he utter the comment about having the plantation of the isle than Antonio (Prospero's usurping brother) intrudes with the sarcastic line, "He'd sow't with nettle-seed". Sebastian, the conspiring brother of the King of Naples, adds, "Or docks, or mallows" (2.1.142). Gonzalo, the ostensible statesman, would in fact cover his plantation with weeds. Even as ideality is being projected, the enemies or forces against it remain in place, moved by cynicism and destructiveness.

We have reason to regard what Gonzalo says as at least well intentioned, particularly because Prospero has already identified him as a figure of virtue. When Prospero recounts to his daughter his usurpation in Milan a dozen years before, when she was a child, he remarks that Gonzalo intervened to save their lives (he provided Prospero with his magic books after the duke and his daughter were expelled by his brother in a leaky boat deliberately to have them drown (1.2.160–68). Gonzalo proceeds with his description undeterred by these interjections. He enumerates a series of "contraries" that would characterize his new plantation:

> for no kind of traffic
> Would I admit; no name of magistrate;
> Letters should not be known; riches, poverty,
> And use of service, none; contract, succession,
> Bourn, bound of land, tilth, vineyard, none;
> No use of metal, corn, or wine, or oil;

No occupation; all men idle, all;
And women too, but innocent and pure;
No sovereignty; —

...

All things in common nature should produce
Without sweat or endeavour. Treason, felony,
Sword, pike, knife, gun, or need of any engine,
Would I not have, but nature should bring forth,
Of it [*sic*] own kind, all foison, all abundance,
To feed my innocent people.

(2.1.146–54; 157–62)

This series of negatives conjures something positive. Nature would provide for mankind, relieving residents of the burden of labour, while the idleness conferred by this condition – he quickly reassures us – would be attended by innocence, not corruption. True to form, Sebastian and Antonio make further objections along the way, noting the contradiction inherent in Gonzalo appointing himself governor of a place without a sovereign, and observing that the promotion of idle pleasure would render the inhabitants "whores and knaves", not innocents (2.1.155–56; 163–64).

Gonzalo's vision remains a fantasy, like much of the play's magical incarnations, not a practical utopia set out with ostensibly plausible political and administrative arrangements. He closes his speech by proclaiming that

I would with such perfection govern, sir,
To excel the golden age.

(2.1.165–66)

The line reminds us of the often shared space between utopia and classical as well as Biblical traditions.[4] Here the transformation required to effect this new arrangement entails a return to an age of innocence now cast into the future.

Gonzalo's vision represents a comprehensive political reform, however backward looking in inspiration. Questions of reform more generally are very much at stake in *The Tempest*, a play that wonders whether such changes can indeed take place. The key individual here is Caliban, the indigenous figure (although born to non-native parents) who enabled Prospero and Miranda to survive after their unanticipated arrival but who now finds himself in a position of brutalized service. Miranda, once the object of Caliban's planned sexual assault, condemns him as an "Abhorrèd slave, / Which any print of goodness wilt not take", and Prospero famously asserts that his is a nature on which "nurture never stuck" (1.2.350–51). Both of them speak in terms of imprinting and the resistance to it, calling into doubt whether moral transformation can occur. Caliban does achieve some moral recognition in the end, promising to "be wise hereafter, / And sue for grace" (5.1.294–95). But we have little reason to believe that the noble conspirators, Antonio and Sebastian, have learned any lessons. They notably remain silent in the closing scene, chastened only by Prospero's whispered threat to reveal their plots. Seen in this light, Gonzalo's projection is all the more illusory.

Much has been made of Gonzalo's lines, including Shakespeare's debt to Montaigne's essay "Of the Cannibals" in John Florio's 1603 translation of the *Essayes*,[5] and his insight into colonialism as a means of rewriting European experience. But they frequently overlook the more obvious tradition he invokes, namely the golden age as described by Ovid in the first book of *Metamorphoses*. In Arthur Golding's 1565 translation, the lines read:

> Then sprang vp first the golden age, which of it selfe maintainde,
> The truth and right of euery thing vnforst and vnconstrainde.
> There was no feare of punishment, there was no threatning lawe
> In brazen tables nayled vp, to cronch[6] or créepe to Iudge with cap in hand,
> They liued safe without a Iudge, in euerie Realme and lande...
> No horne nor trumpet was in vse, no sword nor helmet worne,
> The worlde was suche, that souldiers helpe might easly be forborne.
> The fertile earth as yet was frée, vntoucht of spade or plough,
> And yet it yéelded of it selfe of euery things [*sic*] inough.[7]

As in Ovid, the vision offered by Gonzalo is effectively pre-political – the absence of law and property means that no government is required. Thus it constitutes, conceptually, a state of nature.

Thomas Hobbes, alert to this tradition, used a contradiction of it to rationalize his argument for an all-powerful state in *Leviathan* (1651). For those who might imagine the state of nature as a condition of ease and fulfilment, Hobbes pointed out the opposite. He is best known for characterizing this situation as "poore, nasty, brutish, and short", but the preamble to this conclusion receives less attention:

> there is no place for Industry; because the fruit thereof is uncertain: and consequently no Culture of the Earth; no Navigation,[8] nor use of the commodities that may be imported by Sea; no commodious Building; no Instruments of moving, and removing such things as require much force; no Knowledge of the face of the Earth; no account of Time; no Arts; no Letters; no Society; and which is worst of all, continuall feare, and danger of violent death; And the life of man, solitary, poore, nasty, brutish, and short.[9]

So much for perpetual spring imagined by Ovid. While *The Tempest*'s exploration of inverted political order concentrates on patterns of usurpation, as well as invoking a golden-age utopia, Hobbes plays with a different inversion. His negated age of innocence is the ultimate scenario of conflict: it is so overwhelming that it requires an austere remedy, with the individual resigning authority to the sovereign power to escape from it. Hobbes reveals the utopian condition aligned with the golden age as being, in fact, a dystopia. For him, the proper golden age evidently corresponded not to the state of nature but to an earlier period in *civil* society when sovereign power was revered, a time that had given way to rebellion against monarchs in the era of the civil war.[10]

Travel is essential to the plot of *The Tempest* in a number of ways. On the one hand, the island offers a potential sanctuary from the world of the court, a site of would-be pastoral which might redeem the courtiers – yet pastoral remains at the level of fantasy. Ferdinand's effort to clear logs – a form of labour imposed on him by Prospero and lamented by Miranda – suggests that any opening of spaces will be hard won (3.1.9–11), while the rest of the island carries associations of an unfamiliar New World. The plot is shaped to allow Prospero the opportunity for return, but he places the power to realize this form of comic closure in the hands of the audience. Shakespeare ironizes his own position in the end; Prospero's power mirrors that of the dramatist in its forms of magic and control of the plot, but Shakespeare understands that the audience's willingness to complete the action through wish-fulfilment remains the ultimate source of control.

We are confronted then by a number of realizations: first, that utopia is born out of conflict; that the escape provided by travel, while potentially overturning everything, only proves temporary; and that utopia exists in a condition of irony. This ironic orientation on utopia appears from the beginning in Thomas More, although few would develop its full potential between his time and *The Tempest*.

Some of the lessons in irony may well have come via Montaigne and the essay "Des Cannibales" dating from 1580. In a famous discussion, Montaigne draws some paradoxical conclusions based on reports of the French encounter in Brazil in mid-century (a colonial project itself mired in conflict between Catholic and Protestant interests that preoccupied Montaigne at home). We move of course from the locus of drama to one of philosophical contemplation supposedly contingent on the "real". At the same time, the rhetorical complexity of Montaigne's response requires us to navigate between his sources of testimony, webs of classical allusion, and highly deflected analyses of the contemporary moment which only become clear in the end.

Montaigne tells us that his information on the New World derives from a servant of his who had been in Brazil for ten or twelve years, a man he describes as "a simple, crude fellow".[11] But his humble status and rusticity does not disqualify him as a vehicle of testimony. On the contrary, he represents a valuable resource precisely because he lacks sophistication and remains incapable of theory, unlike the clever who observe more carefully but also interpret what they witness (a failing in which Montaigne must implicate himself). Indeed, Montaigne immediately takes up this opportunity to "interpret", pronouncing himself satisfied that nothing barbarous or savage characterizes the Brazilian nation – those who condemn them merely indulge a preference for their own ways and (in an inverted pattern of utopia) declare their own location to constitute "the perfect religion, the perfect government, the perfect and accomplished manners in all things".[12] Travel in this context serves to endorse the domestic, not to undermine it. Montaigne inverts the order of priorities by praising nature and simplicity above artfulness and cultivation. He wishes, in fact, that Plato could have known of these people when framing his laws and commonwealth. Had this been possible, he would have seen something that surpasses what the poets imagined in an idealized golden age: "This is a nation, I should say to Plato, in which there is no sort of traffic, no knowledge of letters, no science of numbers, no name for a magistrate or for political superiority, no custom of servitude, ... The very words that signify lying, treachery, dissimulation, avarice, envy, ... unheard of".[13] A kind of interplay is going on here that would only be clarified much later by Schiller in his account of relationship between the naive and sentimental:[14] Montaigne takes naive testimony of a people who are themselves naive, living according to nature, wild and yet superior, and in a sentimental turn transforms them into an unreachable paradigm lost to the civilized. Yet at the same time, Montaigne serves as both analyst and witness, reporting that he not only possesses artefacts from this place and examples of their song and poetry to hand, but he adds a personal account of a substance they use in place of bread, which he compares to preserved coriander, "I have tried it; it tastes sweet and a little flat".[15] In this respect he is a hybrid figure with something in common with Thomas More's traveller Raphael – Montaigne is outside the paradise he describes and inside it, able to secure the narrative of a distant place with his own form of participation. As he engages in debate over what is superior and what is inferior, what should be reformed and what preserved, any advantage that home might have in the argument disappears. True, their practices in war are harsh, and when enemies are taken, they suffer cannibalization for purposes of revenge. But the Portuguese presence there merely taught them new and more disturbing methods of dispatching captives that they proceeded to adopt. Events at home in the midst of religious conflict provide too great a confirmation of how we surpass that other world in barbarity, performed on the living and not the dead. Their wars are fought to demonstrate valour, not to make conquests and seize property.

The conclusion of the essay brings together the force of the utopian challenge. At last the New World arrives in the Old to announce its critique: three representatives from Brazil came (allegedly) to Rouen to attend the late French king, Charles IX.[16] Asked for their opinion on what amazed them most, they remarked on the strangeness of grown men in arms submitting themselves to a child rather than enjoying their own command;[17] second, they observed during their visit how some men gorged themselves while others stood at their doors, "emaciated by hunger and poverty".[18] The political message was made clear by these visitors from elsewhere – how could the needy endure this condition and not seize the prosperous by the throat and torch their houses? The radical challenge of utopia becomes apparent. Montaigne reinforces the lesson by engaging one of the men in conversation, although he ironizes the scenario with a remark on the incompetence of the translator who botched the exchange.[19] We must infer, then, that as much as his writing depends on travel and the exotic to make its case, it remains a domestic discourse of self-interrogation. We learn that all the superiority this native individual enjoyed as a war captain came down to taking the foremost position in battle; afterwards, his sole privilege was to have his path cleared for him when visiting villages dependent on him. The image of a golden world, free from exaggerated distinction of rank and forms of sovereignty, remains intact and now places itself in immediate comparison.

While Shakespeare creates an ironic structure in relation to the utopian prospect by introducing discordant voices and denying straightforward identification, Montaigne achieves it by undermining the authority of his audience and substituting the savage for the civil. The potential for an ironized understanding of utopia was in fact set in place by Thomas More from the beginning. More's *Utopia* combines a productive relationship to travel, the scope for critique afforded by locating an alternative polity; the impossibility of it all (utopia is, after all, *ou topos*, no place); and the uncertain relationship forged with the reader – all realized in his account of what is offered, ostensibly, as the best state of the commonwealth, *de optimo reipublicae statu*, the work's Latin title. The two-book format is vital in attaining this effect, but the interpretive instability is enhanced by More's placement of himself in the text as an auditor who receives the narrative of the distant country from a returned traveller with great attention and interest, yet without clarifying his view:

> I can hardly agree with everything he said (though he is a man of unquestionable learning and enormous experience of human affairs), yet I freely confess that in the Utopian commonwealth there are many features that in our own societies I would wish rather than expect to see.[20]

Initially, More draws on the power of circumstantial detail to engage the reader. The report on Utopia comes from an invented Portuguese traveller, Raphael Hythlodaeus, who supposedly accompanied Amerigo Vespucci on the last three of his four voyages before obtaining leave to embark on his own travels. More appears under the name "Morus" as a character and narrator, relating an encounter with Raphael in Flanders during an actual diplomatic mission undertaken by More. Impressed by Raphael's extensive experience, Morus and his friend Peter Giles press him to engage in service to some European prince, but Raphael responds by denying that he could have any effect if he did so, given the absence of virtue among monarchs; the ruthless competition among courtiers; and range of problems requiring redress (including enclosure, the fate of returned soldiers, and military conflict among rulers). This outpouring of objections from Raphael leads to an invitation for him to describe the contrasting policy of Utopia, which occupies Book 2.

By splitting the contexts of description between Europe and Utopia, and leaving the narration and advocacy to Raphael, More indicates, subtly, Raphael's blindness to the contradictions in his tale. In Book 1, Raphael imagines himself at the French court where the subject under consideration is how to retain Milan, regain Naples, and add further conquests. The assembled counsellors variously recommend hiring German mercenaries and paying the Swiss to remain neutral; buying off the Holy Roman Emperor's opposition with gold; and keeping the English in check by stirring up the Scots, as well as encouraging a pretender to the throne to cause trouble. If Raphael were present and raised his voice in opposition, pointing out that the sovereign should concentrate on managing his own kingdom instead of pursuing these schemes, what kind of reception would he receive? Yet in Book 2, after a long detour of laudatory description of Utopia, Raphael tells us that the Utopians employ similar tactics in wartime. They use secret agents to infiltrate their opponents' state and offer rewards for killing the king or assassinating named people, they pay bounties for betraying the prince in vast sums of gold, they encourage the brother of the enemy's king to scheme for the crown, they entice neighbours to antagonize them (like the Scots at home), and they hire mercenaries to commit disgraceful crimes. The parallel evidently eludes Raphael, making his endorsement of their resolve never to allow "foreign auxiliaries onto their island" ring hollow.[21] What is ideal and worthy of adoption remains open to debate. Despite its well-ordered state and contented people, free from internal conflict (due not least to holding property in common), Utopia succumbs to contradictions.

Travel and utopia

The medium of travel remains a constant in these narratives. The productive relationship between "real world" travel and utopia begins with the transfer of descriptive technique. Morus asks Raphael for an account Utopia's "fields, rivers, towns, people, manners, institutions, laws" – essentially a chorography.[22] Structured advice on what to observe in the midst of travel, adopting headings of this kind, would become formalized over the course of the sixteenth century and beyond in a series of contributions to the genre of the *ars apodemica*.[23] This approach could be adapted in utopian writing to generate a framework of description in a realist vein, corresponding to readers' expectations for receiving accounts of distant places. Francis Bacon remarked in his utopia *New Atlantis* (1627) that the person who has visited foreign lands may "commonly know more by the eye than he that stayeth at home can by relation of the traveler", but the two together create what he calls "a mutual knowledge".[24]

The purpose of gathering this information was not merely to become better informed but to adopt what was meritorious in policy or moral practice. Raphael remarks that the Utopians took this approach, gaining the benefits of shipwrecked Romans who arrived 1200 years earlier by following their example and developing it, although he remained doubtful about how quickly his European peers, informed of the fruits of his expeditions, would "adopt any institutions of theirs which are better than ours".[25] Advice literature on early modern travel similarly approved the process of identifying and imitating laudable practices in government, the arts, or "manners" maintained elsewhere. As Bacon put it in his essay "Of Travaile" (1626), the encounter with other countries offered the chance to "prick in some Flowers" from customs established abroad.[26] The opportunity for imitation was typically confined in this period to the social practices of the "polite". More's *Utopia* proceeds on this assumption. "Morus" comments that he was unable to record everything that Raphael had to say of value: "Perhaps on another occasion we shall tell more about these things, especially those that it would be useful not to be ignorant of – above all, the wise and prudent provisions that he observed among the civilised nations".[27] Of course, Montaigne later overturns the pattern by positioning the cannibals as corrective critics. (Bacon, for his part, thought that visitors from America would have the scales fall from their eyes, recognizing the superiority of the metropole.)[28]

Thus we find embedded in the conventions of travel an expectation of imitation and adoption, a premise developed in utopian writing as a means of reform. In accounts of "authentic" travel, the emphasis typically fell on a selective commendation of customs rather than on wholesale endorsement of another way of life. Outright negation of home and chastisement of the mother country remained the preserve of utopian writing, for the most part. We can find occasional works of travel that bring the two possibilities of adoption and critique together, but they are instructive for their relative rarity. Edward Terry's *A Voyage to East-India* (1655) provides one such example. This volume was a much-extended description of a prior version included by Samuel Purchas in *Purchas his Pilgrimes* thirty years earlier. Terry, an Anglican priest educated in Oxford, embarked in 1616 as an East India Company chaplain. On his arrival in India, he became chaplain to Sir Thomas Roe, English ambassador to the Mughal court of the emperor Jahangir, spending twenty-seven months there before entering the church on his return.[29] Although Terry maintained his staunchly critical view of the Muslim faith in his description of India, he continually appropriates the occasion of his travels as an opportunity to reprove his readers for their moral and religious failings and to set them in contrast with what he observed abroad. In fact he devotes an entire section to "the most excellent moralities, which are to be observed amongst the people of those Nations".[30] He commends Indian virtues of temperance; justice; "unwearied *devotion*" and prayerfulness; respect for parents; and their refusal to indulge a taste in fashion, preferring traditional dress. As for their military practice, they had "not well learned that horrid bloody *art of war,* as the *Europeans* have; and wanting Commanders, and other Officers to manage their great Companies, are not so *skilfull to destroy*, as otherwise they might be", anticipating a theme on which Swift would expand in *Gulliver's Travels*.[31] The sensibility of the satirist (Terry is drawn to Juvenal) and the sermonizer appear throughout, no doubt speaking to his condition as a disaffected royalist in the revolutionary period and situation as an aging cleric.

A prefatory poem to Terry's 1655 edition by Edward Waterhouse picks up on the adjacency of the work to utopia but he seeks to distance it from association with the unreal. Terry pleases us, Waterhouse says,

> Not with *Utopian* fancies, nor with vain
> Delusions, brought unto us from the main
> Invention, backt with boldnesse, so set out
> As if we must believe, not dare to doubt.[32]

Waterhouse dwells on the story's credibility, but his insistence would have been unnecessary if Terry had not engaged with the precedent set by utopian writing in offering us a distant ideal, bound up with an often swingeing critique.

Terry pitches his argument in a reformist mode, consistent with his role as minister of the word. The other polarity in narratives of travel that border on utopia (as we see in *The Tempest*) is that of fantasy and wish-fulfilment. New World promotional tracts drew on these associations, depicting a place of abundant resources and welcoming natives, heightening expectations of a golden age of ease and contentment. Arthur Barlowe made this explicit in 1584 in his account of the settlement attempted on Roanoke Island in a letter to the mission's sponsor, Walter Ralegh:

> We were entertained with all loue, and kindnes, and with as much bountie, after their manner, as they could possibly deuise. Wee found the people most gentle louing, and faithful, void of all guile, and treason, and such as liued after the manner of the golden age. The earth bringeth foorth all things in aboundance, as in the first creation, without toile or labour.[33]

This happy prospect would quickly dissolve, as subsequent reports from Ralph Lane and others demonstrated (Richard Hakluyt printed them in sequence in his *Principall Navigations* in 1589). Not only did lethal conflict ensue, but also starvation in this fecund territory. Nonetheless, Barlowe's remark further demonstrates the ready availability of an idiom apparent in both Montaigne and Shakespeare, deployed here to capture the experience of the actual.

Nor was this discourse confined to the New World. In 1571, Sir Thomas Smith (author of *De republica Anglorum*) and his son (also Thomas) received a grant of sizeable lands in Ireland in the Ards Peninsula in Co. Down to establish a private colony. He published an anonymous *Letter* describing the handsome possibilities associated with this tract that he presented as full of utopian potential. He asks in the work, "How say you now, haue I not set forth to you another *Eutopia*?"[34] Here utopia (no place) is transformed into a happy or good place – *eu-topia*, situating it in a literal landscape.[35] Like the later Roanoke project, the enterprise of plantation in the Ards, led by the younger Thomas Smith, ended in total failure and he was killed in 1573 by an Irishman under his command.[36] When fantasy and utopian aspiration became grafted onto the actual, the result was usually profound disillusionment.

Among colonial tracts engaging with utopian prospects, an anonymous work issued in 1610 by order of the Council of Virginia to support the fledgling colony (re-founded at Jamestown three years before) made a more honest assessment. The author openly addressed the incredulous response of those obliged to reconcile tales of the territory's merits in foodstuffs and climate with stories of disease and starvation:

> If any man shall accuse these reports of partiall falshood, supposing them to be but Vtopian, and legendarie fables, because he cannot conceiue, that plentie and famine, a temperate climate, and distempered bodies, felicitie, and miserie can be reconciled together, let him now reade with judgement, but let him not judge before he hath read.[37]

To explain the contradictions, he had to describe the ungoverned condition of the colony, which came about when the new leader of Virginia, Sir Thomas Gates, encountered a violent storm during his journey to take up his governorship (the narrative of which had provided Shakespeare with a source). The remainder of the fleet, arriving on the mainland, contended over mastery: in this "tempest of dissention", "euery man ouervaluing his worth, would be a Commander: every man vnderprising an others value, denied to be commanded". To make matters worse, the settlers, a band of "mutinous loiterers",[38] failed to sow or manure the land. Covetous mariners, meanwhile, embezzled goods, inflated the market in trifles, and drove up the price of corn and food. Nets for fishing were spoiled due to men's idleness. Thus, utopia was transformed needlessly into dystopia. Observing the situation, Powhatan (the Algonquian leader) exploited the misrule and massacred 30 settlers in an ambush. When Gates arrived after his ten-month detour in Bermuda, he resolved in fact to abandon the Virginia colony and set sail for England, dissuaded only by the timely appearance of Lord De La Warr with a relief convoy. Rescuing utopia from reality was no easy task.

Utopian wish-fulfilment

Utopian fictions provided a more resilient medium for wish-fulfilment, often on the premise of locating spaces in which conflict disappears. In Margaret Cavendish's *The Blazing World* (1666), the attributes that make it utopian are not so much a perfected code of laws or set of institutions but its political order. The narrative describes the arrival in the Blazing World of a beautiful visitor who marries the emperor and assumes the role of sovereign. Under her rule, the country "was governed without secret and deceiving policy; neither was there any ambition, factions, malicious detractions, civil dissensions or home-bred quarrels, divisions in religion, foreign wars, etc. but all the people lived in a peaceful society, united tranquility, and religious conformity";[39] it has no rebels or enemies.[40] When contentions grow among the different schools of naturalists active in the country, she promptly disbands them. The empress acknowledges that some difficulties have emerged since she began her reign, resulting from her decision to alter the form of government (a move she attributes to "the nature of women, being much delighted with change and variety").[41] The Duchess of Newcastle (here Margaret Cavendish inserts herself into the narrative as a companion and advisor to the protagonist) responds by encouraging her to return to the original conception, based on "one sovereign, one religion, one law, and one language, so that all the world might be as one united family, without divisions".[42] The "restoration" of an original authority makes the historical moment of the text abundantly clear.

For all its authoritarian political structure, *The Blazing World* offers a refreshing alternative to patriarchal wish-fulfilment that elsewhere characterizes the utopian genre,[43] to the extent that Cavendish places a female figure in charge. The more common approach was to sustain patriarchy and idealized gender relations in such narratives. Bacon, for example, devotes a substantial part of *New Atlantis* to describing the custom of the "Feast of the Family" in Bensalem, which "shews that nation to be compounded of all goodness".[44] The feast is accorded to any man who has thirty living descendants above three years of age. At the feast, the father (called a Tirsan) appears and processes with his lineage, first the males, then the females. "If there be a mother from whose body the whole lineage is descended", she is present in a sort of artificial chamber but is not seen. During the feast itself, the Tirsan's male children "perform unto him all service of the table upon the knee", with female offspring standing around him, "leaning against the wall".[45]

Idealized obedience to male authority appears from the start of the tradition: in More's *Utopia*, we learn that "Wives act as servants to their husbands, children to their parents, and generally the younger to their elders". As part of their religious observances, Utopian wives kneel before their husbands, and children before their parents, to ask for forgiveness.[46] The enforcement of obedience in *The Tempest* is one of the crucial functions of Prospero's "rough magic" – he threatens his servant Ariel with the same punishment inflicted on the spirit by the wicked Sycorax before Prospero's arrival (confinement for a twelve-year term within a cloven pine). His spell puts his daughter Miranda to sleep so that she does not witness these exchanges, and she later avows that she has never seen him "touch'd with anger" (4.1.144–45).

Perfected social and domestic relations described in utopian writing include provisions for regulating sexuality. The "Lunars" visited by the protagonist of Francis Godwin's *Man in the Moone* (1638) have known no "Whoremonger" in a thousand years; after being with a woman once, a man there finds that by a "secret disposition of nature…[he] never desireth any other".[47] The Utopians are similarly monogamous, although divorce is allowed under certain conditions, hinting at a territory of discord. No brothels exist on the island.[48] Bacon's visitor to Bensalem likewise declares that "there are no stews, no dissolute houses, no courtesans" and adds that, "as for masculine love, they have no touch of it".[49]

Religion was the other great context in which the fantasy of an unconflicted space could flourish in utopianism. Gabriel Plattes, a member of the Hartlib Circle, composed a short utopia called *Macaria*, published in 1641 after the easing of Laudian censorship in the lead-up to the Long Parliament, which it was clearly designed to influence. Composed as a dialogue, *Macaria* features a "Traveller" who informs an English "Scholar" about an ideal state he has visited, which lives in "plenty, prosperitie, health, peace, and happinesse".[50] Macaria on his account is a Christian country, but when the Traveller is asked whether Catholic or Protestant, he replies, "Their Religion consists not in taking notice of severall opinions and sects, but is made up of infallible tenets…[which] will abide the grand test of extreme dispute", allowing them to evade the difficulty of "Schismes and Heresies".[51] The state punishes divines with death if they publish their opinions to the common people. As a result, there is "no diversitie of opinions amongst them". If they contrive a new conception, they can present it to the "Great Councell" for debate but not to the wider public. The Anglican divine and participant in the Royal Society, Joseph Glanvill, wrote a continuation of Bacon's *New Atlantis*, where he used the better part of his text to lambast dissenters who had plunged the country into civil war, employing allegory to fierce effect.[52] In *Utopia*, religious stability occurs through measures of toleration, but other narratives achieved it on the basis of uniformity. Bacon has his Bensalemites converted twenty years after the crucifixion (St Bartholomew imparted Scripture to them, via an ark containing the Bible and an explanatory letter; helpfully, the revealed text contained books of the New Testament not yet written).[53] The miraculous event was certified by a member of Salomon's House, showing the compatibility between science, piety, and true religion. In *The Blazing World*, Cavendish's visitor-empress simply converts everyone to her religion, without a hint of further trouble.[54] As these examples demonstrate, for all of the reversals made possible by utopianism, Christianity itself could hardly be negated.

Ironic reflection on religion was nonetheless possible, facilitated by exploring the contrast between practices in the utopian state and those maintained at home. In this way, utopianism could serve as a prospective vehicle of reform. The 1573 edition of William Bullein's *Dialogue against the Fever Pestilence* expands the storytelling of the comic traveller Mendax to include his description of a distant city Metonoyae, a paragon of reformation, in a land which he reports to be "our Antipodie, foote against foote", called *Taerg Natrib* – Great Britain in reverse. Metonoyae is thus a metonym or substitution. Bullein was a cleric who resigned his living at the outset of Mary's reign, and traces of his sharp perspective appear in the midst of the ironized account. The reformist accomplishments of the model city include observances of the sabbath, with proper church attendance; the closing of taverns; and dedication to prayer and fasting. These practices improved from an earlier time when dens of iniquity, called *Seiabba* (abbeys), still existed. Each church has two ministers, but no pluralities – that is, no holders of benefices allowing them to employ others to perform their roles. They have no "mingled doctrine" or "tromperie of Papistrie" but only the purity of the word. The embrace of virtue is accompanied by severe punishments for vice, suitable to the crime (the "double handed Lawier" lands in prison and has his goods seized, the extortioner becomes a beggar, and Catholics are burned for treason). Fortunately no crimes have been committed for twenty-one years – perhaps a reference to the Act of Uniformity of 1552, enacted under Edward VI, the same length of time before the appearance of the second edition of the *Fever Pestilence*. Should we interpret the text as implying that the reverse is true of Britain? Bullein complicates matters by having Mendax explain before he embarks on this account that his own name in the "*Ethiope* tongue" signifies the name of a great city, Emor, with manners like those of "*Modos* & *Romog*" – that is, Rome and the customs of Sodom and Gomor. In short, the description of an alien place, committed to the Reformation, which is the antipode of England, comes from a liar aligned with Rome.[55]

Concluding questions: Crusoe and Gulliver

The utopian prospect features in different ways in the two great fictional works of travel of the eighteenth century – *Robinson Crusoe* (1719) and *Gulliver's Travels* (1726). In the case of Defoe's novel, this moment occurs late on in the narrative, following the arrival of Friday during Crusoe's twenty-fifth year on the island. The community subsequently expands to include Friday's father and a Spanish captive, both redeemed, like Friday, from imminent slaughter. What had initially been a mock kingdom takes on the form of a model commonwealth in which Crusoe exercises exclusive authority ("I was absolute Lord and Law-giver") and extends religious toleration to his "Subjects", Protestant, pagan, and Catholic alike.[56] The utopian dimension appears not only in the creation of a well-regulated colony, with Crusoe as uncontested ruler to whom the others owe their lives and service in gratitude,[57] but also in the deeply patriarchal conception of his role. Reconciliation with the will of the father – above all, the divine father – represents the great lesson of his experience of shipwreck and spiritual regeneration. Crusoe's whole history unfolds as a result of ignoring his own father's sensible direction to him as a young man to adhere to the middle station of life; he impatiently urges on his fortune with disastrous consequences. The retrieval of the Bible from the ship and Crusoe's new-found awareness of a providential shaping of his life leads to a religious awakening that reinstates a proper (patriarchal) order.

The commitment to patriarchal authority continues in the closing of the island sequence of the book when Crusoe restores an English captain to his command after his crew has mutinied – an episode that resonates in interesting ways with *The Tempest*. The ship arrives in the harbour and sends a boat towards the island, containing three captives, including the captain, along with a number of the rebel sailors who intend to abandon them there. Crusoe finds out later that the boatswain was the chief conspirator, the first hint of a connection with *The Tempest*'s inversion of political authority.[58] The malefactors separate to search the area, allowing Crusoe to present himself to the beleaguered captain. Crusoe tells him his story, "a whole Collection of Wonders",[59] before plotting how to address the predicament. When a new search party comes to the island after the boat fails to return, Crusoe instructs Friday and the captain's mate to halloo to the visitors in order to draw them into the island, always keeping some distance ahead until they fatigue the men searching for the source of the noise. This stratagem picks up on Ariel's use of music and strange sounds to lead the noblemen in exhaustion across the island. Crusoe describes the reaction of the tormented mariners: "We could hear them call to one another in a most lamentable Manner, telling one another, they were gotten into an inchanted Island; that either there were Inhabitants in it, and they should all be murther'd, or else there were Devils and Spirits".[60] This narrative is ultimately one restoring rightful power, with the captain regaining his direction of the vessel, and the retaken culprits saying they would "own him for a Father to them as long as they liv'd".[61] Crusoe, having concealed himself from these enemies (as Prospero does until the end), is named governor by the captain, and the ship declared "*all yours, and so are we and all that belong to her*".[62]

The other great travel fiction of the century, *Gulliver's Travels*, came from an author whose perspective on human nature made him unlikely to embrace utopian possibilities. Instead we find a deeply ironic relationship to the form in Swift's satirical vision. This attitude emerges most clearly in the voyage to Brobingnag and Gulliver's reporting of his conversation with the king, a man of "excellent Understanding", who interrogates Gulliver on the social, economic, and legal practices of his homeland. Gulliver repeats a familiar pattern of insisting on the virtues of his place of origin; he laments only that he lacked the tongue of Demosthenes or Cicero to "celebrate the Praise of my own dear native Country in a Style equal to its Merits and Felicity". He launches into an extended discourse that occupied five audiences of several hours each, during which he endorsed the country's natural resources, political and religious system, trade, and military strength. The king thanks him for his "admirable Panygyrick" but concludes that he had succeeded only in proving that in England "Ignorance, Idleness, and Vice are the proper Ingredients for qualifying a Legislator"; those responsible for interpreting the laws merely subverted them. He did allow some traces of a system which, in its original form, might have been "tolerable", but corruption had blotted it out. Point by point, he negates all of Gulliver's endorsements, ending with the famous observation that the populace consisted of "the most pernicious Race of little odious Vermin that Nature ever suffered to crawl upon the Surface of the Earth".[63]

What are we to make of this exchange? We can situate Gulliver's discourse in various genres, including panegyric, as the king suggests. But we can also see him as offering up his home country as utopia, describing it in the language suitable to that form. Swift arguably revisits the ironized relationship between Book 1 and Book 2 of More's *Utopia* here: the person who thinks he is living in a utopia is actually living in a dystopia. Raphael, conversely, knows he is living in a dystopia in Europe but fails to see the limits of Utopia itself.

In the succeeding chapter (chapter 7 of Part II), Gulliver admits his resentment of this abuse of his beloved country. He attributes the king's contemptuous attitude not to any perception of real demerits of England but to his confined experience. To prove the point, Gulliver describes an occasion when he told the king about the invention of gunpowder, offering an enthusiastic account of its capacity to wreak violent destruction.[64] We are in reach here of Edward Terry's sanction of the Mughals for failing to emulate Western warfare. Gulliver attributes the king's horrified reaction to his description to "*narrow Principles* and *short Views*!" In another conversation, Gulliver pointed out the existence of thousands of books on the art of government produced at home, but the king found cause to dispraise this too, despising "*Mystery*, *Refinement*, and *Intrigue*...in a Prince or a Minister" and the cultivation of secrets of state. In his view, governing consists of the exercise of common sense and reason and rapid resolution of civil and criminal cases. Brobingnagian laws are expressed in no more than twenty-two words (corresponding the number of letters in their alphabet), using plain and simple language; engagement in commentary on the laws represents a capital crime. Among their other values, they content themselves with a modest citizen army. And they regard a person who can make two ears of corn grow where one appeared before as doing more to serve the country "than the whole Race of Politicians put together".[65] Gulliver considers the learning of the Brobdingnagians as "very defective" since it consists only of morality, history, poetry, and mathematics, even if they excelled in these domains. Unlike the Academy of Lagado described in Part III of *Gulliver's Travels*, with its ruinous impracticality, Brobdingnagians apply themselves to useful things, such as agriculture and mechanical arts, rather than indulging in abstractions. Once more, we have an ironic reversal in which the traveller fails to recognize that he has encountered a kind of utopia.[66]

Gulliver does give himself over to the utopian impulse in the fourth voyage, which takes him to the Houyhnhnms, the noble horses who adhere to reason and truth, unable to understand the concept of lying.[67] His identification with them mirrors Raphael's in *Utopia* – both travellers wish to remain in the ideal country they have located. After his unwanted departure, Gulliver returns home disaffected, unable to integrate into his own family and left to commune with the sadly degenerate horses in his stable. Swift leaves us in an ironic predicament of non-identification – not just with Gulliver as a narrator since his account cannot be trusted, but with Gulliver as advocate for the Houyhnhnms. His endorsement depends on distancing himself from the Yahoos, his own species, and overlooking the rational Houyhnhnms' readiness to exterminate the race. Only under this threat does he leave, embarking in a canoe he makes for himself by covering it in Yahoo skins and using Yahoo tallow to make it sound. The world he upholds is literally inhuman.

The deep connection between travel and utopia in the early modern period made possible a series of powerful explorations of identity and difference, wish-fulfilment and critique, at once resolving conflict and embedding it in the form as a locus of contradiction. Those who ironized utopia, from More to Montaigne, Shakespeare, and Swift, recognized not only the fixation with alternative worlds but also the incapacity of human beings to realize them.

Notes

1 For further discussion, see Daniel Carey, "Utopian Fiction", in *The Oxford History of the Novel in English*, vol. 1: *Prose Fiction in English from the Origins of Print to 1750*, ed. Thomas Keymer (Oxford: Oxford University Press, 2017), 227–42. I have drawn occasionally on this essay for my discussion here.

2 William Shakespeare, *The Tempest*, ed. Stephen Orgel (Oxford: Oxford University Press, 1994), 2.1.141, 143, 145–46.

3 Interruptions are important elsewhere in play; notably, Prospero's spell is broken by his own involuntary interruption when he stages a pastoral fantasy to celebrate the engagement of Miranda and the recovered Ferdinand. Shakespeare, 4.1.126–27, 139.

4 For discussion of different modalities, see J.C. Davis, *Utopia and the Ideal Society: A Study of English Utopian Writing, 1516–1700* (Cambridge: Cambridge University Press, 1981).

5 See in particular Kenji Go, "Montaigne's 'Cannibals' and *The Tempest* Revisited," *Studies in Philology* 109 (2012): 455–73.

6 *sic*: crouch.

7 Ovid, *The. xv. bookes of P. Ouidius Naso, entytuled Metamorphosis*, trans. Arthur Golding (London, 1567), B2r–v.

8 The reference to navigation also responds to Ovid, I would argue, whose account of the golden age reports, in Golding's version, "The loftie Pynetrée was not hewen from mountaines where it stood, / In séeking straunge and forren landes, to roue vpon the flood. / Men knew none other countries yet, than where themselues did kéepe" (B2v).

9 Thomas Hobbes, *Leviathan*, ed. Richard Tuck, rev. ed. (Cambridge: Cambridge University Press, 1996), 89. In answer to the suggestion that such a time never existed, Hobbes goes on to introduce anthropological support, "the savage people in many places of *America*, except the government of small Families, the concord whereof dependeth on naturall lust, have no government at all; and live at this day in that brutish manner, as I said before" (89).

10 In the English translation of *De cive* (*Philosophicall Rudiments Concerning Government and Society* (1651), Hobbes comments, "*in truth, the simplicity of those times was not yet capable of so learned a piece of folly* [as opposing the authority of the sovereign]. *Wherefore was peace, and a golden age, which ended not before that* Saturn being *expelled, it was taught lawfull to take up arms against Kings*". *De Cive: The English Version*, ed. Howard Warrender (Oxford: Clarendon Press, 1983), 31. On this passage see Richard Tuck, who does not make a connection between *Leviathan* and Ovid. "The Utopianism of *Leviathan*," in *Leviathan after 350 Years*, ed. Tom Sorell and Luc Foisneau (Oxford: Clarendon Press, 2004), 125–38 (127). John Locke issues a reply, I would argue, in the Second Treatise of Government. In the Golden Age "there was then *no stretching Prerogative* on the one side to oppress the People; *nor* consequently on the other any *Dispute about Priviledge*, to lessen or restrain the Power of the Magistrate". The problem begins with overextended power, in other words; in a note to the passage he clarifies that "*They saw, that to live by one Man's Will, became the cause of all Mens misery*". *Two Treatises of Government*, ed. Peter Laslett (Cambridge: Cambridge University Press, 1988), 342–43 (§111).

11 Montaigne supplements this source by reports from sailors and merchants. Michel de Montaigne, *The Complete Essays of Montaigne*, trans. Donald Frame (Stanford: Stanford University Press, 1965), 152. This length of stay in Brazil is somewhat implausible; Philippe Desan points out that this was twice the duration of the entire short-lived Antarctic France. *Montaigne: A Biography*, trans. Steven Rendall and Lisa Neal (Princeton: Princeton University Press, 2017), 162.

12 Montaigne, 152.

13 *Ibid.*, 153.

14 Friedrich von Schiller, "On Naive and Sentimental Poetry," in Hugh Barr Nisbet, ed., *German Aesthetic and Literary Criticism: Winckelmann, Lessing, Hamann, Herder, Schiller, Goethe* (Cambridge: Cambridge University Press, 1985), 180–232.

15 Montaigne, 154.

16 Desan, 167–75, suggests that this encounter happened not in Rouen but in Bordeaux in 1565 when Montaigne was involved in organizing a parade of Brazilians on the occasion of Charles IX's entrance into the city.

17 If this alleged encounter took place in 1565 in Bordeaux, as Desan suggests, then the king was then aged 15.

18 Montaigne, 159.

19 For discussions of the role of the translator and the relationship between translation, mistranslation, and cultural misunderstandings, see essays by Marie-Christine Gomez-Géraud and Céline Bonnotte-Hoover in this collection.

20 Thomas More, *Utopia*, ed. George M. Logan and trans. Robert M. Adams, 3rd ed. (Cambridge: Cambridge University Press, 2016), 113.
21 *Ibid.*, 97.
22 *Ibid.*, 40.
23 On this literature, see Daniel Carey, "Advice on the Art of Travel," in *The Cambridge History of Travel Writing*, eds. Nandini Das and Tim Youngs (Cambridge: Cambridge University Press, 2019), 392–407; Juliette Morice, *Le Monde ou la bibliothèque: voyage et éducation à l'âge classique* (Paris: Les Belles Lettres, 2016). For further considerations on aspects of this literature, see Natalya Din-Kariuki, Eva Johanna Holmberg, and Gábor Gelléri's essays in this volume.
24 Francis Bacon, *The Advancement of Learning and New Atlantis*, ed. Arthur Johnston (Oxford: Clarendon Press, 1974), 224.
25 More, 42.
26 Francis Bacon, "Of Travaile," in *The Essayes or Counsels, Civill and Morall*, ed. Michael Kiernan (Oxford: Clarendon Press, 1985), 58.
27 More, 12.
28 Bacon, "Of Plantations," in *Essayes*, 108.
29 See Daniel Carey, "Edward Terry's *A Voyage to East-India* (1655): A Chaplain's Narrative of the Mughal World," *Études anglaises* 70 (2017): 187–208.
30 Edward Terry, *A Voyage to East-India* (London, 1655), R3v–R8v.
31 *Ibid.*, A6r.
32 *Ibid.*, A7r.
33 David Beers Quinn, ed., *The Roanoke Voyages 1584-1590*, 2 vols (London: Hakluyt Society, 1955), 1:108.
34 Anon, *A Letter sent by I.B. Gentleman* (London, [1572]), E1r.
35 Cathy Shrank, *Writing the Nation in Reformation England 1530–1580* (Oxford: Oxford University Press, 2004), 176.
36 Christopher Maginn, "Smith, Thomas (1547–1573), Colonial Adventurer." *Oxford Dictionary of National Biography* (Oxford: Oxford University Press, 2008), https://www.oxforddnb.com/view/10.1093/ref:odnb/9780198614128.001.0001/odnb-9780198614128-e-69171; on the Ards project, see Hiram Morgan, "The Colonial Venture of Sir Thomas Smith in Ulster, 1571–1575," *Historical Journal* 28 (1985): 261–78; Nicholas Canny, *Making Ireland British 1580–1650* (Oxford: Oxford University Press, 2001), 120–21.
37 Anon, *A True Declaration of the estate of the Colonie in Virginia* (London, 1610), 33–34.
38 *Ibid.*, 34, 36.
39 Margaret Cavendish, "The Description of a New World, Called the Blazing World," in *Political Writings*, ed. Susan James (Cambridge: Cambridge University Press, 2003), 1–119 (75–76).
40 *Ibid.*, 107.
41 *Ibid.*, 87. The narrator reports that even before the empress ascends to power, the polity was free from conflict, "for there was but one language in all that world, nor no more than one Emperor, to whom they all submitted with the greatest duty and obedience, which made them live in a continued peace and happiness, not acquainted with other foreign wars, or home-bred insurrections" (13).
42 *Ibid.*, 87.
43 On the development of other female utopias, see Alessa Johns, *Women's Utopias of the Eighteenth Century* (Urbana: University of Illinois Press, 2003); Nicole Pohl, *Women, Space, and Utopia, 1600–1800* (Aldershot: Ashgate, 2006).

44 Bacon, *New Atlantis*, 231.
45 *Ibid.*, 232, 233.
46 More, 57, 107. The country does, however, have female priests, and women engage in military exercises (94, 104).
47 Francis Godwin, *The Man in the Moone*, ed. William Poole (Peterborough, ON: Broadview, 2009), 112. For a brief overview of lunar narratives and representations of a harmonious society, see Rachel Willie's essay in this volume.
48 More, 62, 84. Montaigne reports the Tupinambá to be polygamous, but rather than this situation leading to conflict, women in this social system strive for their men to have more wives as a demonstration of their husbands' honour (*Complete Essays*, 158).
49 Bacon, *New Atlantis*, 235, 236.
50 Gabriel Plattes, "A Description of the Famous Kingdome of Macaria," in Charles Webster (ed.), *Samuel Hartlib and the Advancement of Learning* (Cambridge: Cambridge University Press, 1970), 79–90 (81).
51 *Ibid.*, 85.
52 Joseph Glanvill, "Anti-fanatical Religion and Free Philosophy. In a Continuation of the New Atlantis," in Glanvill, *Essays on Several Important Subjects in Philosophy and Religion* (London, 1676), 3–47. In the 1720 utopia attributed to Ambrose Phillips, no disputes or dissension over religion occur. "The Fortunate Shipwreck, or a Description of New Athens," first published in *Miscellanea Aurea; or The Golden Medley* (London, 1720), 80–117, in *Utopias of the British Enlightenment*, ed. Gregory Claeys (Cambridge: Cambridge University Press, 2004), 27–53 (43).
53 Bacon, "New Atlantis," 233.
54 Cavendish, 49.
55 William Bullein, *A Dialogue against the Feuer Pestilence*, ed. Mark W. Bullen and A.H. Bullen (London: N. Trübner & Co., 1888), 105–11.
56 Daniel Defoe, *Robinson Crusoe*, eds. Thomas Keymer and James Kelly (Oxford: Oxford University Press, 2007), 203. There is no atheist among them to test the limits of toleration. In Utopia they exempt non-believers from the benefits of toleration (98).
57 When Crusoe returns to his island in the *Farther Adventures* (1719), he discovers that this experiment has undergone various trials and difficulties, but he leaves it in a condition of order with an elaborated political system to ensure its survival.
58 Defoe, 219, 224.
59 *Ibid.*, 217. This may also recall *The Tempest*, whether the naming of Miranda or the dilemma of imparting travellers' tales mentioned in Act 3, Scene 3.
60 Defoe, 224.
61 *Ibid.*, 227.
62 *Ibid.*, 229.
63 Jonathan Swift, *Gulliver's Travels*, ed. David Womersley (Cambridge: Cambridge University Press, 2012), 178, 179, 188, 189.
64 Swift may possibly respond to the unironic narrative of a European visitor helpfully describing gunpowder and demonstrating the "wonderful Force" of it to the utopian New Athenians in "The Fortunate Shipwreck," 35–36.
65 Swift, 193, 194.
66 The end of the chapter does reveal that Brobdingnag has had its political problems, experiencing the same difficulties as the rest of mankind in the contention for power by the nobility, the common people's demand for liberty, and the king's desire for "absolute dominion" (Swift, 201). Civil wars had taken place two generations before (paralleling Gulliver's home country), ended by the

king's grandfather. Swift was hardly likely to present an unalloyed utopian vision.

67 In the "Letter from Capt. Gulliver" added to the 1735 edition, the protagonist defends his truthfulness against those who "are so bold as to think my Book of Travels a meer Fiction ... and have gone so far as to drop Hints, that the *Houyhnhnms*, and *Yahoos*, have no more Existence than the Inhabitants of *Utopia*" (Swift, 13). Thomas More is one of the worthies called up in Part 3 during Gulliver's visit to the island of magicians, Glubbdubdrib (292).

Bibliography

1572 Anon. *A Letter Sent by I.B. Gentleman*. London, 1572.

1610 Anon. *A True Declaration of the Estate of the Colonie in Virginia*. London, 1610.

Bacon, Francis. *The Advancement of Learning and New Atlantis*. Edited by Arthur Johnston. Oxford: Clarendon Press, 1974.

Bacon, Francis. *The Essayes or Counsels, Civill and Morall*. Edited by Michael Kiernan. Oxford: Clarendon Press, 1985.

Bullein, William. *A Dialogue against the Feuer Pestilence*. Edited by Mark W. Bullen and A.H. Bullen. London: N. Trübner & Co., 1888.

Canny, Nicholas. *Making Ireland British 1580–1650*. Oxford: Oxford University Press, 2001.

Carey, Daniel. "Edward Terry's *A Voyage to East-India* (1655): A Chaplain's Narrative of the Mughal World." *Études anglaises 70* (2017): 187–208.

Carey, Daniel. "Utopian Fiction." In *The Oxford History of the Novel in English, vol. 1: Prose Fiction in English from the Origins of Print to 1750*. Edited by Thomas Keymer, 227–42. Oxford: Oxford University Press, 2017.

Carey, Daniel. "Advice on the Art of Travel." In *The Cambridge History of Travel Writing*. Edited by Nandini Das and Tim Youngs, 392–407. Cambridge: Cambridge University Press, 2019.

Cavendish, Margaret. "The Description of a New World, Called the Blazing World." In *Political Writings*. Edited by Susan James, 1–109. Cambridge: Cambridge University Press, 2003.

Claeys, Gregory, ed. *Utopias of the British Enlightenment*. Cambridge: Cambridge University Press, 2004.

Davis, J.C. *Utopia and the Ideal Society: A Study of English Utopian Writing, 1516–1700*. Cambridge: Cambridge University Press, 1981.

Defoe, Daniel. *Robinson Crusoe*. Edited by Thomas Keymer and James Kelly. Oxford: Oxford University Press, 2007.

Desan, Philippe. *Montaigne: A Biography*. Translated by Steven Rendall and Lisa Neal. Princeton: Princeton University Press, 2017.

Glanvill, Joseph. *Essays on Several Important Subjects in Philosophy and Religion*. London, 1676.

Go, Kenji. "Montaigne's 'Cannibals' and *The Tempest* Revisited." *Studies in Philology 109* (2012): 455–73.

Godwin, Francis. *The Man in the Moone*. Edited by William Poole. Peterborough, ON: Broadview, 2009.

Hobbes, Thomas. *De Cive: The English Version*. Edited by Howard Warrender. Oxford: Clarendon Press, 1983.

Hobbes, Thomas. *Leviathan*. Edited by Richard Tuck, rev. ed. Cambridge: Cambridge University Press, 1996.

Johns, Alessa. *Women's Utopias of the Eighteenth Century*. Urbana: University of Illinois Press, 2003.

Locke, John. *Two Treatises of Government*, ed. Peter Laslett. Cambridge: Cambridge University Press, 1988.

Maginn, Christopher "Smith, Thomas (1547–1573), Colonial Adventurer." *Oxford Dictionary of National Biography*. Oxford: Oxford University Press, 2008. https://www.oxforddnb.com/view/10.1093/ref:odnb/9780198614128.001.0001/odnb-9780198614128-e-69171.

Montaigne, Michel de. *The Complete Essays of Montaigne*. Translated by Donald Frame. Stanford: Stanford University Press, 1965.

More, Thomas. *Utopia*. Edited by George M. Logan and translated by Robert M. Adams, 3rd ed. Cambridge: Cambridge University Press, 2016.

Morgan, Hiram. "The Colonial Venture of Sir Thomas Smith in Ulster, 1571–1575." *Historical Journal* 28 (1985): 261–78.

Morice, Juliette. *Le Monde ou la bibliothèque: voyage et éducation à l'âge classique*. Paris: Les Belles Lettres, 2016.

Ovid. *The. xv. bookes of P. Ouidius Naso, entytuled Metamorphosis*. Translated by Arthur Golding. London, 1567.

Plattes, Gabriel. "A Description of the Famous Kingdome of Macaria." In *Samuel Hartlib and the Advancement of Learning*. Edited by Charles Webster, 79–90. Cambridge: Cambridge University Press, 1970.

Pohl, Nicole. *Women, Space, and Utopia, 1600–1800*. Aldershot: Ashgate, 2006.

Quinn, Davis Beers, ed. *The Roanoke Voyages 1584–1590*. London: Hakluyt Society, 1955.

Schiller, Friedrich von. "On Naive and Sentimental Poetry." In *German Aesthetic and Literary Criticism: Winckelmann, Lessing, Hamann, Herder, Schiller, Goethe*. Edited by Hugh Barr Nisbet, 180–232. Cambridge: Cambridge University Press, 1985.

Shakespeare, William. *The Tempest*. Edited by Stephen Orgel. Oxford: Oxford University Press, 1994.

Shrank, Cathy. *Writing the Nation in Reformation England 1530–1580*. Oxford: Oxford University Press, 2004.

Swift, Jonathan. *Gulliver's Travels*. Edited by David Womersley. Cambridge: Cambridge University Press, 2012.

Terry, Edward. *A Voyage to East-India*. London, 1655.

Tuck, Richard. "The Utopianism of *Leviathan*." In *Leviathan after 350 Years*. Edited by Tom Sorell and Luc Foisneau, 125–38. Oxford: Clarendon Press, 2004.

12 Lunar travel and lunacy

Reading conflict in Aphra Behn's *The Emperor of the Moon* (1687)

Rachel Willie

Early modern travel writing had its infancy in the accounts written by travellers and published by stationers keen to cash in on public interest in tales about the world. Medieval travel texts such as *The Book of Sir John Mandeville* (*c.* 1356) made claims to being accounts of physical travel that remained fantastical, unverifiable and located within a Christian framework that viewed the globe as fundamentally sacred.[1] Later texts, such as Thomas Coryate's *Crudities* (1611), explored locales that were familiar to the rich but perhaps equally intriguing to those less able to travel abroad, whereas the emerging genre of *ars apodemica* instructed consumers on the art of travel. Although by no means forming a single, clearly defined genre, literature on and about travel proliferated in the sixteenth, seventeenth and eighteenth centuries and, as with the examples of Aphra Behn's *Oroonoko* (1688) and Jonathan Swift's *Gulliver's Travels* (1726), these texts influenced the development of the novel and often blurred the boundaries between the real and the fictional, veracity and verisimilitude.[2] Early modern astronomy also sat at the threshold of what could be imagined and what was real. As I go on to argue below, the planets, once construed as a site of harmonious movement, became a space of travel and conflict: Behn utilizes these tensions to question perceived truisms concerning gender, spectacle, and bodily cognition.

This essay explores the treatment of travel writing as a way of falsifying knowledge of the cosmos to counter poor reading practices and reconcile familial conflict in Behn's *The Emperor of the Moon*. This play utilized the visual culture of *commedia dell'arte* and was first staged to great success and published in 1687.[3] It has received limited critical attention: recent scholarship has sought to locate Behn's text within a nexus of royalist propaganda and cultural memory or else has attempted to mine the text for further evidence of Behn's complicated Toryism and politics in the play's very specific use of spectacle, gaze, and natural philosophy.[4] Conversely, Judy A. Hayden has argued that Behn employs *commedia dell'arte* to ridicule the notion of plurality of worlds.[5] Claire Preston has considered how the sites of scientific endeavour are offstage and, instead, the audience is presented with the folly of faux-scientific spectacle.[6] In all of these readings, visual culture is brought to the fore. Scholars have tended to consider reading and performance as separate events, but, as criticism increasingly makes apparent, reading aloud blurs the distinctions between performance and print.[7] Not all acts of reading happened in private closets, and reading aloud presented its own forms of public performance and spectacle. This essay will therefore build on Katherine Mannheimer's suggestion that *The Emperor of the Moon* establishes a symbiotic relationship between private reading practices and publicly staged spectacle, to argue that reading and misreading travel writing is fundamental to the spectacle of performance in the play. The opposing epistemological approaches of private, individualized modes of reading and public, communal acts of performance align partly through reading being an embodied act.[8]

Travel itself was often used as a metaphor for the experience of reading: whether through space or through text, movement was perceived as having a cognitive, emotional, and physical transformation upon the subject. Indeed, the pervasiveness of travel as a metaphor for critical thought has led Georges Van Den Abbeele to comment upon the "banality" and paradoxical nature of the trope even as he acknowledges its centrality to intellectual endeavour.[9] Voyages necessitate movement, but thinking tends to be perceived as a static activity. Yet thinking requires a metaphorical journey through ideas to stimulate conclusions. Central to travel narratives is the credulity of the reader and the extent to which they are willing to suspend disbelief and accept the veracity of the account of travel. Narratives of travel bore many similarities to the cognate genre of utopian fiction, casting doubt on the kinds of empirical knowledge that can be garnered from tales of the wider world.[10] Literature of travel made claims to truth that were often untenable and the narratives deconstruct themselves through making bold and spurious claims. However, these fabrications are not necessarily indicative of a deceitful writer attempting to dupe a gullible reader. Instead, they may offer broader philosophical truths about far-off lands from which the reader may draw and make comparisons with their lived experiences.

In considering reading as performative, we can gain some insights into the material conditions of enacting space and the relationships between reading, spectacle, the cosmos, and travel. Consequently, after this brief overview of the relationship between travel and reading, this essay will go on to consider early modern conceptions of travel, natural philosophy, and wonder to show how the play uses spectacle and space (both fabricated and performed) as a way to castigate reading practices that lead to misinterpretation. First, I will turn to early modern notions of the universe before moving on to the travel narrative to show how the cosmos and this slippery genre each feed into the comedy of the play. What emerges is that both reading and scientific observation were considered as ways of acquiring knowledge, but this knowledge could be dangerous and unverifiable.[11]

From geocentrism to heliocentrism

The basic plot of the *The Emperor of the Moon* centres on the need to dupe the moon-obsessed but apparently otherwise learned Dr Baliardo into believing that two gallants, Cinthio and Charmante, are the Emperor of the Moon and his brother, the Duke of Thunderland, respectively. Such an enterprise is deemed necessary, not only for the gallants to achieve their designs to wed Baliardo's daughter Elaria and his niece Bellemante but also to cure Baliardo of his malady. Baliardo, it would seem, has been rendered lunatic from reading too many fantastical tales about lunar travel and this has caused him to believe that terrestrial beings are inferior to their neighbouring moon-dwellers. Baliardo's malady reflects not only debates over what to read and how to read it but also exposes radical changes in how the cosmos was conceived. In 1609, Galileo Galilei invented the first telescope strong enough to view in detail celestial objects; viewing the motions of the planets and four of Jupiter's moons led to a growing rejection of geocentric Ptolemaic cosmography and a reconfiguring of the universe.[12]

The technology supporting reading and the technology to see the stars were closely intertwined: optical lenses had their genesis in spectacles developed to aid sight and to enable the reader or writer to see the page. Although reading glasses had been invented *c.* 1300, it was another three hundred years before magnifying lenses had advanced sufficiently to enable Galileo and his contemporaries to make their discoveries. Experiments with optical lenses in Italy towards the end of the sixteenth century led to the beginnings of the telescope, but it was not until late 1608 that the first spyglasses were revealed in The Netherlands. This invention instantly provoked much interest across Europe, but the magnification was limited. By late 1609, Galileo had developed a telescope that had a magnification of x20. It was this telescope that Galileo used in his astronomical observations that are documented in his *Sidereus Nuncius* (1610, commonly translated as *The Sidereal Messenger*).[13] At least two other astronomers had turned the telescope to the night sky before Galileo, but he had the strongest instrument and consequently observed the heavens more clearly. Galileo discovered that, contrary to conventional wisdom, the moon was not a perfect, smooth, celestial orb but had craters. More importantly in terms of advancing heliocentrism, Galileo also discovered four of the moons of Jupiter.

The discovery that Jupiter had satellites was of major scientific importance: Galileo provided empirical evidence that there was more than one fixed point in the cosmos and orbs wandered around orbs. Planetary motion thus presents travel and conflict as being at the centre of scientific inquiry in the seventeenth century. As we will see, it is the attempt to reconcile heliocentrism with geocentrism and misunderstanding the epistemologies of travel writing that drive Baliardo's malady and underpins the plot of *The Emperor of the Moon.*

Galileo's discoveries revolutionized ways of knowing the wider world, the cosmos and humanity's status in creation. It also led to renewed fascination in the relationship between humanity and the universe. Yet Galileo was also part of a wider movement that reevaluated the heavenly spheres. Texts as diverse as Robert Burton's *Anatomy of Melancholy* (first published 1621, then going through numerous revised and expanding editions throughout the seventeenth century) and Johannes Kepler's *Somnium* (*c.* 1611, published posthumously in 1634) turned to imaginative, dream-like visions to comprehend the space of the moon, the cosmos's relationship to the body, and to make the case for heliocentric Copernican theory.[14] The Copernican revolution also gave rise to early science fiction as writers and readers tried to make sense of the new cosmological order.[15] Yet not all were ready to embrace the heliocentric universe. As Edward, Viscount Conway, crudely observed in a letter to his daughter-in-law, the earth lacked the agility to travel around the sun:

> for the Earth a heavy dull grosse body to move and the heaven and Starres who are light to stand still is as if a Prince should upon a festival day appoint all the old and fat men an woemen to dance and all the yonge men and woemen of sixteen and twenty to sit still.[16]

In the geocentric universe, base, mutable matter and death occupied the sublunary sphere. Ironically and paradoxically, perhaps, given that some of its symbolism aligned it to both mutability and chastity, the moon becomes gatekeeper to the celestial or heavenly spheres, where the immutable planets move in their orbits. By reclassifying the earth as a planet, this sense of the two-tiered universe, with base, gross matter at its centre, disappears. However, Conway's comments demonstrate the ongoing insistence that the earth is a "heavy dull gross body" in comparison to the lithe celestial planets. Rather than placing the earth in a privileged place at the centre of the universe, as in Aristotelian and Ptolemaic cosmology, this positioning becomes representative of the earth's coarse bulkiness.

"Gross body" not only invokes celestial bodies but also imbalance within the body and within the globe. In medicine, "gross humours" signifies a thickening of the humours, leading to disease. In discussing gout, Philemon Holland underscores the connection between gross humour, heaviness, and disease:

> *Gout* is a griefe of the feet, occasioned by some distemperature, or irregular humor ... A griefe (as concerning this purpose) is occasioned after foure sorts: Either by way of oppressing a part, when as a grosse humor weighing downe a part ponderously, causeth it to greeve in bearing the burden thereof: such kinde of griefe proceedeth of a phlegmatick, and melancholick humor, because they are heavy and weighty.[17]

Grossness in the macrocosm and in the microcosm, then, points to plethoric, lethargic and weighty bodies, but the analogy of a prince instructing lithe young men and women to sit still as the older couples dance implies a grotesque, Saturnalian and carnivalesque element to Copernican theory. A prince may have the authority to temporarily invert social structures on feast days but such inversion only emphasizes the natural order of things.[18] For Conway, the apparent absurdity of Copernican theory exposes its own impossibility and becomes a means by which geocentric models are reaffirmed.

Conway's analogy not only asserts an assumed natural order with regards to class and age but also assumes particular roles for men and for women and how they ought to move when young and when old. This gendering of movement can also be seen in how Ptolemaic cosmology was refashioned by Christianity. Since the early Christian church, Ptolemaic cosmology had been absorbed into Christian narratives of creation, lending a more theologically fraught subtext to Conway's dismissal of Copernican theory. Within a Christian framework, as a consequence of Adam and Eve being banished from the Garden of Eden, humanity has been relegated to dwelling in the basest part of the cosmos: Eve was subjugated to Adam and both were subjugated to the harshness of the land. Unlike the bountiful Garden of Eden, earth was a space of toil and tribulation. Medieval writers conceived earth as the dull, fixed, base centre of the universe: Albertus Magnus went as far as to assert, "among the simple bodies, the earth is like an excrement".[19] Far from privileging the earth and humanity, the geocentric universe effectively marks humanity as the dung beetles dwelling in a heap of cosmic faeces.

Conway's comments focus upon the grossness of matter and he uses the simile of withered, gross, fat earthly bodies to emphasize this sense that the decayed matter of the earth cannot be a celestial body. Yet his views were already deemed outmoded by those who elevated the earth beyond the negative connotations of it being the fixed, dull centre of the universe.[20] What this emphasizes is how older notions regarding cosmic movement continued to be known, held, and understood.

Thirty years after Conway made these wry observations, Behn presents in Baliardo a fictional man of learning who believes, paradoxically, that terrestrial creatures are too base for his earth-dwelling charges. In focusing upon elevated lunar creatures, Baliardo appears to be harkening back to a geocentric view of the universe, but the very mentioning of a lunar society hints at heliocentrism. In 1638, John Wilkins speculated over the possibility of a lunar world:

> Very many others, both English and French, all who affirmed our Earth to bee one of the Planets, and the Sunne to be the Centre of all, about which the heavenly bodies did move, and how horid soever this may seeme at the first, yet is it likely enough to be true, nor is there any maxime or observation in Opticks (saith *Pena*) that can disprove it.

> Now if our earth were one of the Planets (as it is according to them) then why may not another of the Planets be an earth?

> Thus have I shewed you the truth of this proposition: Before I proceede farther, 'tis requisite that I informe the Reader, what method I shall follow in the proving of this chiefe assertion, that there is a World in the Moone.[21]

For Wilkins, the very reordering of the universe makes possible the plurality of worlds and the existence of aliens; since the earth is re-categorized as a planet, other planets can be inhabited like earth and there can be a society ("a World") dwelling on the moon. Wilkins points to the novelty and to the horror of acknowledging that the sun – not the earth – is the fixed point of the universe, demonstrating the extent to which conflicting accounts of how the planets wandered in space was a site of discord. This anxiety connects closely to perceptions of how God ordered the cosmos and what this ordering implied about humanity, salvation, and an individual's relationship with God. In acknowledging that the earth was a planet, not only was earth reconsidered, but also other planets. William Gilbert, who was amongst the handful of people to accept Copernican theory in the sixteenth century, concluded the moon must have continents, oceans, and a terrain that is similar to earth.[22] Such conjecture could be contentious within a Christian framework.

Wilkins was a cleric and a founding member of the Royal Society, illustrating how science and reconciling astrological discoveries with God concerned some Christians. In particular, theories regarding the plurality of worlds were controversial and texts such as Bernard le Bovier de Fontenelle's *Entretiens sur la pluralité des mondes* ("Conversations on the plurality of worlds" 1686) were blacklisted by the Roman Catholic Church: Fontenelle's text continued to be listed on the Catholic Church's index of heretical works until the nineteenth century.[23] However, at this point in his discussions, Wilkins is less concerned with reconciling readers with the "horrid" notion that the earth is a planet and, instead, focuses upon optics. Observation, especially through the telescope, provides the viewer with empirical truths about the nature of matter and how it travels through space.

The telescope enables the viewer to see what is being perceived more clearly, but hidden within this recognition of the telescope as an instrument for scientific enquiry is an acceptance of the instrument's strategic importance. When Galileo presented his telescope with a magnification of x8 to the Doge of Venice and the Venetian Senate, he emphasized its military relevance:

> This is a thing of inestimable benefit for all transactions and undertakings, maritime or terrestrial, allowing us at sea to discover at a much greater distance than usual the hulls and sails of the enemy.[24]

The telescope had strategic importance from its inception, particularly in times of conflict and as an aid to assist safe passage across treacherous seas. As such, the telescope could be construed as a potent weapon in the quest for knowledge acquisition, whether this knowledge was military or astronomic. Through observation and magnification, forms of knowledge previously hidden to the human eye could be revealed.

Observation and optics, as Wilkins maintains, thus have the potential to expand knowledge. However, seventeenth century telescopes were difficult to handle and the fields of vision were very small: it was not easy to keep celestial objects in sight.[25] Although later examination of the moon through more powerful telescopes showed Galileo's plotting of lunar craters to be surprisingly accurate, the small sight lines could expose the drawings to doubt and to suspicion.[26] The seventeenth century may have witnessed the decisive rejection of geocentrism, but the possibility of doubting what the eye sees meant that Edward, Viscount Conway, could scoff at the absurdity of heliocentrism: common sense told us that the earth was static and did not move. Nevertheless, Galileo's invention revolutionized how the cosmos was conceived and his discussions became popularized in print. Readers could engage with astronomy through consuming texts that imagined new extraterrestrial lands, and it is this fascination with the planets that Behn draws from in fashioning the spectacle of her play.

Reading and performing the early modern moon

Behn took a keen interest in developments in scientific method and was one of the earliest translators into English of Fontenelle.[27] Despite – or perhaps because of – its controversy, Fontenelle's text eventually became a European best seller. As Line Cottegnies has argued, Behn does not produce a faithful word for word or sense for sense translation of Fontenelle but instead distances herself from the most controversial elements of the text through how she frames it in her preface and from her loose translation of certain parts.[28] Fontenelle's narrative is structured around a series of nocturnal conversations between a philosopher and his host, a marquise, as they wander around her garden observing the night sky, and, as Cottegnies has argued, Behn makes women more prominent in the narrative and less passive agents in the philosophical discourse.[29] This echoes and challenges Elaria and Bellemante's active participation in the fabrication of a lunar society in *The Emperor of the Moon*: whereas Behn's translation presents women as engaged in philosophical discussions regarding the order of the cosmos, *The Emperor of the Moon* presents women as energetically challenging cosmic philosophy through using visual culture to cast doubt on scientific method.

Although Fontenelle wrote in the vernacular for a wide readership, Behn's interjections, as Cottegnies observes, "show Behn's awareness of herself as a critic and ... a scrupulous mediator between a philosophical/scientific text and a public of non-specialists".[30] In the preface to her translation, Behn playfully contests, distorts, and renders ludicrous Fontenelle's most contentious arguments – especially with regards to the existence of other inhabited planets, asserting that Fontenelle "failed in his design" due to his endeavours to make astronomy accessible:

> I must tell you freely, he hath failed in his Design; for endeavouring to render this part of Natural Philosophy familiar, he hath turned it into Ridicule; he hath pushed his wild Notion of the *Plurality of Worlds* to that heighth of Extravagancy, that he most certainly will confound those Readers, who have not Iudgment and Wit to distinguish between what is truly solid (or, at least, probable) and what is trifling and airy: and there is no less Skill and Vnderstanding required in this, than in comprehending the whole Subject he treats of.[31]

These critiques of Fontenelle might seem at odds with Behn's interventions in the text that are aimed at making it accessible, but it casts light on reading and the reading strategies needed to engage with scientific learning. Behn presents herself as the mediator of a controversial text, enabling the reader to engage "appropriately" with the more provocative elements of Fontenelle's argument. When Behn's comments regarding Fontenelle are considered in relation to her play and reading, however, we see that Behn consistently returns to questions of reading and how the reader consumes and digests works that are concerned with deciphering the secrets of the universe.

In the play, Baliardo is presented as a learned natural philosopher who is obsessed with the moon. He conflates heliocentric cosmography with some aspects of geocentrism: for Baliardo, the moon continues to be the threshold to the celestial sphere, even though the earth is a planet. Convinced that the moon is inhabited, Baliardo privileges lunar dwellers and asserts that the lunar society is superior to terrestrial civilizations. The moon thus becomes a dominant force, not by governing human physiology as some early modern medics believed, but instead by dominating Baliardo psychologically.[32] Baliardo's interest places impediments upon his household and prevents him from engaging with his physical surroundings: it is only through calling a halt to Baliardo's fixation with the moon that order will be restored. In the opening scene, we are told Baliardo's malady was triggered by reading lunar travel narratives:

SCARAMOUCH: lunatic we may call him ... for he is always travelling to the moon.

ELARIA: And so religiously he believes there is a world there, that he discourses gravely of the people, their government, institutions, laws, manners, religion and constitution, as if he had been bred a Machiavel there.

SCARAMOUCH: How came he thus infected first?

ELARIA: With reading foolish books, Lucian's *Dialogue of Icaromenipus*, who flew up to the moon and thence to heaven; and heroic business called *The Man in the Moon*, if you'll believe a Spaniard, who was carried thither upon an engine drawn by wild geese; with another philosophical piece, *A Discourse of the World in the Moon*; with a thousand other ridiculous volumes too hard to name.

SCARAMOUCH: Aye, this reading of books is a pernicious thing. I was like to have run mad once, reading Sir John Mandeville.[33]

Imaginary travel, the servant Scaramouch asserts, is at the root of Baliardo's malady, but the punning use of "lunatic" highlights how, since antiquity, madness and the moon were perceived as closely intertwined. Humoural theory asserted that the balance of the humours was governed by the phases of the moon – women, with their weaker bodies and reasoning, were more susceptible to sudden changes in emotion and, inevitably, were more likely to be influenced by the moon.[34] In imagining himself to have an affinity with lunar dwellers, Baliardo renders himself lunatic and effeminate. In setting up this premise, the text becomes a means to refute the misogyny inherent within forms of medical theory that continued to hold sway in the seventeenth century. The moon-infected Baliardo is cured of his lunacy by his daughter and his niece, and this act enables the wards to marry the men of their choosing. Far from being governed by the mutable moon (as previous ways of considering the relationship between the moon and the body attested), female agency is the catalyst for conflict resolution.

Whereas Behn celebrated female agency, others viewed it as a disruptive force, especially if excited through reading: romances, especially, were believed to affect the senses and encourage hysteria. These views continued into the eighteenth century, where the novel was feared to be detrimental reading matter for women in particular.[35] Baliardo's imagination has deceived him, but this deception has been brought on by the material practices of reading. Aristotelian theories regarding the faculties of the mind presented cognition as being reached through sensory information being conveyed to the brain by the animal spirits. The imagination converted this data into images that were stored and (after being stored) continued to be perceived when the external object was no longer in sight. These images were then transmitted to reason and to memory: the internal senses thus made sense of data gathered by the external senses.[36] This not only shows how interior and exterior knowledge was conceived as closely intertwined, but it also highlights the belief in the porous, leaky quality of the human body and mind. Right reason could enter the body through sensory experience, but so too could false knowledge: Charles Gobinet instructed the young Christian reader to avoid "unchast Books ... as the Plague of the Soul", demonstrating ongoing anxieties regarding what to read and how to read.[37] Women, perceived as being especially leaky vessels, were believed to be particularly susceptible to absorbing data that would be cognitively and physiologically detrimental. What women read and how they engaged with fiction were therefore a cause for concern to moralists. By presenting Baliardo as falling prey to delusion, Behn not only draws from fears regarding what ought to be read and how to read, she also inverts gendered concerns regarding reading. Elaria dismisses the "foolish books" that led to her father's faculties being perplexed: far from being deceived by reading pernicious books, the female characters in Behn's play understand the importance of careful and judicious reading.

Behn realigns false reading practices to be a malady of men who pursue scientific enquiry with little understanding of scientific method. Elaria's criticism of her father centres on how Baliardo has placed trust in there being a perfect lunar society. Baliardo has moved beyond the parameters of what can be known empirically through plotting information received through the telescope, and, instead, he subscribes to metaphysical speculation that can be neither proved nor disproved. Baliardo has strayed from scientific method and instead considers what kind of society might inhabit the moon. In so doing, he entered the realms of governance. Machiavelli, whose complex theories of statecraft were contentious and caricatured throughout the early modern period, then becomes the influential force as Baliardo fabricates his lunar society.[38] Baliardo ceases to have right reasoning and instead suffers from a Machiavellian infection.

Through reading, Baliardo is thus deceived into accepting as authoritative the hypothesis that there are other inhabited worlds. In so doing, doubt is cast on Baliardo's erudition: instead of being a learned doctor, Baliardo has shown himself to be a reader who lacks the judgement and critical acumen to distinguish between the "truly solid" and the trifling and airy. Behn's criticism of Fontenelle materializes upon the stage as the relationship between reading, matter, imagination, travel, the performance of travel, and staged spectacle is explored and deemed a site of conflict.

Reading, travel, and the performance of deception

Elaria dismisses Baliardo's reading of "ridiculous volumes" relating to the moon, but the servant Scaramouch's observation that he almost went insane as a consequence of reading the medieval travel romance by John Mandeville is telling. By referencing Mandeville's travels, Scaramouch is not only establishing a connection between lunar exploration and terrestrial travel, he is also identifying the prehistory to ideas of exploration and how the epistemological precepts that underpin travel were gradually changing. Terrestrial travel and lunar travel are both causes for anxiety. At the time of the play's first performance, John Harrison was still fifty years from discovering a means of accurately calculating longitude and the world still presented the wayward traveller with undiscovered countries.[39] Travel narrative in the late

1600s therefore could (and did) encompass the fantastical as well as the factual. The discussion between Elaria and Scaramouch highlights tensions between the impulses towards modernity, the birth of more scientific ways of knowing the wider world, and previous discourses of travel and wonder. Elaria's dismissal of the lunar narratives as "foolish" and Scaramouch's assumption that they are "pernicious books" that lead the reader into the vice of lunacy highlight each character's inability to comprehend the rhetorical mechanisms of early modern travel writing.

Natural philosophy and experimental philosophy may have sought to establish truths through observation, but Jonathan Sell has drawn attention to the fact that the travel narrative demands to know what sort of truths are under discussion. Rather than offering specific empirical truths about the wider world – or even the moon, in the case of authors on lunar travel – travel writing is less concerned with scientific reasoning and instead offers metaphors and parallels that may be unpicked by the reader. Since language itself can cloud and obfuscate meaning, any relaying of the exotic and foreign is flawed: verbal signs could transform a rhinoceros, as Marco Polo found when he attempted to describe the beast to untravelled Europeans by naming it a unicorn.[40]

The rhetoric of early modern travel, therefore, is fraught with difficulty if it were to be taken literally. Instead, early modern travel could offer a way of knowing that is largely based upon perceived universal truisms. One of the "pernicious books" identified by Elaria neatly encapsulates the metaphorical nature of travel writing and the fictionality of utopian tales. *The Man in the Moone*, by Francis Godwin, was printed posthumously in 1638. It is supposedly the autobiographical account of Domingo Gonsales, a Spaniard whose adventures take him to the moon on a contraption he invented and harnessed to a flock of birds (Fig. 12.1). Gonsales' discoveries offer parallels to earthly societies. On the moon, he finds a refined lunar society. This society has a natural accord with Christianity as demonstrated by the moon inhabitants bowing at the name of Jesus – even if they appear to have no concept of Christianity as an organized faith.

Such a representation of moon-dwellers would seem to suggest that there is an inherent harmony within creation and this harmony is something to which Baliardo aspires. In denying his daughter and his niece access to terrestrial lovers, Baliardo sets his sights (if not the sights of his wards) upon higher beings. Advancements in science would seem to locate these discussions within a nexus of scientific learning predicated upon discoveries chartered through knowledge gained by looking through the telescope. However, telescopic insights lack viability in the context of lunar travel narratives. Many of these lunar explorations are less about the moon as a real space and more about the imagined societies that could inhabit it. Perhaps most notably as far as the play is concerned, Godwin's text blends travel writing with the picaresque and utopian fiction.

Figure 12.1 Francis Godwin, *The Man in the Moone* (London, 1638), C5v. Ih G549 638, General Collection, Beinecke Rare Book and Manuscript Library, Yale University

Utopian literature signalled its own fictionality in its very construction: the lack of clear coordinates enabled writers to use the ill-defined space of the far-off land to imagine other ways of ordering society. Central to utopian literature is travel and, as Chloë Houston notes, Godwin's appropriation of forms associated with travel encourages comparison of lunar and terrestrial life even if Godwin makes no suggestions with regards to how the latter may be reformed.[41] Travel writing and utopian literature raise questions regarding authenticity and the reader's credulity: Houston argues that the mid-seventeenth century marks the point where the idealism of utopianism is replaced by comic ridicule. Despite this, the form maintained vitality across many genres and the seventeenth century witnessed a proliferation in texts that appropriate utopian forms.[42] Utopian literature thus informed ideas of travel and constructions of ideal societies, even if readers during a period of continued revolution, political uncertainty, and religious reform failed to invest in the idealist precepts that underpinned the genre.

The moon, imagination, and natural philosophy

Utopian fiction implies that the moon could offer a space to reflect upon earthly matters and humanity's status in the wider cosmos. Central to these discussions were not only ways of understanding the heliocentric universe but also endeavours to comprehend changing attitudes to religion – perhaps especially soteriology and whether or not aliens could receive salvation. As David Cressy writes:

> Like the Utopian tradition to which it was related, the literature of lunar voyaging was part parodic but mostly sober, combining earnestness and jest. Imagining a world on the moon was perhaps a response to a shuffled world, a world turned upside down, in which systems of hierarchy, authority, religion, and gender, as well as planetary revolutions, were called into question.[43]

Lunar travel narratives thus become a means to question and destabilize early modern truisms: Baliardo's foolishness stems from his taking a positivist approach to these narratives and making an inductive leap of faith in the absence of empirical evidence. Rather than seeing lunar civilization as a cognitive exercise and acknowledging an indebtedness to utopian writings and fantastical discourses regarding an ideal society, Baliardo believes the lunar community to be real. The stage directions of Behn's play, such as the one that opens 1.2, emphasize this absurdity:

> *Enter* Doctor, *with all manner of Mathematical Instruments, hanging at his Girdle;* Scaramouch *bearing a Telescope twenty (or more) Foot long.*

The telescope is presented as a preposterously phallic object for comic and visual effect as Baliardo seeks to view the moon-dwellers, "the most beautiful of all the sons and daughters of the Universe" (1.2.55). However, twenty-foot and longer telescopes were not unusual in the seventeenth century due to the smallness of the field of vision and how the lenses needed to be positioned. The 202 feet tall Monument, completed in 1677 to commemorate the 1666 Great Fire of London, was designed to be a giant and powerful telescope: although this experiment was abandoned due to vibrations from passing traffic affecting the sensitivity of the instrument, it underscores the eagerness with which the new technology was replicated and enhanced.[44] Behn's comic telescope thus has a basis in astronomical fact, but it also highlights the need to imagine and construct various elaborate theatrical and verbal spectacles in order to continue the ruse. Much of the deception exercised in the play is predicated upon the way in which Baliardo's imagination can be abused.

As noted previously, imagination was central to Aristotelian ideas regarding sensory experience and cognition. However, imagination can also refer to what is fanciful, likely, expected, or anticipated.[45] The lovers and their accomplices develop Baliardo's fancy by deceiving his imagination. They expand upon Baliardo's belief that the moon is inhabited and encourage him to believe that the two principle moon-dwellers are in love with his daughter and with his ward.

Credulity and imagination thus go hand in glove, but the text establishes tensions between Baliardo's scientific method and the Naples that he inhabits. His scientific instruments are dismissed as "devils" before the decision is taken to act out a farce to cure him of his lunar obsession. Here, the telescope is not presented as a means by which empirical truths can be established, but instead it is viewed as an instrument of visual deception. Folly is to be cured by folly, but the play's metadramatic concern with performance and spectacle does not stop at the impositions placed upon Baliardo. Scientific learning and the instruments associated with it are rejected, but lest we fall into the trap of believing the narrative will unravel into a war between piety and science, other modes of performativity are introduced. In 1.1, Bellemante is witnessed clutching a prayer book and has the appearance of a woman who has returned from church; the sensory gratification gained from reading her prayer book (and in attending public forms of worship) thus provides nourishment for her soul.[46] Yet it becomes apparent that her attendance at chapel was less to feast her eyes upon God and more to gain carnal gratification through assuming the role of voyeur. Her sole purpose in attending church worship is to ogle men and especially her lover Charmante.

Appearances and the realities of the situations presented distort the various narrative strategies employed, showing staged spectacle to be as pernicious a thing as the play purports reading to be. Behn appropriates the rhetoric of celestial orbs and their relationship to God as a way of satirically implying that the church is a place to see and to be seen. Later, Baliardo assumes the role of a chaste voyeur when, as part of the deception meted out on him, a disguised Scaramouch implores the doctor to clear his mind of all vices as only a virtuous man can see the spirits who inhabit the heavens. While Baliardo is praying, Scaramouch affixes a lens onto the telescope that depicts a beautiful celestial nymph sleeping. Private repose is thus scrutinized by the male gaze and there is a tension between the insistence of Baliardo's inherent virtue permitting him to view the sleeping nymph and the sense of perversion generated through the act of illicitly watching a private scene.

The boundaries between public and private – so hotly debated by later critics – become ever more porous as the play establishes more and more elaborate modes of disguise, surveillance, and misunderstandings that require resolution. The various binaries that the play seeks to establish become destabilized through the various strategies used to create the elaborate spectacles that were in vogue with theatre audiences at the time of the play's first performance. Private meditations and public devotions, private repose and public action, masculine and feminine forms, terrestrial and extra-terrestrial activities, interiority and exteriority, static reading and active spectacle, travel and remaining stationary are all called to question as these conflicts remain unresolved. This might be due partly to the creative impulses that underpin the development of staged spectacle colliding with the parallel concerns of travel writing. The velocity of the narrative collapses these binaries as imagination feeds both the staged spectacles that the play appears to celebrate and the travel writing precepts that it seems to attack.

Wonder

How the imagination can be abused brings us to the notion of wonder – in terms of the emotion that is caused by perceiving something novel, inexplicable, and astonishing; the curiosity, perception, and bewilderment that the sensation elicits; and the cognitive methods this emotion may generate. The play uses wonder to establish various deceptions for comic effect and to ridicule the easily duped. In the subplot, Cinthio's servant, Harelequin, disguises himself as a farmer in an attempt to win the hand of the wards' companion, Mopsophil, while his rival for her affections, Scaramouch, assumes the appearance of an apothecary. Mopsophil is not deceived, but Harlequin adopts further disguises: in order to escape paying a tax upon entering the city, he transforms his appearance from a gentleman farmer driving a calash to a baker steering a cart. The various stage directions emphasize the relationship between wonder and spectacle:

> [The Officer] goes to the entrance to call the clerk; [in] the meantime Harlequin whips a frock over himself, and puts down the hind part of the chariot, and then'tis a cart. Enter Clerk (3.1)

This visual trick is present in Behn's source text, *Arlequim, empereur dans la lune*, performed by an Italian troupe in Paris in 1684, yet this trick has little bearing on the plots of either drama.[47] The entire scene is a comedic device where a city official is duped into believing that he must be drunk to have mistaken a baker delivering goods to the city for a gentleman farmer. Consequently, its purpose seems purely to provide visual pleasure to an audience privy to the joke. However, the scene also draws attention to the way in which seeing, perceiving, and cognitive blindness operate within the play. Through Harlequin's duplicity, the official cannot rely upon his sensory perceptions to establish the empirical truth of the matter. Instead, he is left in a state of bewildered wonder.

While the spectacle may point to wonder being a source of deception, wonder is also inherent in early modern travel writing. Upon returning to Europe, the traveller could regale their audiences with astonishing stories of remote and far-off lands and the peoples who inhabit them. Audiences to these narratives both require the capacity to imagine the details described and the ability to marvel at them. An unknown physical object thus becomes a source of wonder, thereby anchoring abstract sensations in an unknown material world. Drawing from the humanist tradition and in accord with practices of knowledge acquisition in the early modern period, Baliardo's scientific learning blends astrology, astronomy, alchemy, medicine, cosmography, and Rosicrucian spiritualism as a way to establish empirical truths about the moon and the wider world. Yet wonder, as asserted by Socrates and later by Francis Bacon, is at the root of all knowledge. Philosophy and science begin with bewilderment. In *The Advancement of Learning* (1605), Bacon refers to wonder as being both "broken knowledge" and "the seed of knowledge". Establishing a difference between the dangerous and "proud knowledge of good and evil" that brought humanity to its postlapsarian state and virtuous knowledge that permitted humanity to name the animals, Bacon asserts the need for wonder as the starting point for all cognitive processes that bring about reasoning and empirical truths.[48]

Wonder, therefore, is at the heart of early modern scientific method and learning, but how does this connect to a play that seems determined to lampoon the folly of placing too much faith in perception and appears dismissive of lunar exploration? In the final scene of the play, spectacle, deception, wonder, and gender conjoin before Baliardo is cured when he learns the spectacle is fabricated. Baliardo is led to believe that he, Elaria, Bellemante, and Mopsophil might have been transported to the moon:

BELLEMANTE:	Are you sure on't, sir? Are we not, think you, in that world above I often heard you speak of? In the moon, sir?
DOCTOR [BALIARDO]:	(*aside*) How shall I resolve her? For ought I know, we are.
ELARIA:	Sure, sir,'tis some enchantment.
DOCTOR:	Let not thy female ignorance profane the highest mysteries of natural philosophy. To fools it seems enchantment, but I've a sense can reach it: sit, and expect the event. Hark! [Aside], I am amazed, but must conceal my wonder, that joy of fools, and appear wise in gravity.

(3.3.6–13)

Rather than perceiving wonder to be the beginning of knowledge, Baliardo dismisses the sensation as the "joy of fools". Baliardo aligns wonder with curiosity and in so doing rejects all scientific learning. Unable to see the distinctions between good and bad curiosity, legitimate and illegitimate learning, Baliardo is exposed as a fool precisely because he is unable to publicly succumb to the sensation of wonder or to admit the limits of his knowledge. The scene culminates in further spectacles as the lovers marry, before it is abruptly asserted that Baliardo is cured of his lunacy. Upon learning that he has been the subject of an elaborate hoax, Baliardo decides to destroy his library:

> DOCTOR: Burn all my books, and let my study blaze;
> Burn all to ashes, and be sure the wind
> Scatter the vile contagious monstrous lies.
> – Most noble youths, you've honoured me with your alliance, and you, and all your friends, assistances in this glorious miracle, I invite tonight to revel with me. – Come, all, and see my happy recantation of all follies fables have inspired till now ... I see there's nothing in philosophy. – of all that writ, he was the wisest bard, who spoke this mighty truth:
> He that knew all that ever learning writ,
> Knew only this: that he knew nothing yet.
>
> (3.3.222–28 and 230–33)

Baliardo thus abjures philosophy, but this decision only emphasizes that he misses the point. The rhetoric employed is loaded with contradictions. The vile contagious books are to be scattered to the wind, but this spreads the "contagion" rather than containing it. To endorse the view that "there is nothing in philosophy", Baliardo cites a maxim first propounded by Socrates to test the limits of knowledge and not to dismiss all learning.[49] Rather than demonstrating reading to be a bad exercise that clouds judgement, we are presented with a doctor who lacks the erudition to understand the knowledge he has acquired. The false reading and interpretative practices that lead Baliardo to conclude that a philosopher is against philosophical learning also encourages him to misunderstand the tenets of early modern travel writing. His reading of the texts is predicated upon the false assumption that the texts address empirical truths. Instead, the barriers between real and imagined space and the beings that inhabit these spaces become fractured sites of conflict. The binary divisions between the earth and the moon that Baliardo assumes to be in place are not supported by the play, yet the oracular distinctions between reading and viewing spectacle are also open to speculation. Despite its allusions to travel narrative, its geneses in *commedia dell'arte,* and its appropriation of the fashion for staged spectacles as a means of conflict resolution, ultimately, Behn's play seems more concerned with the epistemologies that underpin reading.

Notes

1 For a discussion of how Columbus set sail to find the Far East and the Garden of Eden, and Cartier's Pentecostal approach to the New World, see Marie-Christine Gomez-Géraud's essay in this current volume.

2 For a succinct overview of the complexities of travel writing as a genre, see Carl Thompson's *Travel Writing* (Oxford: Routledge, 2011), 9–33. On the relationship between travel writing and the novel, see Percy G. Adams, *Travel Literature and the Evolution of the Novel* (Lexington: The University Press of Kentucky, 1983).

3 Derek Hughes, "Aphra Behn and the Restoration Theatre," in *The Cambridge Companion to Aphra Behn*, ed. Derek Hughes and Janet Todd (Cambridge: Cambridge University Press, 2004), 29–45 (35).

4 Daniel Gustafson, "Cultural Memory and the Royalist Political Aesthetic in Aphra Behn's Later Works," *Restoration* 36 (2012): 1–22; Al Coppola, "Retraining the Virtuoso's Gaze: Behn's *Emperor of the Moon*, The Royal Society, and the Spectacles of Science and Politics," *Eighteenth-Century Studies* 41 (2008): 481–506.

5 Judy A. Hayden, "Harlequin Science: Aphra Behn's *Emperor of the Moon* and the Plurality of Worlds," *English* 64 (2015): 167–82.

6 Claire Preston, *The Poetics of Scientific Investigation in Seventeenth-Century England* (Oxford: Oxford University Press, 2015), 107–08.

7 See Elspeth Jajdelska, *Speech, Print and Decorum in Britain, 1600–1750:Studies in Social Rank and Communication* (Oxford: Routledge, 2016); Jennifer Richards, *Voices and Books in the English Renaissance: A New History of Reading* (Oxford: Oxford University Press, 2019); in recent years, there has been a proliferation of interest in the relationship between literate and oral cultures, influenced by earlier studies such as Bruce Smith's *The Acoustic World of Early Modern England: Attending to the O-Factor* (Chicago: University of Chicago Press, 1999) and Adam Fox's *Oral and Literate Culture in England, 1500–1700* (Oxford: Oxford University Press, 2001). For a discussion of the performance of print, see my "Viewing the Paper Stage: Civil War, Print, Theatre and the Public Sphere,"in *Making Space Public in Early Modern Europe: Performance, Geography, Privacy* ed. Angela Vanhaelen and Joseph Ward (New York and London: Routledge, 2013), 54–75.

8 Katherine Manheimer, "Celestial Bodies: Readerly Rapture as Theatrical Spectacle in Aphra Behn's *Emperor of the Moon*," *Restoration* 31 (2011): 39–60.

9 See Georges Van Den Abbeele, *Travel As Metaphor: From Montaigne to Rousseau* (Minneapolis: University of Minnesota Press, 1991), xiii–xiv.

10 For a discussion of the conflictual nature of utopian fiction, see Daniel Carey's essay in this volume.

11 Elizabeth Spiller, *Science, Reading, and Renaissance Literature: the Art of Making Knowledge* (Cambridge: Cambridge University Press, 2004), 102.

12 David Wootton, *The Invention of Science: A New History of the Scientific Revolution* (London: Allen Lane, 2015), 214–34.

13 Galileo Galilei, *Sidereus Nuncius or The Sidereal Messenger,* trans. Albert Van Helden, 2nd ed. (Chicago and London: The University of Chicago Press, 2015).

14 David Cressy, "Early Modern Space Travel and the English Man in the Moon," *The American Historical Review* 111 (2006): 961–82.

15 The earliest work of science fiction in the English language is often credited as being Francis Godwin's *The Man in the Moone* (pub. 1638). See William Poole, "The Origins of Francis Godwin's *The Man in the Moone* (1638)," *Philogical Quarterly* 84 (2005): 189–210.

16 Edward, Viscount Conway, letter to his daughter-in-law Anne Conway, July 22, 1651, *Conway Letters: The Correspondence of Anne, Viscountess Conway, Henry More, and their Friends, 1642–1684*, ed. Marjorie Hope Nicolson (Oxford: Oxford University Press, 1930), 34.
17 Philemon Holland, *Gutta podrica: a treatise of the gout The severall sorts thereof. VVhat diet is good for such as are troubled therewith. And some approved medicines and remedies for the same* (London, 1633), A4v-B1r.
18 For more on the Carnivalesque, see Mikhail Bakhtin, *Rabelais and his World* trans. Hélène Iswolsky (Bloomington: Indiana University Press, 1984).
19 Albertus Magnus, *De Caelo et mundo*, cited in Dennis Danielson, *"Paradise Lost" and the Cosmological Revolution* (Cambridge: Cambridge University Press, 2014), 158. For a detailed discussion of the kind of space Earth is conceived as being, see Danielson, chapter 7.
20 Danielson, 159.
21 John Wilkins, *The discovery of a world in the moone, or, A discourse tending to prove, that 'tis probable there may be another habitable world in that planet* (London, 1638), G7r–G7v. For a sustained discussion of this text and how its connections to Galileo and Christianity, see Howard Marchitello, "Telescopic Voyages: Galileo and the Invention of Lunar Cartography" in *Travel Narratives, the New Science, and Literary Discourse, 1569–1750*, ed. Judy A. Hayden (Farnham: Ashgate, 2012), 161–78.
22 Wootton, 157–58.
23 Line Cottegnies, "The Translator as Critic: Aphra Behn's Translation of Fontenelle's *Discovery of New Worlds*," *Restoration* 27 (2003): 23–38 (25).
24 Cited in Van Helden's introduction to *Sidereal Messenger*, 8.
25 Van Helden, 14.
26 Spiller, 101. Anna Marie E. Roos notes that in 1609, Thomas Harriot made telescopic observations of the moon and, after viewing Galileo's wash drawings of the moon, drafted more precise lunar sketches (Roos, *Luminaries in the Natural World: The Sun and the Moon in England, 1400–1720* (New York: Peter Lang, 2001), 94).
27 Behn's translation was printed under the title *A Discovery of New Worlds* in 1688. Two further translations in English appeared in 1687 and 1689 (Cottegnies, 23).
28 *Ibid.*, 30.
29 *Ibid.*, 26.
30 *Ibid.*, 33.
31 Bernard de Fontenelle, *A discovery of new worlds from the French, made English by A. Behn* (London, 1688), A8r.
32 See Anna Marie Roos on how, after the Restoration, mechanistic explanations over the physiological effects of the moon on the body replaced occult rationalizations.
33 Aphra Behn, "The Emperor of the Moon," in *The Rover and Other Plays*, ed. Jane Spencer (Oxford: Oxford University Press, 1995). 1.1.87–102.
34 Keith Thomas, *Religion and the Decline of Magic* (London: Penguin, 1971), 351–52; Dolly MacKinnon, "'Poor Senseless Bess, Clothed in her Rags and Folly': Early Modern Women, Madness, and Song in Seventeenth-Century England," *Parergon* 18 (2001): 119–51 (139). For a discussion of women as "weaker vessels" who are forbidden to travel except through reading, see Gábor Gelléri's chapter in this current volume.
35 On attitudes to women readers, see, for example, Jacqueline Pearson, "Women Reading, Reading Women," in *Women and Literature in Britain, 1500–1700*, ed. Helen Wilcox (Cambridge: Cambridge University Press, 1996), 80–99; Helen

Hackett, *Women and Romance Fiction in the English Renaissance* (Cambridge: Cambridge University Press, 2000); Barbara M. Benedict, "'I Never Read Novels': The Gender of Readers and Reading," in *The Oxford Handbook of the Eighteenth-Century Novel*, ed. J. A. Downie (Oxford: Oxford University Press, 2016), 355–71.

36 Thibaut Maus de Rolley, "The Devil. Delusions and Early Modern Cognition," in *Cognitive Confusions: Dreams, Delusions and Illusions in Early Modern Culture*, ed. Ita Mac Carthy, Kirsti Sellevold and Olivia Smith (Cambridge: Legenda, 2016), 71–88 (73–74).

37 Charles Gobinet, *The instruction of youth in Christian piety taken out of the sacred Scriptures, and Holy Fathers; divided into five parts. ... The last edition in French, now render'd into English.* (London, 1687), Q7v.

38 Erica Benner has offered an overview of Machiavelli's early reception history before going on to make a detailed, revisionist account of his most controversial text. See *Machiavelli's "Prince": A New Reading* (Oxford: Oxford University Press, 2013).

39 Dava Sobel, *Longitude: The True Story of a Lone Genius who Solved the Greatest Scientific Problem of His Time* (London: Fourth Estate, 1996).

40 Jonathan P.A. Sell, *Rhetoric and Wonder in English Travel Writing, 1560–1613* (Aldershot: Ashgate, 2006), 2–3.

41 Chloë Houston, *The Renaissance Utopia: Dialogue, Travel and the Ideal Society* (Farnham: Ashgate, 2014), 144.

42 *Ibid.*, 162–63.

43 Cressy, 981–82.

44 J.A. Bennett, *The Mathematical Science of Christopher Wren* (Cambridge: Cambridge University Press, 1982), 42.

45 "imagination, n." OED.

46 On private devotion, see, for examples, Jessica Martin and Alec Ryrie, eds., *Private and Domestic Devotion in Early Modern Britain* (Farnham: Ashgate, 2012). On women reading and piety, see Edith Snook, *Women, Reading, and the Cultural Politics of Early Modern England* (Aldershot: Ashgate, 2005). On congregational engagement with sermon culture, see Arnold Hunt, *The Art of Hearing: English Preachers and Their Audiences, 1590–1640* (Cambridge: Cambridge University Press, 2010).

47 Christa Knellwolf King, *Faustus and the Promise of the New Science, c. 1580–1730: From the Chaptbooks to Harlequin Faustus* (Farnham: Ashgate, 2008), 173.

48 Sell develops this point, 4.

49 Apology 23b3–4. For a study of Socrates and knowledge, see Hugh H. Benson, *Socratic Wisdom: The Model of Knowledge in Plato's Early Dialogues* (Oxford: Oxford University Press, 2000).

Bibliography

Adams, Percy G. *Travel Literature and the Evolution of the Novel*. Lexington: The University Press of Kentucky, 1983.

Bakhtin, Mikhail. *Rabelais and his World*. Translated by Hélène Iswolsky. Bloomington: Indiana University Press, 1984.

Behn, Aphra. *The Rover and Other Plays*. Edited by Jane Spencer. Oxford: Oxford University Press, 1995.

Benedict, Barbara M. "'I Never Read Novels': The Gender of Readers and Reading." In *The Oxford Handbook of the Eighteenth-Century Novel*. Edited by J.A. Downie, 355–71. Oxford: Oxford University Press, 2016.

Benner, Erica. *Machiavelli's "Prince": A New Reading*. Oxford: Oxford University Press, 2013.

Bennett, J.A. *The Mathematical Science of Christopher Wren*. Cambridge: Cambridge University Press, 1982.

Benson, Hugh H. *Socratic Wisdom: The Model of Knowledge in Plato's Early Dialogues*. Oxford: Oxford University Press, 2000.

Cressy, David. "Early Modern Space Travel and the English Man in the Moon." *The American Historical Review 111* (2006): 961–82.

Coppola, Al. "Retraining the Virtuoso's Gaze: Behn's *Emperor of the Moon*, The Royal Society, and the Spectacles of Science and Politics," *Eighteenth-Century Studies 41* (2008): 481–506.

Cottegnies, Line. "The Translator as Critic: Aphra Behn's Translation of Fontenelle's *Discovery of New Worlds*." *Restoration 27* (2003): 23–38.

Danielson, Dennis. *"Paradise Lost" and the Cosmological Revolution*. Cambridge: Cambridge University Press, 2014.

Fontenelle, Bernard de. *A Discovery of New Worlds from the French, Made English by A. Behn*. London, 1688.

Fox, Adam. *Oral and Literate Culture in England, 1500–1700*. Oxford: Oxford University Press, 2001.

Galilei, Galileo. *Sidereus Nuncius or The Sidereal Messenger*. Translated by Albert Van Helden, 2nd ed. Chicago and London: The University of Chicago Press, 2015

Gobinet, Charles. *The instruction of youth in Christian Piety Taken out of the Sacred Scriptures, and Holy Fathers; Divided into Five Parts. ... The Last Edition in French, Now Render'd into English*. London, 1687.

Godwin, Francis. *The Man in the Moone*. Edited by William Poole. Peterborough, ON: Broadview, 2009.

Gustafson, Daniel. "Cultural Memory and the Royalist Political Aesthetic in Aphra Behn's Later Works." *Restoration 36* (2012): 1–22.

Hackett, Helen. *Women and Romance Fiction in the English Renaissance*. Cambridge: Cambridge University Press, 2000.

Hayden, Judy A. "Harlequin Science: Aphra Behn's *Emperor of the Moon* and the Plurality of Worlds." *English 64* (2015): 167–82.

Holland, Philemon. *Gutta podrica: a treatise of the gout The severall sorts thereof. VVhat diet is good for such as are troubled therewith. And some approved medicines and remedies for the same*. London, 1633.

Hughes, Derek. "Aphra Behn and the Restoration Theatre." In *The Cambridge Companion to Aphra Behn*. Edited by Derek Hughes and Janet Todd, 29–45. Cambridge: Cambridge University Press, 2004.

Houston, Chloë. *The Renaissance Utopia: Dialogue, Travel and the Ideal Society*. Farnham: Ashgate, 2014.

Hunt, Arnold. *The Art of Hearing: English Preachers and Their Audiences, 1590–1640*. Cambridge: Cambridge University Press, 2010.

2016 Jajdelska, Elspeth. *Speech, Print and Decorum in Britain, 1600–1750: Studies in Social Rank and Communication*. Oxford: Routledge, 2016.

Knellwolf King, Christa. *Faustus and the Promise of the New Science, c. 1580–1730: From the Chaptbooks to Harlequin Faustus*. Farnham: Ashgate, 2008.

MacKinnon, Dolly. "'Poor Senseless Bess, Clothed in her Rags and Folly': Early Modern Women, Madness, and Song in Seventeenth-Century England." *Parergon 18* (2001): 119–51.

Manheimer, Katherine. "Celestial Bodies: Readerly Rapture as Theatrical Spectacle in Aphra Behn's *Emperor of the Moon*." *Restoration 31* (2011): 39–60.

Marchitello, Howard. "Telescopic Voyages: Galileo and the Invention of Lunar Cartography." In *Travel Narratives, the New Science, and Literary Discourse, 1569–1750*. Edited by Judy A. Hayden, 161–78. Farnham: Ashgate, 2012.

Martin, Jessica and Alec Ryrie, eds. *Private and Domestic Devotion in Early Modern Britain*. Farnham: Ashgate, 2012.

Nicolson, Majorie Hope, ed. *Conway Letters: The Correspondence of Anne, Viscountess Conway, Henry More, and their Friends, 1642 1684*. Oxford: Oxford University Press, 1930.

Pearson, Jacqueline. "Women Reading, Reading Women." In *Women and Literature in Britain, 1500–1700*. Edited by Helen Wilcox, 80–99. Cambridge: Cambridge University Press, 1996.

Poole, William "The Origins of Francis Godwin's *The Man in the Moone* (1638)." *Philogical Quarterly 84* (2005): 189–210.

Preston, Clair. *The Poetics of Scientific Investigation in Seventeenth-Century England*. Oxford: Oxford University Press, 2015.

Richards, Jennifer. *Voices and Books in the English Renaissance: A New History of Reading*. Oxford: Oxford University Press, 2019.

Rolley, Thibaut Maus de. "The Devil. Delusions and Early Modern Cognition." In *Cognitive Confusions: Dreams, Delusions and Illusions in Early Modern Culture*. Edited by Ita Mac Carthy, Kirsti Sellevold and Olivia Smith, 71–88. Cambridge: Legenda, 2016.

Roos, Anna Marie E. *Luminaries in the Natural World: The Sun and the Moon in England, 1400–1720*. New York: Peter Lang, 2001.

Sell, Jonathan P.A. *Rhetoric and Wonder in English Travel Writing, 1560–1613*. Aldershot: Ashgate, 2006.

Smith, Bruce. *The Acoustic World of Early Modern England: Attending to the O-Factor*. Chicago: University of Chicago Press, 1999.

Snook, Edith. *Women, Reading, and the Cultural Politics of Early Modern England*. Aldershot: Ashgate, 2005.

Sobel, Dava. *Longitude: The True Story of a Lone Genius who Solved the Greatest Scientific Problem of His Time*. London: Fourth Estate, 1996.

Spiller, Elizabeth. *Science, Reading, and Renaissance Literature: The Art of Making Knowledge*. Cambridge: Cambridge University Press, 2004.

Thomas, Keith. *Religion and the Decline of Magic*. London: Penguin, 1971.

Thompson, Carl. *Travel Writing*. Oxford: Routledge, 2011.

Van Den Abbeele, Georges. *Travel As Metaphor: From Montaigne to Rousseau*. Minneapolis: University of Minnesota Press, 1991.

Wilkins, John. *The Discovery of a World in the Moone, or, A Discourse Tending to Prove, that 'tis Probable There May Be Another Habitable World in That Planet*. London, 1638.

Willie, Rachel. "Viewing the Paper Stage: Civil War, Print, Theatre and the Public Sphere." In *Making Space Public in Early Modern Europe: Performance, Geography, Privacy*. Edited by Angela Vanhaelen and Joseph Ward, 54–75. New York and London: Routledge, 2013.

Wootton, David. *The Invention of Science: A New History of the Scientific Revolution*. London: Allen Lane, 2015.

Index

9780367524210